Information Theory for Data Science

Changho Suh

now
the essence of knowledge

Published, sold and distributed by:
now Publishers Inc.
PO Box 1024
Hanover, MA 02339
United States
Tel. +1-781-985-4510
www.nowpublishers.com
sales@nowpublishers.com

Outside North America:
now Publishers Inc.
PO Box 179
2600 AD Delft
The Netherlands
Tel. +31-6-51115274

ISBN: 978-1-63828-114-6
E-ISBN: 978-1-63828-115-3
DOI: 10.1561/9781638281153

Suggested citation: Changho Suh. (2023). *Information Theory for Data Science*. Boston–Delft: Now Publishers

This work was supported by Institute of Information & Communications Technology Planning & Evaluation (IITP) grant funded by the Korea government (MSIT) (2020-0-00626, Ensuring high AI learning performance with only a small amount of training data).

Table of Contents

To my family, Yuni Kim, Hyun Seung Suh, and Ian Suh

Acknowledgement

We would like to extend our heartfelt appreciation to Hyun Seung Suh, who has generously offered numerous insightful comments and valuable feedback on the organization of the book, the writing style, and the overall accessibility of the content to those who are new to the subject.

This work was supported by the 2022 Google Research Award; and Institute of Information & Communications Technology Planning & Evaluation (IITP) grant funded by the Korea government (MSIT) (2020-0-00626, Ensuring high AI learning performance with only a small amount of training data).

Preface

Features of the book The writing of this book was prompted by the surge of research activities in data science, and the role of information theory in the field. This forms the motivation for this book, enabling three key features.

The first feature is the demonstration of principles and tools of information theory in the context of data science applications, such as social networks, DNA sequencing, search engine, and artificial intelligence (AI). Information theory is a fundamental field that have made foundational impacts upon a wide spectrum of domains in science and engineering. It was established by Claude Shannon in 1948 and deals with mathematical laws that govern the flow, representation and transmission of information. The most significant achievement of the field is the invention of digital communication which forms the basis of our daily-life digital products such as smart phones, laptops and Internet of Things (IoT) devices. While the field was founded in communication, it has since expanded beyond its original domain, contributing to a widening array of contexts, including networks, computational biology, quantum science, economics, finance, and even gambling. Therefore, several books on information theory have been published over the past few decades, covering a broad range of subjects (Gallager, 1968; Cover, 1999; MacKay, 2003; Yeung, 2008; Csiszár and Körner, 2011; El Gamal and Kim, 2011; Gray, 2011; Gleick, 2011; Pierce, 2012; Wilde, 2013). However, this book focuses on a single field: *data science*. Out of the vast content, we emphasize the information-theoretic concepts and tools related to data science applications. These applications include: community detection in social networks, DNA sequencing in biological networks, ranking in search engine, supervised learning, unsupervised learning and social AI.

Secondly, this book is written in a lecture-style format. Most books on this subject cover numerous mathematical concepts and theories, as well as various applications in diverse domains. The concepts and relevant theories are presented in a dictionary-style organization, with topics listed in a sequential order. Although this dictionary-style organization makes it easy to find specific material, it often lacks a cohesive narrative that can engage and motivate readers. This book aims to engage and motivate those who are interested in data science and its interconnections with other disciplines. Our aim is to create a compelling narrative that emphasizes the significance of fundamentals in the field. To achieve this, we have adopted a lecture-style format, with each section serving as notes for a lecture lasting approximately 80 minutes. A consistent connection is established across sections through themes and concepts. To ensure a smooth transition from one section to the next, we have included two paragraphs: (i) the "recap" paragraph that summarizes what has been covered and motivates the contents of the current section; and (ii) the "look ahead" paragraph that introduces the upcoming contents by linking it to previous material.

The final feature of this book is the inclusion of many programming exercises via two software languages: (i) Python; and (ii) TensorFlow. While C++ and MATLAB are widely used in traditional fields, Python has become a key software in data science. Given the breadth of data science applications covered in the book, we have selected Python as our primary platform. To implement machine learning and deep learning algorithms, we utilize TensorFlow, one of the most popular deep learning frameworks. TensorFlow provides many built-in functions for performing many important procedures in deep learning, and its integration with Keras, a high-level library that emphasizes fast user experimentation. With Keras, we can easily transition from an idea to implementation with minimal steps.

Structure of the book This book consists of course materials developed at KAIST over the past decade: (i) EE623 Information Theory (offered from Fall 2012 to 2016 and in 2018 and in 2019); (ii) EE326 Introduction to Information Theory and Coding (offered in Spring 2016 and 2017); (iii) EE321 Communication Engineering (offered from Spring 2013 to 2015 and in 2022); (iv) EE523 Convex Optimization (Spring 2019); and (v) EE424 Introduction to Optimization (Fall 2020 and 2021). It is structured into three parts, each consisting of many sections. Each section covers the material from a single lecture, which lasted approximately 80 minutes. Problem sets, which served as homework in the courses, are included every three or four sections. The detailed contents are summarized as below.

I. *Source coding (9 sections and 3 problem sets):* A brief history of information theory; in-depth study of key notions and Python exercise (entropy, joint entropy, mutual information, Kullback-Leibler (KL) divergence); role of entropy in source coding theorem; prefix-free codes; Kraft's inequality; typical sequences and the asymptotic equipartition property; entropy rate; Huffman code and Python implementation.

II. *Channel coding (9 sections and 3 problem sets):* Role of mutual information in channel coding theorem; capacity of binary erasure channels (BECs), binary symmetric channels (BSCs), and discrete memoryless channels (DMCs); random coding; maximum a posteriori probability (MAP) decoding, maximum likelihood (ML) decoding; jointly typical sequences; union bound; Fano's inequality; data processing inequality; polar code and Python implementation.

III. *Data science applications (20 sections and 6 problem sets):* Community detection in social networks; the achievability and converse proofs of the fundamental limits; the spectral algorithm and Python implementation; Haplotype phasing in computational biology; the achievability and converse proofs of the fundamental limits; an advanced two-staged algorithm and Python implementation; top-K ranking in search engine; a variant of PageRank, an advanced two-staged algorithm and their Python implementation; supervised learning, and the role of cross entropy in logistic regression and deep learning; gradient descent; TensorFlow implementation of a digit classifier; unsupervised learning, and the role of the KL divergence in Generative Adversarial Networks (GANs); alternating gradient descent; TensorFlow implementation of a digit image GAN; fair machine learning, and the role of mutual information in the design of a fair classifier; TensorFlow implementation of a recidivism predictor.

In terms of data science applications, several sections are adapted from the author's previous books: (i) "Convex Optimization for Machine Learning" (Suh, 2022); and (ii) "Communication Principles for Data Science" (Suh, 2023). The contents has been tailored to fit the theme of this book, which focuses on the role of information-theoretic concepts and tools. The book also includes three appendices: two providing brief tutorials on the programming languages used (Python and TensorFlow); and one offering guidance on how to conduct research (primarily aimed at student readers). These tutorials have been adapted from (Suh, 2022, 2023), with appropriate modifications to suit the focused topics. At the end of the book, a list of references relevant to the discussed content is provided, but these are not explained in detail, as we do not aim to exhaust the extensive research literature.

How to use this book This book is written as a textbook for a senior-level undergraduate course and is also suitable for a first-year graduate course. The expected background includes solid undergraduate courses in probability and random processes, as well as basic familiarity with Python.

For students and interested readers, we provide the following guidelines:

1. *Study one section per day and two sections per week:* This is recommended, as each section is designed for a single lecture and two lectures are typical per week in a course offering.
2. *Complete the content in Parts I and II:* One of the most important concepts in information theory is phase transition, together with the achievabilty and converse proofs. Also the following three notions are crucial: entropy, mutual information and the KL divergence. If you are already familiar with these, you can quickly review Parts I and II before proceeding to Part III. However, if you are not familiar with these concepts, it is recommended to read Parts I and II in a sequential manner. The sections are arranged in a way that builds a motivating storyline, and appropriate exercise problems are interspersed throughout to enhance your understanding and motivation.
3. *Explore Part III as per your interest:* Part III focuses on applications and can be partially read, if desired. However, it is structured so that each section builds upon the previous one, assuming a sequential reading. One of the key aspects of Part III is the implementation of algorithms in Python and TensorFlow. With the guidance provided in the main text, problem sets, and appendices, you should be able to implement the algorithms covered.
4. *Solve four to five basic problems in each problem set:* Over 130 problems (including more than 280 subproblems) are provided. Most of them elaborate on the concepts discussed in the main text. The exercises cover basics in probability and random processes, relatively simple derivations of results from the main text, in-depth exploration of non-trivial concepts not fully explained in the main text, and implementation through Python or TensorFlow. The problems are closely tied to the established storyline, so it is essential to work on at least some of them to fully understand the material.

In the course offerings at KAIST, we have covered most of the materials in Parts I and II, but only a limited number of applications in Part III. Based on the students' backgrounds, interests, and available time, there are several ways to structure a course utilizing this book. For example:

1. *Semester-based course (24–26 lectures):* This option would entail covering all the sections in Parts I and II, as well as two to three selected applications from Part III, e.g., (i) community detection and supervised learning, (ii) Haplotype phasing and GANs, or (iii) top-K ranking and fair machine learning.
2. *Quarter-long course (18–20 lectures):* This option would encompass almost all the materials in Parts I and II, excluding certain topics like Huffman coding,

source-channel separation theorem, the role of feedback, and polar codes. Investigate two applications picked up from Part III.

3. *A graduate-level course for students with prior knowledge of information theory:* This option would provide a brief review of the contents in Parts I and II, taking approximately 6–8 lectures. The focus would be on covering as many of the materials in Part III as possible.

Programming exercises can be included as homework assignments to enhance the in-class learning experience.

DOI: 10.1561/9781638281153.ch1

Chapter 1

Source Coding

1.1 Overview of the Book

Outline In this section, we will cover two basic stuffs. Firstly, we will discuss the logistics of the book, providing details on its organization. Secondly, we will provide a brief overview of the book, including the history of how information theory was developed and what will be covered throughout the book.

Prerequisite A basic understanding of probability and random processes is required before proceeding with this book. This can be achieved by taking introductory-level courses on the topics, typically offered in the Department of Electrical Engineering. If you have taken equivalent courses, this is also acceptable. The importance of probability in this book is rooted in the fact that information theory was developed in the context of communication, where the relationship between information theory and probability is evident.

Communication is the transfer of information from one end (called the *transmitter*) to the other (called the *receiver*), over a physical medium (like an air) between the two ends. The physical medium is called the *channel*. The channel links the concept of probability to communication. If you think about how the channel behaves, you can easily see why. The channel can be interpreted as a *system* (in other words, a function) that inputs a transmitted signal and outputs a received signal. However, the channel is not a deterministic function, as it is subject to random elements, also known as noise, that are added to the system. Typically, the noise is additive, meaning that the received signal is the sum of the transmitted signal and the noise. In mathematics or statistics, such a random quantity is referred to as a random variable or random process, which is based on probability. This is why a comprehensive understanding of probability is crucial for understanding this book. If you have taken a basic course on probability but are not familiar with random processes, don't be concerned. Whenever the topic of random processes arises, we will provide detailed explanations and exercises to help you comprehend the material.

There is another important course that can aid in understanding the material in this book, such as a course on random processes, e.g., EE528 at KAIST or EE226 at UC Berkeley. This is a graduate-level course that delves deeper into probability, encompassing many crucial concepts related to random processes. If you have the passion and time, we strongly recommend taking this course while reading this book, though it is not a prerequisite.

Problem sets Problem sets are provided every three to four sections, with a total of 12 problem sets. We encourage working together with other peers if available, as problem sets serve as opportunities for learning, and any method that enhances your learning is encouraged, including discussion, teaching others, and learning from others. Solutions will be made available only to instructors upon request. Some problems may require the use of programming tools such as Python and TensorFlow. We will be using Jupyter notebook. For further information, please refer to the installation guide in Appendix A.1 or seek assistance from:

https : //jupyter.readthedocs.io/en/latest/install.html

We provide tutorials for the programming tools in appendices: (i) Appendix A for Python; and (ii) Appendix B for TensorFlow.

History of communication We will explore the establishment and evolution of information theory within the field of communication. We will begin with a recount of the communication industry's role in the establishment of information theory. Afterwards, we will delve into the in-depth topics that we will cover throughout the book.

Communication is the transfer of information from one end (the transmitter) to the other (the receiver). In between lies a physical medium, known as the channel. The history of communication dates back to the beginning of civilization, where people communicated through dialogue. However, this form of communication has no relation to the electronic communication systems prevalent today. There was a major breakthrough in the history of communication with the invention of the telegraph by Samuel Morse (Beauchamp, 2001). Morse code[1] was the first instance of a simple transmission system used in the telegraph. The invention was based on the discovery in physics that electrical signals, such as voltage or current signals, could be transmitted over wires, such as copper lines. This was the first communication system to use electrical signals and is the reason that communication systems are studied in electrical engineering. Over time, this technology was improved and Alexander Graham Bell invented the telephone (Coe, 1995).

Later advancements in communication systems were made based on another discovery in physics, that electrical signals could be transmitted wirelessly through electromagnetic waves, known as radio waves. This discovery inspired Guglielmo Marconi to develop a wireless version of the telegraph, known as wireless telegraphy (Bondyopadhyay, 1995). Over time, this technology was further developed, leading to the invention of radio and television.

The state of affairs in the early 20th century and Claude E. Shannon In the early 20th century, several communication systems emerged, such as telegraphs, telephones, wireless telegraphs, radios, and televisions. Claude E. Shannon, known as the father of information theory, made a noteworthy observation about these systems during this time. He pointed out that the engineering designs of these systems were customized and specific to each application, resulting in varying design principles for different signals.

Shannon was discontent with this ad-hoc approach and felt that a general framework was necessary to unify these different communication systems. With this in mind, he formulated three questions aimed at integrating the fragmented approach.

Shannon's questions The first question is the most fundamental in terms of the possibility of unification in communication systems.

Question 1: Is there a general unified methodology for designing communication systems?

The second question is a logical follow-up to the first and is aimed at addressing it. Shannon believed that if unification was possible, there could be a common currency (such as the dollar in economics) with respect to information. In communication systems, there are a variety of information sources, such as text, voice,

1. The term "code" should not be mistaken for computer programming languages, such as C++ and Python, as they have no relationship to it. In the communication literature, "code" refers to a transmission scheme.

video, and images. The second question addresses the existence of such a common currency.

Question 2: Is there a common currency of information that can represent different information sources?

The last question pertains to the communication process itself:

Question 3: Is there a limit on the speed of communication?

By addressing these questions, Shannon was able to develop a single theory, which would later be known as *information theory*.

What Shannon did What Shannon did can be divided into three parts. Firstly, he demonstrated that the answer to the second question was affirmative, and he devised a common currency of information that could represent different types of information sources. With this common currency, he then addressed the first question and developed a single comprehensive framework that could unify all the various communication systems. Under this unified framework, he answered the third question by showing that there is a limit to the amount of information that can be communicated, expressed in terms of the common currency. He also characterized this limit.

Interestingly, during the process of characterizing the limit, Shannon made an important observation. The limit is solely dependent on the channel, regardless of any transmission and reception strategy. This means that for a given channel, there is a fundamental limit to the amount of information that can be transmitted, and beyond this limit, communication becomes impossible, no matter what. This quantity does not change, regardless of the actions of the transmitter and receiver. It is like a fundamental law dictated by nature. Shannon theorized this law in a mathematical framework and referred to it as "a mathematical theory of communication" in his landmark paper (Shannon, 2001). Later, this theory became known as information theory or the Shannon theory.

A communication architecture Next, we will explain how Shannon accomplished these tasks, and then we will outline the specific topics that will be covered in this book.

To begin, we will introduce an additional term, in addition to the three terms of transmitter, receiver, and channel. This new term is called "information source," and it refers to the information that one wishes to transmit, such as text, voice, or image pixels. According to Shannon, there must be a process that transforms the information source before it is transmitted. He envisioned this process as a black box, which he called an encoder. At the receiver, there must also be a process that

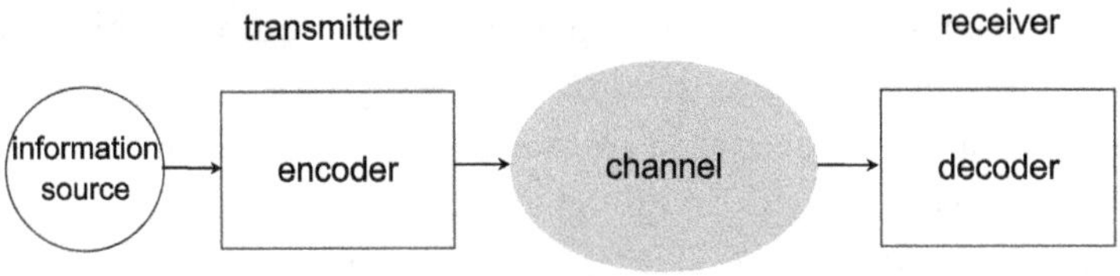

Figure 1.1. A basic communication architecture.

tries to recover the information source from the received signals, which Shannon referred to as a decoder. This was the first block diagram that Shannon imagined for a communication architecture, as shown in Fig. 1.1. From Shannon's perspective, a communication system is simply a collection of an encoder and a decoder, and designing a communication system involves creating an appropriate pair of encoder and decoder.

Representation of an information source With the basic architecture (Fig. 1.1) in mind, Shannon sought to unify the various communication systems that existed at the time. Many engineers at the time transmitted information sources without significant modification, despite the variations in the information sources based on the application. Shannon believed that this was the reason behind the existence of multiple communication systems.

To achieve unification, Shannon believed that there had to be a common representation that could be used to describe different information sources. His work on his Master's thesis at MIT (Shannon, 1938) was pivotal in finding a universal way of representing information sources. He used Boolean algebra to demonstrate in his thesis that logical relationships in circuit systems could be represented using binary strings, represented by the 0/1 logic.

Encouraged by this, Shannon theorized that the same approach could be applied to communication systems, meaning that any type of information source could be represented using a binary string. He proved that this was indeed possible by demonstrating that binary strings, known as "bits," could represent the meaning of information. For instance, in the case of an English text consisting of multiple letters, how could each letter be represented using a binary string? One key realization was that there is a finite number of possibilities for each letter. This number refers to the total number of letters in the English alphabet, which is 26, excluding any special characters like spaces. From this observation, it can be deduced that $\lceil \log_2 26 \rceil = 5$ number of bits suffices to represent each letter.

A two-stage architecture This realization led Shannon to propose bits as a standard unit of information. He proposed a two-stage architecture where the encoder was divided into two parts. The first part, known as the "source encoder," was responsible for converting the information source into bits. The second part, known as the "channel encoder," was responsible for converting the bits into a signal that could be transmitted over a channel.

Similarly, the receiver operates in two stages, but in reverse order. The received signals are first converted into bits through the channel decoder, and then the information source is reconstructed from the bits through the source decoder. The source decoder should have a one-to-one mapping, or there will be no way to recreate the original information source.

The portion of the system that extends from the channel encoder, through the channel, to the channel decoder is referred to as the "digital interface." This digital interface is universal and agnostic to the type of information source, as the input to the digital interface is always bits, regardless of the source. In this sense, it provides a unified communication architecture.

Two questions on the fundamental limits Keeping the two-stage architecture (Fig. 1.2) in his mind, Shannon tried to address the third question: Is there a limit on how fast one can communicate? Shannon discovered the importance of having bits as a standard unit of information. In his proposed two-stage architecture, the source encoder is responsible for converting the information source into bits, before the channel encoder converts the bits into a signal that can be transmitted over a channel.

In order to maximize the amount of information transmitted, Shannon considered the efficiency of the source encoder. To do this, he split his third question into two sub-questions: the first focused on finding the minimum number of bits needed to represent the information source, and the second focused on determining the maximum number of bits that can be transmitted over a channel successfully.

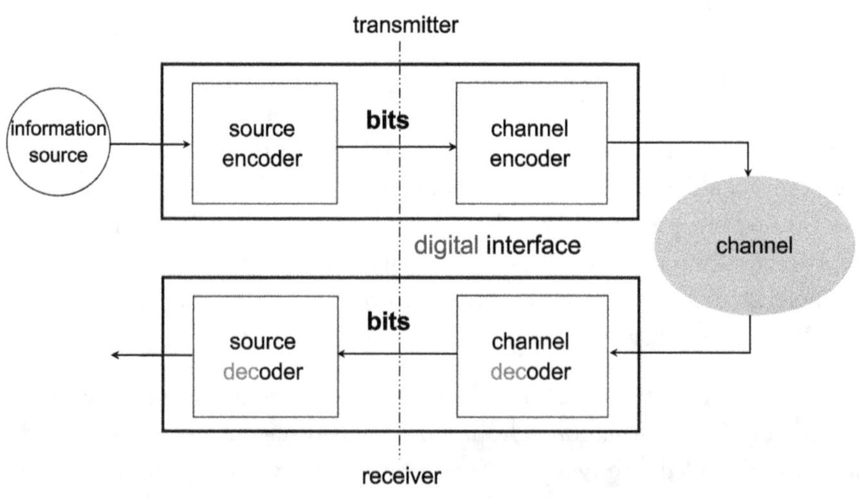

Figure 1.2. A two-stage communication architecture.

Shannon developed two theorems to answer these sub-questions. The first, called the "source coding theorem," determined the minimum number of bits needed to represent the information source. The second, called the "channel coding theorem," characterized the maximum transmission capability.

Source coding theorem Let's first discuss the source coding theorem. An information source can be represented as a sequence of elementary components, such as a sequence of English alphabets, audio signals, or image pixels, represented as $S_1, S_2, S_3, \ldots$. Shannon viewed this sequence as a random process, as the information source is unknown to the receiver.

The minimum number of bits needed to represent the information source depends on the probabilistic properties of the random process, specifically its joint distribution. For example, when the random process has a simple joint distribution, such as when the individual variables are independent and identically distributed (i.i.d.), it is straightforward to state the source coding theorem.

In this context, each individual random variable is referred to as a "symbol." The minimum number of bits needed to represent the information source is related to the concept of entropy, which is a measure of disorder or randomness. This measure plays a crucial role in formulating the source coding theorem, which states that the minimum number of bits needed to represent the information source is proportional to the entropy of the random process.

Theorem 1.1 (Source coding theorem in the i.i.d. case). *The minimum number of bits that can represent the source per symbol is the entropy of the random variable S, denoted by*

$$H(S) := \sum_{s \in \mathcal{S}} \mathbb{P}_S(s) \log_2 \frac{1}{\mathbb{P}_S(s)} \tag{1.1}$$

where $\mathbb{P}_S(s)$ denotes the probability distribution[2] of S, and $\mathcal{S}$ (that we call "caligraphy S") indicates the range (the set of all possible values that S can take on).

Source code example To gain a clearer understanding of the source coding theorem, let's examine a practical example. The objective in source coding is to establish a functional relationship, denoted by f, between the input sequence S and the output from the source encoder, referred to as the "codeword". For instance, consider a DNA sequence where each symbol S can take on one of four values: A, C, T, G. To make things simple, let's consider an unrealistic yet straightforward

2. It is a probability mass function, simply called pmf, for the case where S is a discrete random variable. It is often denoted by $p(s)$ for brevity.

scenario where the random process, represented by $\{S_i\}$, is independent and identically distributed (i.i.d.). In this case, each symbol in the sequence is distributed as follows:

$$S = \begin{cases} A, & \text{with probability (w.p.) } \frac{1}{2}; \\ C, & \text{w.p. } \frac{1}{4}; \\ T, & \text{w.p. } \frac{1}{8}; \\ G, & \text{w.p. } \frac{1}{8}. \end{cases}$$

How can we design the functional relationship $f(S)$ in order to minimize the average length of the codeword, represented by $\mathbb{E}[f(S)]$, and reduce the number of bits needed to represent S? With a total of four letters, it is sufficient to use two bits per symbol. A simple approach might be to assign A to 00, C to 01, T to 10, and G to 11. This would result in an average of 2 bits per symbol. However, according to the source coding theorem, it is possible to attain better results. The limit promised is:

$$H(S) = \frac{1}{2} \cdot 1 + \frac{1}{4} \cdot 2 + \frac{1}{8} \cdot 3 + \frac{1}{8} \cdot 3 = 1.75. \tag{1.2}$$

The existence of a code that achieves the desired limit has been confirmed. The code is based on the following observations: (i) the letter A occurs more frequently than the other letters, and (ii) the length of the codeword does not have to be fixed. These observations lead to a natural idea. Assigning short codewords to frequent letters and long codewords to less frequent letters. To implement this idea, a "binary code tree" is introduced to facilitate the mapping from S to $f(S)$.

A binary code tree: A binary code tree is a tree structure where every internal node has only two branches, and a node without any branches is referred to as a leaf. An example of this can be seen in Fig. 1.3. The binary tree is related to a code by assigning a symbol to a leaf and defining the functional relationship by specifying the pattern of the corresponding codeword. This is done by following the sequence of binary labels (associated with branches) from the root to the leaf. For instance, if an upper branch is labeled 0 and a lower branch is labeled 1, and the symbol A is assigned to the top leaf, then $f(A) = 0$, as there is only one branch (labeled 0) linking the root to the leaf. Similarly, $f(C) = 10$, as there are two branches (labeled 1 and 0, respectively) connecting the root to the leaf assigned to C.

How to implement an optimal mapping rule that achieves 1.75 bits per symbol using a binary code tree? As previously noted, the goal is to assign short codewords to frequent letters. The most frequent letter, A, should be assigned to the top leaf, as it has the shortest codeword length. This is evident. However, what about the second most frequent letter, C? It may seem like a good idea to assign it to the internal node marked with a blue square in Fig. 1.3, but this is not a valid solution.

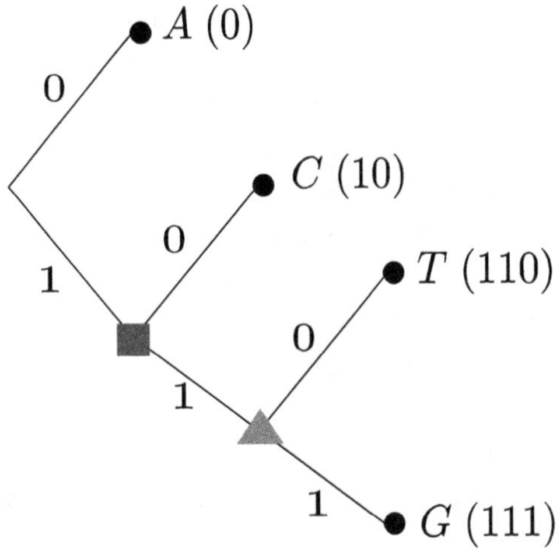

Figure 1.3. Representation of an optimal source code via a binary code tree.

The reason for this is that the codeword pattern would end. This is problematic, as there are only two leaves available, but four are required in total. Another two branches need to be generated from the internal node. C can be assigned to the second top leaf with a codeword length of 2. Similarly, the next frequent letter, T (or G), cannot be assigned to the internal node marked with a red triangle in Fig. 1.3. The node should have another set of two branches. The remaining letters T and G are assigned to the two remaining leaves. With this mapping rule, we achieve:

$$\begin{aligned}\mathbb{E}[\text{length}(f(S))] &= \mathbb{P}(S=A)\text{length}(f(A)) + \mathbb{P}(S=C)\text{length}(f(C)) \\ &\quad + \mathbb{P}(S=T)\text{length}(f(T)) + \mathbb{P}(S=G)\text{length}(f(G)) \\ &= \frac{1}{2}\cdot 1 + \frac{1}{4}\cdot 2 + \frac{1}{8}\cdot 3 + \frac{1}{8}\cdot 3 = 1.75 = H(S).\end{aligned}$$

Channel coding theorem The channel coding theorem states that the maximum number of bits that can be transmitted over a channel is its capacity, represented by C. There is a mathematical definition for C, which involves several important concepts and notions that we will need to study. One of these important concepts is "mutual information." We will delve into the definitions of these concepts later.

Book outline The two theorems are at the core of the material covered in this book. The book is divided into three parts. In Part I, we will study the basic principles of the field, including (i) entropy, which is an essential component in establishing the source coding theorem; (ii) mutual information, which is critical for the channel coding theorem; and (iii) Kullback-Leibler (KL) divergence, a powerful concept that has similar functions to mutual information and is widely used in fields like mathematics, statistics, and machine learning. By using entropy, we will demonstrate the source coding theorem. In Part II, we will prove the channel coding theorem with the use of mutual information and present a code that satisfies the fundamental limit set by the channel coding theorem.

Information theory, and these three crucial concepts in particular, form the foundation for solving numerous important problems across various fields such as communication, social networks, computational biology, machine learning, and deep learning. In Part III, we will examine the applications of information theory in the field of data science, with a specific emphasis on six major examples that highlight the fundamental limit and key concepts of information theory.

The first of these applications is *community detection* (Girvan and Newman, 2002; Fortunato, 2010; Abbe, 2017), a well-researched problem in data science with applications in social networks (such as Facebook, LinkedIn, and Twitter) and biological networks. The second is *Haplotype phasing*, one of the significant DNA sequencing problems that shares a similar structure with community detection (Browning and Browning, 2011; Das and Vikalo, 2015; Chen *et al.*, 2016a; Si *et al.*, 2014). The third is a *ranking problem* which forms the basis of search engines. Google's PageRank (Page *et al.*, 1999) is a famous ranking algorithm that serves as the backbone of Google's website search engine.

In these three problems, we will demonstrate that the concept of fundamental limits plays a crucial role in addressing the problems and in the development of optimal algorithms. The last three applications are related to machine learning and deep learning: (i) supervised learning, one of the most widely used machine learning methods; (ii) Generative Adversarial Networks (GANs) (Goodfellow *et al.*, 2014), a groundbreaking model for unsupervised learning; and fair classifiers (Larson *et al.*, 2016; Zafar *et al.*, 2017; Cho *et al.*, 2020), a timely and socially significant topic in machine learning. In particular, we will emphasize: (i) the central role of entropy and KL divergence in the design of a loss function for optimization in supervised learning; (ii) the fundamental role of KL divergence in the design of GANs; and (iii) the recently discovered role of mutual information in the design of fair classifiers.

1.2 Entropy and Python Exercise

Recap In the previous section, we recounted out the story of Shannon's founding of information theory. Motivated by the desire to unify the disparate communication systems of the early 20th century, Shannon proposed the use of bits to represent different information sources and established a unified two-stage architecture. The first stage transforms an information source into bits, while the second stage generates a signal that can be transmitted over a channel. With this framework in place, Shannon determined the limit on the amount of information that can be communicated, leading to the formulation of two fundamental theorems. The first of these theorems, the source coding theorem, defines the maximum compression rate of an information source, while the second theorem, the channel coding theorem, defines the maximum number of bits that can be transmitted reliably.

Outline Before delving into the two landmark theorems, we will first examine three crucial concepts that play central roles in the theorems: (i) entropy; (ii) mutual information; and (iii) Kullback-Leibler (KL) divergence. These concepts are essential in addressing many critical issues across a range of disciplines, including statistics, physics, computational biology, and machine learning. Therefore, it is advisable to familiarize oneself with the detailed properties of these concepts for various purposes.

In this section, we will focus on the first concept: entropy. Our tasks will include: (i) reviewing the definition of entropy, providing intuitive explanations for the meaning of entropy to better understand why the maximum compression rate must be entropy, as stated in the source coding theorem; (ii) studying key properties of entropy that are useful in various contexts; and (iii) completing a Python exercise to compute entropy. In a later section, we will see how entropy factors into the proof of the source coding theorem.

Definition of entropy The entropy is defined in relation to a random quantity. Specifically, it deals with the probability distribution of a random variable (for scalar random quantities) or a random process (for vector quantities). For simplicity, we will begin by examining the case of a random variable, and then move on to cover the general case of a random process later.

More precisely, the entropy is defined w.r.t. a *discrete*[3] random variable. Let X be a discrete random variable and $\mathbb{P}_X(x)$ be its probability mass function (pmf). For

3. We say that a random variable is *discrete* if its range (the set of values that the random variable can take on) is finite or at most countably infinite.

brevity, we employ a simpler notation $p(x)$ to indicate $\mathbb{P}_X(x)$. Let $\mathcal{X}$ (that we call "caligraphy ex") be its range: the set of values that X can take on. The entropy is defined as:

$$H(X) := \sum_{x \in \mathcal{X}} p(x) \log_2 \frac{1}{p(x)} \qquad \text{bits.} \tag{1.3}$$

Throughout this book, the logarithmic function most commonly used is the base 2 logarithm. However, to simplify the notation, we will omit the "2" and use the logarithm function without any base specification.

We introduce an alternative expression for the entropy formula. It is much simpler and thus easy to remember. In addition, it serves to simplify the proof of some important properties that we are going to investigate. Observe in (1.3) that the entropy is a weighted sum of $\log \frac{1}{p(x)}$ for different values of x's. So it can be represented as:

$$H(X) := \mathbb{E}\left[\log \frac{1}{p(X)}\right] \tag{1.4}$$

where the expectation is taken over the distribution $p(x)$ of X.

Interpretation #1 We will highlight two commonly recognized interpretations of entropy that are intuitive and can provide some understanding of why entropy is associated with the maximum compression rate. The first is:

Entropy is a measure of the uncertainty of a random quantity.

This interpretation is supported by a tangible instance, as demonstrated below. Consider two experiments: (i) tossing a fair coin; and (ii) rolling a fair dice. One simple random variable that one can think of for the first experiment is a function that maps the head (or tail) event to 0 (or 1). Since the coin is fair, we have:

$$X = \begin{cases} 0, & \text{w.p. } \frac{1}{2}; \\ 1, & \text{w.p. } \frac{1}{2}. \end{cases}$$

The abbreviation "w.p." stands for "with probability". On the other hand, a natural random variable in the second experiment is a function that maps a dice result to the same number:

$$X = \begin{cases} 1, & \text{w.p. } \frac{1}{6}; \\ 2, & \text{w.p. } \frac{1}{6}; \\ 3, & \text{w.p. } \frac{1}{6}; \\ 4, & \text{w.p. } \frac{1}{6}; \\ 5, & \text{w.p. } \frac{1}{6}; \\ 6, & \text{w.p. } \frac{1}{6}. \end{cases}$$

One may inquire about which random variable is more unpredictable. Intuitively, it appears that the second random variable is more uncertain. The entropy provides precise numerical evidence to confirm this intuition. Note that $H(X) = 1$ in the first experiment while $H(X) = \log 6 > 1$ in the latter. The entropy plays a role to quantify such uncertainty.

Let us give you another example. Suppose we have a bent coin. Then, it yields a different probability for the head event, say $p \neq \frac{1}{2}$:

$$X = \begin{cases} 0, & \text{w.p. } p; \\ 1, & \text{w.p. } 1-p. \end{cases} \tag{1.5}$$

Can it be inferred that this random variable is more unpredictable than the fair coin scenario? To examine this, let us contemplate an extreme situation in which $p \ll 1$, yielding the following:

$$H(X) = p \log \frac{1}{p} + (1-p) \log \frac{1}{1-p} \approx 0. \tag{1.6}$$

Here we used the fact that $\lim_{p \to 0^+} p \log \frac{1}{p} = 0$. Remember L'Hospital's theorem that you may learn from calculus (Stewart, 2015). According to this theorem, it can be concluded that the bent-coin scenario is unquestionably more certain. This is logically sound since a very small value of $p(\ll 1)$ implies that the tail event occurs almost all the time, making the outcome highly foreseeable.

Interpretation #2 The second interpretation concerns a method for eliminating uncertainty. To clarify this point, consider the following example: Imagine meeting a person for the first time. At this point, the person is entirely unknown to us. However, one can remove this uncertainty by asking questions. With each answer, randomness associated with the person can be eliminated. With enough questions, it is possible to gain complete knowledge about the individual. Therefore, the number of questions needed to obtain comprehensive knowledge reflects the degree of uncertainty: the greater the number of questions required, the more unpredictable the situation. This leads to:

> *Entropy is intimately related to the number of questions required to uncover the value of X.*

In some cases, the number of questions (on average) precisely corresponds to $H(X)$. The following is an example of such a scenario:

$$X = \begin{cases} 1, & \text{w.p. } \frac{1}{2}; \\ 2, & \text{w.p. } \frac{1}{4}; \\ 3, & \text{w.p. } \frac{1}{8}; \\ 4, & \text{w.p. } \frac{1}{8}. \end{cases}$$

A straightforward calculation yields $H(X) = 1.75$. Assume that the questions are binary, requiring a yes or no answer. In this case, the minimum average number of questions needed to ascertain the value of X is $H(X)$. The optimal approach for posing questions to achieve $H(X)$ is as follows: start by asking if X is equal to 1. If the answer is yes, then X is 1; otherwise, ask if X is equal to 2. If the answer is yes, then X is 2; if not, ask if X is equal to 3. Let $f(x)$ represent the number of questions required to determine X when $X = x$. This method results in:

$$\begin{aligned} \mathbb{E}[f(X)] &= \frac{1}{2}f(1) + \frac{1}{4}f(2) + \frac{1}{8}f(3) + \frac{1}{8}f(4) \\ &= \frac{1}{2}\cdot 1 + \frac{1}{4}\cdot 2 + \frac{1}{8}\cdot 3 + \frac{1}{8}\cdot 3 = 1.75. \end{aligned}$$

Some of you may recognize that this is exactly the same as the number that appears in the prior source code example. See (1.2) for details.

Key properties The entropy has several important properties. We can identify them by making some observations.

Recall the bent-coin example. See (1.5) for the distribution of the associated random variable X. Consider the entropy calculated in (1.6). Notice that the entropy is a function of p. Fig. 1.4 illustrates how $H(X)$ behaves as a function of p. One can make two observations here: (i) the minimum entropy is 0; and (ii) the entropy is maximized when $p = \frac{1}{2}$, i.e., X is uniformly distributed.

Consider another example in which $X \in \mathcal{X} = \{1, 2, \ldots, M\}$ and is uniformly distributed:

$$X = \begin{cases} 1, & \text{w.p. } \frac{1}{M}; \\ 2, & \text{w.p. } \frac{1}{M}; \\ \vdots & \\ M, & \text{w.p. } \frac{1}{M}. \end{cases}$$

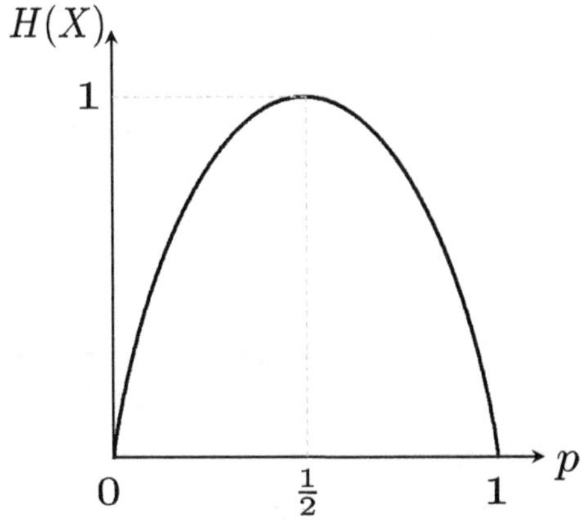

Figure 1.4. The entropy of a binary random variable X with $\mathbb{P}(X=0)=p$ and $\mathbb{P}(X=1)=1-p$.

In this case,

$$H(X)=\sum_{x=1}^{M}\frac{1}{M}\log M=\log M=\log|\mathcal{X}|$$

where $|\mathcal{X}|$ indicates the cardinality of $\mathcal{X}$ (the size of the set). This leads to another observation: the entropy of the uniformly distributed random variable $X\in\mathcal{X}$ is $\log|\mathcal{X}|$.

The above three observations lead us to conjecture the following two: for $X\in\mathcal{X}$,

Property 1: $H(X)\geq 0$.

Property 2: $H(X)\leq\log|\mathcal{X}|$.

These properties hold indeed. The first is easy to prove. Using the definition of entropy and the fact that $p(X)\leq 1$, we get: $H(X)=\mathbb{E}[\log\frac{1}{p(X)}]\geq\mathbb{E}[\log 1]=0$. Using *Jensen's inequality*, it is straightforward to prove the second property. This inequality is a popular and fundamental mathematical concept that underlies many results in information theory. Its formal statement is presented below.

Theorem 1.2 (Jensen's inequality). *For a concave*[4] *function* $f(\cdot)$,

$$\mathbb{E}[f(X)]\leq f(\mathbb{E}[X]).$$

4. We say that a function f is *concave* if for any (x_1,x_2) and $\lambda\in[0,1]$, $\lambda f(x_1)+(1-\lambda)f(x_2)\leq f(\lambda x_1+(1-\lambda)x_2)$. In other words, for a concave function, the weighted sum w.r.t. functions evaluated at two points is less than or equal to the function at the weighted sum of the two points. See Fig. 1.5.

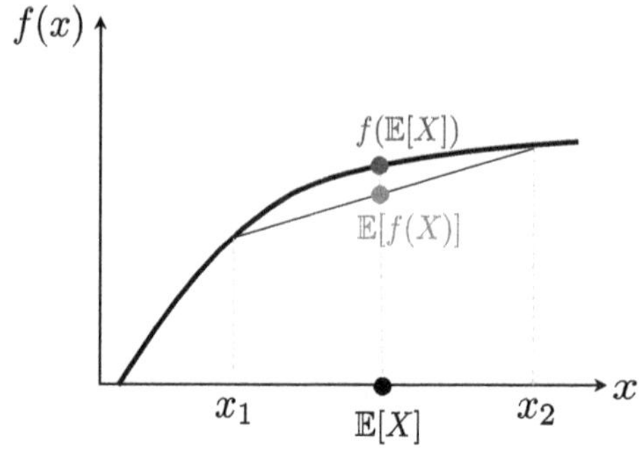

Figure 1.5. Jensen's inequality for a concave function: $\mathbb{E}[f(X)] \leq f(\mathbb{E}[X])$.

Proof. The proof is immediate for a simple binary case, say $X \in \mathcal{X} = \{x_1, x_2\}$. By setting $p(x_1) = \lambda$ and $p(x_2) = 1 - \lambda$, we get: $\mathbb{E}[X] = \lambda x_1 + (1 - \lambda)x_2$ and $\mathbb{E}[f(X)] = \lambda f(x_1) + (1 - \lambda)f(x_2)$. The definition of concavity (see the associated footnote for the definition or Fig. 1.5) completes the proof. The generalization to an arbitrary $\mathcal{X}$ can be done by induction. Try this in Prob 1.5. □

Using the definition of entropy and the fact that $\log(\cdot)$ is a concave function, we get:

$$\begin{aligned} H(X) &= \mathbb{E}\left[\log \frac{1}{p(X)}\right] \\ &\leq \log\left(\mathbb{E}\left[\frac{1}{p(X)}\right]\right) \\ &= \log\left(\sum_{x\in\mathcal{X}} p(x) \cdot \frac{1}{p(x)}\right) \\ &= \log|\mathcal{X}| \end{aligned}$$

where the inequality is due to Jensen's inequality.

Joint entropy We will examine the entropy defined w.r.t. multiple (say two) random variables. Since this calculation involves multiple quantities, it is referred to as *joint* entropy, and is defined as follows: for two discrete random variables, $X \in \mathcal{X}$ and $Y \in \mathcal{Y}$,

$$H(X, Y) := \sum_{x\in\mathcal{X}} \sum_{y\in\mathcal{Y}} \mathbb{P}_{X,Y}(x, y) \log \frac{1}{\mathbb{P}_{X,Y}(x, y)}$$

where $\mathbb{P}_{X,Y}(x, y)$ denotes the joint distribution of (X, Y). Again, we employ a simpler notation $p(x, y)$. The only distinction w.r.t. the single random variable case is that the joint distribution $p(x, y)$ comes into picture. Similarly, an alternative expression reads:

$$H(X, Y) = \mathbb{E}\left[\log \frac{1}{p(X, Y)}\right]$$

where the expectation is taken over $p(x, y)$.

Chain rule We highlight a significant characteristic of joint entropy, known as the chain rule, which demonstrates the correlation between multiple random variables. For the two random variable case, it reads:

$$\textit{Property 3 (chain rule): } H(X, Y) = H(X) + H(Y|X)$$

where $H(Y|X)$ is *conditional entropy* and defined as:

$$H(Y|X) := \sum_{x \in \mathcal{X}} \sum_{y \in \mathcal{Y}} p(x, y) \log \frac{1}{p(y|x)} = \mathbb{E}\left[\frac{1}{p(Y|X)}\right]$$

where the expectation is taken over $p(x, y)$ and $p(y|x)$ denotes a simpler notation of conditional distribution $\mathbb{P}_{Y|X}(y|x)$. The proof of the chain rule is straightforward:

$$\begin{aligned}
H(X, Y) &= \mathbb{E}\left[\log \frac{1}{p(X, Y)}\right] \\
&\overset{(a)}{=} \mathbb{E}\left[\log \frac{1}{p(X)p(Y|X)}\right] \\
&\overset{(b)}{=} \mathbb{E}_{X,Y}\left[\log \frac{1}{p(X)}\right] + \mathbb{E}\left[\log \frac{1}{p(Y|X)}\right] \\
&\overset{(c)}{=} \mathbb{E}_{X}\left[\log \frac{1}{p(X)}\right] + \mathbb{E}\left[\log \frac{1}{p(Y|X)}\right] \\
&= H(X) + H(Y|X)
\end{aligned}$$

where (*a*) follows from the definition of conditional probability ($p(y|x) := \frac{p(x,y)}{p(x)}$); (*b*) follows from the linearity of expectation; and (*c*) follows from $\sum_{y \in \mathcal{Y}} p(x, y) = p(x)$ (total probability law). The last step is due to the definition of entropy and conditional entropy.

We provide an interesting interpretation on the chain rule. Remember that entropy is a measure of uncertainty. Hence, one can interpret *Property 3* as follows. The uncertainty of (X, Y) (reflected in $H(X, Y)$) is the sum of the following two: (i) the uncertainty of X (reflected in $H(X)$); and (ii) residual uncertainty in Y when X is known (reflected in $H(Y|X)$). See Fig. 1.6 for visual illustration.

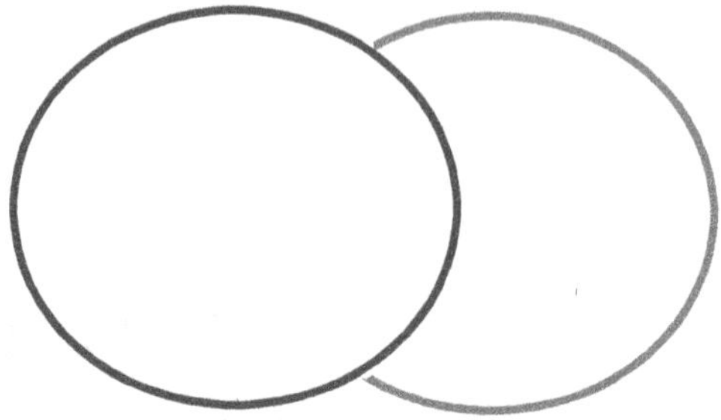

Figure 1.6. A Venn diagram interpretation of the chain rule.

The interpretation of Property 3 becomes clearer when we map the Venn diagram to the amount of uncertainty associated with a random variable. The area of the blue circle represents $H(X)$; the area of the red part represents $H(Y|X)$; and the entire area represents $H(X, Y)$.

A remark on conditional entropy Another way to express conditional entropy is:

$$\begin{aligned} H(Y|X) &= \sum_{x\in\mathcal{X}}\sum_{y\in\mathcal{Y}} p(x,y)\log\frac{1}{p(y|x)} \\ &= \sum_{x\in\mathcal{X}} p(x)\sum_{y\in\mathcal{Y}} p(y|x)\log\frac{1}{p(y|x)} \\ &= \sum_{x\in\mathcal{X}} p(x)H(Y|X=x) \end{aligned}$$

where the last equality is due to the conventional definition of:

$$H(Y|X=x) := \sum_{y\in\mathcal{Y}} p(y|x)\log\frac{1}{p(y|x)}.$$

Here $H(Y|X=x)$ is the entropy defined w.r.t. Y when $X = x$. So the conditional entropy can be interpreted as the weight sum of $H(Y|X=x)$. This interpretation serves to remember the formula as well as provides an easy way to calculate.

Python exercise Let us investigate how to compute entropy, joint entropy and conditional entropy via Python. As per the definition (1.3) of entropy, one can calculate entropy from scratch via the following code.

```
import numpy as np

def entropy(pX):
    return sum(pX*np.log2(1/pX))

# probability distribution of a binary random variable
pX = np.array([1/2, 1/2])
print(entropy(pX))
```

1.0

Alternatively one can use a built-in function in scipy.stats.

```
from scipy.stats import entropy

pX1 = np.array([1/2, 1/2]) # numpy.array
pX2 = [1/2, 1/2]  # list
print(entropy(pX1, base=2))
print(entropy(pX2, base=2))
```

1.0
1.0

As for the input distribution, entropy can take either numpy.array or list.

Using the above entropy function, we draw the entropy of a binary random variable as a function of p, as demonstrated in Fig. 1.4.

```
import matplotlib.pyplot as plt

p = np.arange(0.001,0.999,0.001)
Hp = np.zeros(len(p))
for i,val in enumerate(p):
    pX = np.array([val, 1-val])
    Hp[i] = entropy(pX, base=2)

plt.figure(figsize=(5,5), dpi=150)
plt.plot(p,Hp)
plt.xlabel('p')
plt.ylabel('H(X)')
plt.title('Entropy of a binary random variable')
plt.show()
```

The way to compute joint entropy is the same as that of entropy. The only distinction is that the cardinality of the range set grows. To see this, consider a simple two-random variable example where the joint distribution reads: $p(x, y) = \frac{1}{4}, \frac{1}{4}, \frac{1}{3}, \frac{1}{6}$ for $(x, y) = (0, 0), (0, 1), (1, 0), (1, 1)$, respectively. The joint distribution is then represented as an array like $[\frac{1}{4}, \frac{1}{4}, \frac{1}{3}, \frac{1}{6}]$. This gives the computation of joint entropy as:

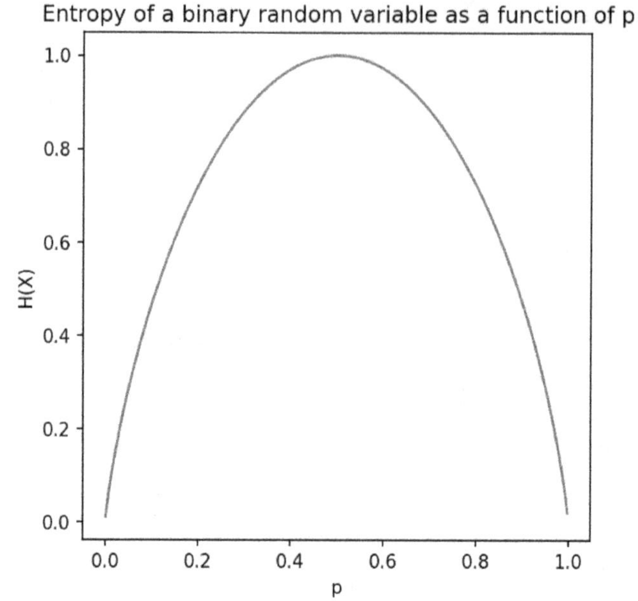

Figure 1.7. Python plotting: Entropy of a binary random variable.

```
pXY = np.array([1/4, 1/4, 1/3, 1/6])
HXY = entropy(pXY, base=2)
print(HXY)
```

1.959147917027245

Next, we compute $H(X)$ and $H(Y|X)$ (conditional entropy) to verify the chain rule.

```
# Compute p(x)
pX = np.array([1/4+1/4, 1/3+1/6])
# Compute p(y|0)
pY_x0 = np.array([1/4, 1/4])/pX[0]
# Compute p(y|1)
pY_x1 = np.array([1/3, 1/6])/pX[1]

# Compute H(X)
HX = entropy(pX, base=2)
# Compute H(Y|X)=\sum p(x)*H(Y|X=x)
HY_X = pX[0]*entropy(pY_x0,base=2) \
       + pX[1]*entropy(pY_x1,base=2)

# Verify the chain rule: H(X,Y) = H(X) + H(Y|X)
print(HX+HY_X)
```

1.9591479170272448

Notice that $H(X) + H(Y|X)$ is the same $H(X, Y)$, taking ≈ 1.9591.

Look ahead We can use the entropy concept to prove the source coding theorem, but not the channel coding theorem. To prove the channel coding theorem, we need to introduce another concept: mutual information. In the next section, we will delve into mutual information, covering its definition and several important properties. These properties not only serve to prove the theorem but also play critical roles in other fields. Additionally, we will examine another essential concept: the KL divergence. The KL divergence, together with entropy, is crucial in proving the source coding theorem. It has also played a significant role in other disciplines, such as serving as a distance measure between distributions in statistics.

1.3 Mutual Information, KL Divergence and Python Exercise

Recap In the preceding section, we gained knowledge about entropy, which is crucial in proving the source coding theorem. Initially, we defined entropy for a single random variable and subsequently extended it to cases where multiple random variables are present. Furthermore, we explored the chain rule, a significant principle that governs the association among multiple random variables. Specifically, for two random variables X and Y, the chain rule can be stated as follows:

$$H(X, Y) = H(X) + H(Y|X) \tag{1.7}$$

where $H(Y|X)$ denotes conditional entropy. Remember the definition of conditional entropy:

$$H(Y|X) := \sum_{x \in \mathcal{X}} p(x) H(Y|X = x) \tag{1.8}$$

where $H(Y|X = x)$ is the entropy w.r.t. $p(y|x)$.

Towards the end, it was highlighted that to prove the channel coding theorem, an understanding of another significant concept, mutual information, is necessary. Additionally, we emphasized the need to delve into another important notion, the Kullback-Leibler (KL) divergence.

Outline This section is dedicated to exploring two important concepts: mutual information and the KL divergence. It is divided into five parts. Firstly, we will begin with the definition of mutual information. Secondly, we will delve into the key properties of mutual information. Thirdly, we will examine the relationship between mutual information and the KL divergence. In the fourth part, we will discuss how mutual information is related to the channel coding theorem. Finally, we will conclude this section with a Python exercise that involves computing mutual information and the KL divergence.

Observation An interesting observation we made w.r.t. the chain rule (the Venn diagram interpretation in Fig. 1.8) brings about a natural definition for mutual information. First recall the interpretation. The randomness of two random variables X and Y (reflected in the total area of two Venn diagrams) is the sum of the randomness of one variable, say X, (reflected in the area of the blue Venn diagram) and the uncertainty that remains about Y conditioned on X (reflected in the crescent-moon-shaped red area). By the chain rule, the crescent-moon-shaped red area can be represented as: $H(Y|X)$.

We see an overlap between the blue and red areas. The area of the overlapped part depends on how large $H(Y|X)$ is: the larger the overlapped area, the smaller

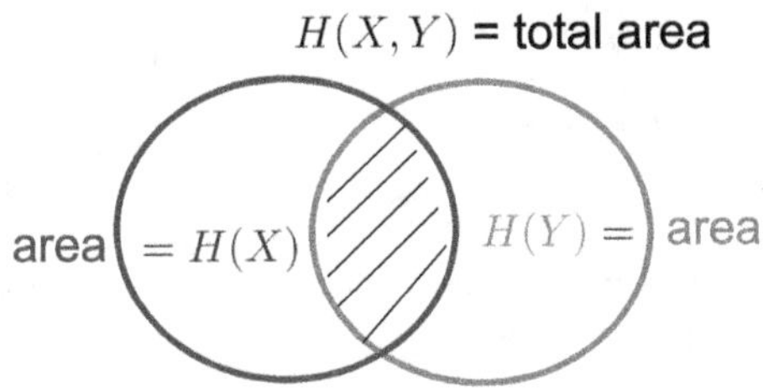

Figure 1.8. A Venn diagram interpretation of the chain rule.

$H(Y|X)$. A low value of $H(Y|X)$ implies a high level of dependence between X and Y. Therefore, the larger the area of overlap between the two Venn diagrams, the greater the dependence between them. Consequently, the overlapping area quantifies the degree of shared information between X and Y.

Definition of mutual information This observation leads to the definition of the overlapped area that captures the shared information:

$$I(X;Y) := H(Y) - H(Y|X). \tag{1.9}$$

In the literature, this notion is called mutual information instead of shared (or common) information.

From the picture in Fig. 1.8, one can define it instead as $I(X;Y) := H(X) - H(X|Y)$ because the alternative indicates the same overlapped area. By convention, we follow the definition of (1.9) though: The entropy of the right-hand-side term inside $I(\cdot;\cdot)$ minus conditional entropy of the right-hand-side term conditioned on the left-hand-side term.

Key properties Remember that entropy respects: (i) the non-negativity $H(X) \geq 0$; and (ii) the cardinality bound $H(X) \leq \log|\mathcal{X}|$. Similarly mutual information exhibits the following properties:

$$\begin{aligned} &\textit{Property 1:} \quad I(X;Y) = I(Y;X); \\ &\textit{Property 2:} \quad I(X;Y) \geq 0; \\ &\textit{Property 3:} \quad I(X;Y) = 0 \Longleftrightarrow X \perp\!\!\!\perp Y. \end{aligned}$$

The first property (named the symmetry property) is obvious from the picture. For rigorousness, we leave the proof as below:

$$\begin{aligned} I(X;Y) &:= H(Y) - H(Y|X) \\ &\overset{(a)}{=} H(Y) - (H(X,Y) - H(X)) \\ &\overset{(b)}{=} H(Y) + H(X) - (H(Y) + H(X|Y)) \end{aligned}$$

$$
\begin{aligned}
&= H(X) - H(X|Y) \\
&\overset{(c)}{=} I(Y;X)
\end{aligned}
$$

where (a) and (b) follow from the chain rule (1.7); and (c) is due to the definition of mutual information (1.9).

The second property is also straightforward, as mutual information captures the overlapped area and therefore it must be non-negative. But the proof is not that simple. It requires a bunch of steps as well as the usage of an important inequality that we learned in the previous section. That is, Jensen's inequality. We will prove the second property in the sequel.

The third property also makes an intuitive sense. Mutual information being 0 means no correlation between X and Y, implying the independence between the two. But the proof is not trivial either. We will provide the proof right after proving the second property.

Proof of $I(X;Y) \geq 0$ & its implication Starting with the definition of mutual information, we obtain:

$$
\begin{aligned}
I(X;Y) &:= H(Y) - H(Y|X) \\
&\overset{(a)}{=} \mathbb{E}_Y\left[\log \frac{1}{p(Y)}\right] - \mathbb{E}_{X,Y}\left[\log \frac{1}{p(Y|X)}\right] \\
&\overset{(b)}{=} \mathbb{E}_{X,Y}\left[\log \frac{1}{p(Y)}\right] - \mathbb{E}_{X,Y}\left[\log \frac{1}{p(Y|X)}\right] \\
&\overset{(c)}{=} \mathbb{E}\left[\log \frac{p(Y|X)}{p(Y)}\right] \\
&\overset{(d)}{=} \mathbb{E}\left[-\log \frac{p(Y)}{p(Y|X)}\right] \\
&\overset{(e)}{\geq} -\log \mathbb{E}\left[\frac{p(Y)}{p(Y|X)}\right] \\
&= -\log\left\{\sum_{x\in\mathcal{X}}\sum_{y\in\mathcal{Y}} p(x,y)\frac{p(y)}{p(y|x)}\right\} \\
&\overset{(f)}{=} -\log\left\{\sum_{x\in\mathcal{X}}\sum_{y\in\mathcal{Y}} p(x)p(y)\right\} \\
&\overset{(g)}{=} -\log 1 = 0
\end{aligned}
$$

where (*a*) follows from the definition of entropy and joint entropy; (*b*) is due to the total probability law $\sum_{x\in\mathcal{X}} p(x,y) = p(y)$; (*c*) is due to the linearity of expectation; (*d*) comes from $\log x = -\log\frac{1}{x}$; (*e*) is due to the fact that $-\log(\cdot)$ is a convex function and applying Jensen's inequality; (*f*) follows from the definition of conditional distribution $p(y|x) := \frac{p(x,y)}{p(x)}$; and (*g*) is due to an axiom of the probability distribution: $\sum_{x\in\mathcal{X}} p(x) = \sum_{y\in\mathcal{Y}} p(y) = 1$.

This non-negativity property has another intuitive implication. Applying the definition of mutual information and then re-arranging the two terms $H(Y)$ and $H(Y|X)$ properly, we get:

$$H(Y) \geq H(Y|X). \tag{1.10}$$

Remember one interpretation of entropy: a measure of uncertainty. So $H(Y)$ can be viewed as the uncertainty of Y, while $H(Y|X)$ being interpreted as the residual uncertainty in Y after X being revealed. Our intuition then says: Given side information like X that is given as conditioning, we know more about Y (the uncertainty is removed further) and therefore, such conditional entropy must be reduced. In short, *conditioning reduces entropy*. The above property proves this intuition.

Some curious readers may want to ask: What if X is realized as a certain value $X = x$? In such a case, does the particular form of conditioning still reduce entropy:

$$H(Y) \geq H(Y|X = x)? \tag{1.11}$$

Please think about it while solving Prob 1.9.

Proof of $I(X;Y) = 0 \Longleftrightarrow X \perp\!\!\!\perp Y$ To prove this, first recall one procedure that we had in the process of proving the second property:

$$\begin{aligned} I(X;Y) &:= H(Y) - H(Y|X) \\ &= \mathbb{E}\left[-\log\frac{p(Y)}{p(Y|X)}\right] \\ &\geq -\log\mathbb{E}\left[\frac{p(Y)}{p(Y|X)}\right]. \end{aligned}$$

Remember that the last inequality is due to Jensen's inequality. As you will figure out while solving Prob 1.5, the sufficient and necessary condition for the equality to hold in the above is:

$$\frac{p(Y)}{p(Y|X)} = c \text{ (constant)}.$$

The condition then implies that

$$p(y) = cp(y|x) \quad \forall x \in \mathcal{X}, \forall y \in \mathcal{Y}.$$

Using the axiom of probability distribution (the sum of the probabilities being 1), we get $c = 1$ and therefore:

$$p(y) = p(y|x) \quad \forall x \in \mathcal{X}, \forall y \in \mathcal{Y}.$$

Due to the definition of independence between two random variables, the above implies that X and Y are independent. Hence, this completes the proof:

$$I(X;Y) = 0 \Longleftrightarrow X \perp\!\!\!\perp Y.$$

Interpretation on $I(X;Y)$ Let us say a few words about $I(X;Y)$. Using the chain rule and the definitions of entropy and joint entropy, one can rewrite $I(X;Y) := H(Y) - H(Y|X)$ as

$$\begin{aligned} I(X;Y) &= H(Y) + H(X) - H(X,Y) \\ &= \mathbb{E}\left[\log \frac{1}{p(Y)}\right] + \mathbb{E}\left[\log \frac{1}{p(X)}\right] - \mathbb{E}\left[\log \frac{1}{p(X,Y)}\right] \\ &= \mathbb{E}\left[\log \frac{p(X,Y)}{p(X)p(Y)}\right]. \end{aligned} \tag{1.12}$$

This leads to the following observation:

$$\begin{aligned} p(X,Y) \text{ close to } p(X)p(Y) &\Longrightarrow I(X;Y) \approx 0; \\ p(X,Y) \text{ far from } p(X)p(Y) &\Longrightarrow I(X;Y) \text{ far above } 0. \end{aligned}$$

This enables us to interpret mutual information as a sort of *distance measure* that captures how far the joint distribution $p(X,Y)$ is from the product distribution $p(X)p(Y)$. In statistics, there is a well-known divergence measure that reflects a distance between two distributions. That is, KL divergence. So mutual information can be represented as the KL divergence. Before detailing the representation, let us first introduce the definition of the KL divergence.[5]

5. The classic book "Elements of Information Theory" by Cover and Thomas (Cover, 1999) employs a different naming for the KL divergence. That is, relative entropy. This naming is popular yet mainly in the information theory literature. It is not the case in other societies; the naming of the KL divergence is more prevalent. Hence, we have chosen the naming of the KL divergence in this book.

Definition of the KL divergence Let $Z \in \mathcal{Z}$ be a discrete random variable. Consider two probability distributions w.r.t. Z: $p(z)$ and $q(z)$ where $z \in \mathcal{Z}$. The KL divergence between the two distributions are defined as:

$$\begin{aligned}\mathsf{KL}(p\|q) &:= \sum_{z\in\mathcal{Z}} p(z)\log\frac{p(z)}{q(z)} \\ &= \mathbb{E}_{p(Z)}\left[\log\frac{p(Z)}{q(Z)}\right].\end{aligned} \tag{1.13}$$

Mutual information in terms of the KL divergence Applying the definition (1.13) to (1.12), we obtain:

$$\begin{aligned}I(X;Y) &= \mathbb{E}\left[\log\frac{p(X,Y)}{p(X)p(Y)}\right] \\ &= \mathbb{E}_{p(X,Y)}\left[\log\frac{p(X,Y)}{p(X)p(Y)}\right] \\ &\overset{(a)}{=} \mathbb{E}_{p(Z)}\left[\log\frac{p(Z)}{q(Z)}\right] \\ &\overset{(b)}{=} \mathsf{KL}(p(Z)\|q(Z)) \\ &= \mathsf{KL}(p(X,Y)\|p(X)p(Y))\end{aligned} \tag{1.14}$$

where (a) comes from our own definition: $Z := (X, Y)$ (note that $p(x)p(y)$ is a valid probability distribution. Why?); and (b) is because of the definition of the KL divergence.

Properties of the KL divergence As mutual information has the three properties, the KL divergence has three similar properties:

Property 1: $\mathsf{KL}(p\|q) \neq \mathsf{KL}(q\|p)$;

Property 2: $\mathsf{KL}(p\|q) \geq 0$;

Property 3: $\mathsf{KL}(p\|q) = 0 \iff p = q$.

The first property of the KL divergence is different from mutual information in that it is not symmetric. The definition of the KL divergence in (1.13) only takes the expectation over the first probability distribution p, which breaks symmetry. The second and third properties, on the other hand, are similar to those of mutual information and their proofs are also similar. Please refer to Prob 1.12 for more details.

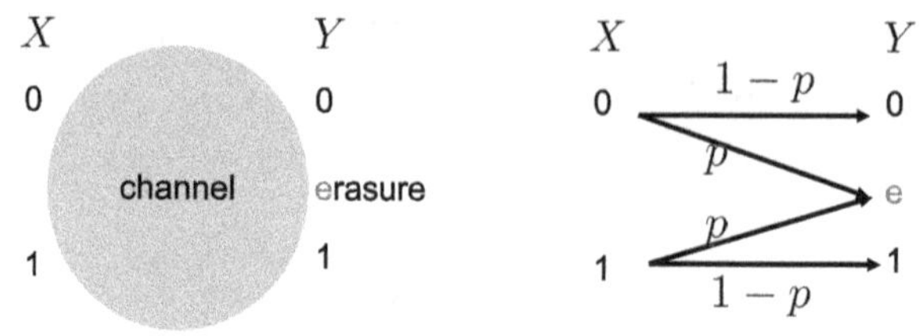

(*a*) binary erasure channel (BEC) (*b*) erasure probability p

Figure 1.9. Binary erasure channel.

Connection between mutual information and channel capacity We establish a connection between mutual information and the channel coding theorem. To illustrate this connection, we consider a concrete exemplary channel called the binary erasure channel (BEC). The BEC is the first toy-example channel that Shannon proposed. It has an input, X, which is binary and takes values 0 or 1. The output, Y, is ternary and takes values 0, 1, or an erasure symbol (denoted by e). As discussed in Section 1.1, the channel introduces uncertainty in the form of noise, which can be characterized by the conditional distribution $p(y|x)$. In the BEC, the output is the same as the input with probability $1-p$, otherwise, it takes an erasure symbol regardless of the value of x. The conditional distribution is given by:

$$p(y|x) = \begin{cases} 1-p, & \text{for } (x,y) = (0,0); \\ p, & \text{for } (x,y) = (0,\mathsf{e}); \\ p, & \text{for } (x,y) = (1,\mathsf{e}); \\ 1-p, & \text{for } (x,y) = (1,1); \\ 0, & \text{otherwise.} \end{cases} \tag{1.15}$$

A pictorial description of the BEC is in Fig. 1.9(*b*). A value placed above each arrow indicates the transition probability for a transition reflected by the arrow.

To see the connection, let us compute mutual information between the input and the output.

$$I(X;Y) = H(Y) - H(Y|X).$$

As you can see, it requires a computation of the entropy $H(Y)$ of a ternary random variable Y. It turns out that $H(Y)$ is a bit complicated to compute, while a simpler calculation comes from an alternative expression:

$$I(X;Y) = H(X) - H(X|Y). \tag{1.16}$$

In order to compute $H(X)$, we need to know about $p(x)$. However, $p(x)$ is not given. So let us make a simple assumption: X is uniformly distributed, i.e., (in other words), $X \sim \mathsf{Bern}(\frac{1}{2})$. Here "$\sim$" refers to "is distributed according to"; Bern denotes the distribution of a binary (or Bernoulli[6]) random variable; and the value inside $\mathsf{Bern}(\cdot)$ indicates the probability that the variable takes 1, simply called the Bernoulli parameter. Assuming $X \sim \mathsf{Bern}(\frac{1}{2})$, the entropy of X is $H(X) = 1$ and the conditional entropy $H(X|Y)$ is calculated as:

$$\begin{aligned} H(X|Y) &\overset{(a)}{=} \mathbb{P}(Y = \mathsf{e})H(X|Y = \mathsf{e}) + \mathbb{P}(Y \neq \mathsf{e})H(X|Y \neq \mathsf{e}) \\ &\overset{(b)}{=} \mathbb{P}(Y = \mathsf{e})H(X|Y = \mathsf{e}) \\ &\overset{(c)}{=} p \end{aligned}$$

where (a) is due to the definition of conditional entropy; (b) follows from the fact that $Y \neq \mathsf{e}$ completely determines X (no randomness) and therefore $H(X|Y \neq \mathsf{e}) = 0$; and (c) follows from the fact that $Y = \mathsf{e}$ does not provide any information about X and hence $X|Y = \mathsf{e}$ has the same distribution as X, so $H(X|Y = \mathsf{e}) = H(X) = 1$. Applying this to (1.16), we get:

$$I(X;Y) = H(X) - H(X|Y) = 1 - p.$$

This is where we can see the connection between mutual information and channel capacity C: the maximum number of bits that can be transmitted over a channel. It turns out that $I(X;Y) = 1 - p$ is the capacity of the BEC. Remember that we assume the distribution of X in computing $I(X;Y)$. For a general channel indicated by an arbitrary $p(y|x)$, such $p(x)$ serves as an optimization variable and the channel capacity is characterized as:

$$C = \max_{p(x)} I(X;Y). \tag{1.17}$$

This is the statement of the channel coding theorem. We see that mutual information indeed characterizes the channel capacity. Later in Part II, we will prove the theorem.

Python exercise Finally we explore how to compute mutual information and the KL divergence via Python. As per the definition of mutual information together

---

6. The binary random variable is named after Jacob Bernoulli, a Swiss mathematician from the 1600s who used this simple random variable to discover one of the foundational laws in mathematics, the *Law of Large Numbers* (LLN). Hence, the binary random variable is commonly referred to as the Bernoulli random variable. Later, we will have an opportunity to explore the LLN in more detail.

with the way to compute entropy and conditional entropy (that we learned in Section 1.2), one can calculate mutual information. Let us do some exercise with the same example introduced in the previous section: $p(x,y) = \frac{1}{4}, \frac{1}{4}, \frac{1}{3}, \frac{1}{6}$ for $(x,y) = (0,0), (0,1), (1,0), (1,1)$. First we compute $I(X;Y) = H(Y) - H(Y|X)$.

```
import numpy as np
from scipy.stats import entropy

# Compute p(y)
pY = np.array([1/4+1/3, 1/4+1/6])
# Compute H(Y)
HY = entropy(pY, base=2)

# Compute p(x)
pX = np.array([1/4+1/4, 1/3+1/6])
# Compute p(y|0)
pY_x0 = np.array([1/4, 1/4])/pX[0]
# Compute p(y|1)
pY_x1 = np.array([1/3, 1/6])/pX[1]
# Compute H(Y|X)=\sum p(x)*H(Y|X=x)
HY_X = pX[0]*entropy(pY_x0,base=2) \
      + pX[1]*entropy(pY_x1,base=2)

# Compute I(X;Y)=H(Y)-H(Y|X)
IXY = HY - HY_X
print(IXY)
```

```
0.020720839623907916
```

We also compute $I(Y;X) = H(X) - H(X|Y)$ to do sanity check for the symmetry property.

```
# Compute p(x)
pX = np.array([1/4+1/4, 1/3+1/6])
# Compute H(X)
HX = entropy(pX, base=2)

# Compute p(y)
pY = np.array([1/4+1/3, 1/4+1/6])
# Compute p(x|0)
pX_y0 = np.array([1/4, 1/3])/pY[0]
# Compute p(x|1)
pX_y1 = np.array([1/4, 1/6])/pY[1]
```

```
# Compute H(X|Y)=\sum p(y)*H(X|Y=y)
HX_Y = pY[0]*entropy(pX_y0,base=2) \
        + pY[1]*entropy(pX_y1,base=2)

# Compute I(Y;X)=H(X)-H(X|Y)
IYX = HX - HX_Y
print(IYX)
```

```
0.02072083962390825
```

Up to a numerical precision error on a computer, we observe that $I(X;Y)$ is equivalent to $I(Y;X)$.

Using the definition (1.13) of the KL divergence, one can implement it from scratch.

```
def kl(p,q):
    return sum(p*np.log2(p/q))
```

We employ this function to verify the relationship (1.14) between mutual information and the KL divergence.

```
# Compute p(x,y)
pXY = np.array([1/4, 1/4, 1/3, 1/6])
# Compute p(x)p(y)
pXpY = np.array([pX[0]*pY[0],pX[0]*pY[1],
                 pX[1]*pY[0],pX[1]*pY[1]])
# Compute KL(pXY||pXpY)
print(kl(pXY,pXpY))
```

```
0.020720839623908215
```

Below we check that the symmetry property does not hold for the KL divergence.

```
print(kl(pXY,pXpY))
print(kl(pXpY,pXY))
```

```
0.020720839623908215
0.020945827042758484
```

For computation of the KL divergence, one can alternatively employ a built-in function rel_entr provided in the scipy.special package. The rel_entr employs the natural logarithm instead of log base 2. It returns a list of all the associated values $p(x)\ln\frac{p(x)}{q(x)}$. Hence, we need a proper conversion.

```
from scipy.special import rel_entr

kl_builtin = rel_entr(pXY,pXpY)
print(sum(kl_builtin))
# To convert into log base 2
```

```
print(sum(kl_builtin)/np.log(2))
print(kl(pXY,pXpY))
```

```
0.014362591564146779
0.020720839623908218
0.020720839623908215
```

Look ahead We have completed our exploration of the key concepts in information theory. Moving forward to the next section, we will begin the process of proving the source coding theorem.

Problem Set 1

Prob 1.1 (Bits) In Section 1.1, we learned that *bits* is a common currency of information that can represent any type of information source. Consider an image signal that consists of many pixels. Each pixel is represented by three real values which indicate the intensity of red, green and blue colors respectively. We assume that each value is quantized, taking one of 256 equal-spaced values in $[0, 1)$, i.e., $\frac{0}{256}, \frac{1}{256}, \frac{2}{256}, \ldots, \frac{254}{256}, \frac{255}{256}$. Explain how bits can represent such an image source.

Prob 1.2 (Digital communication architecture) Draw the digital communication architecture that Shannon came up with. Also, point out the digital interface.

Prob 1.3 (Source coding example) Let S be a discrete random variable with the probability distribution:

$$S = \begin{cases} 1, & \text{w.p. } 0.4; \\ 2, & \text{w.p. } 0.2; \\ 3, & \text{w.p. } 0.2; \\ 4, & \text{w.p. } 0.1; \\ 5, & \text{w.p. } 0.1. \end{cases}$$

Consider a source code that maps $S \in \{1, 2, \ldots, 5\}$ to codeword $f(S)$.

(*a*) Calculate the entropy of the random variable S.
(*b*) Using the binary code tree that we learned in Section 1.1, construct a source code that minimizes the expected codeword length $\mathbb{E}\left[\text{length}(f(S))\right]$.
(*c*) Compare the expected codeword length of your code with $H(S)$. Which one is smaller? Also explain why.

Prob 1.4 (Channel coding example) Consider a binary erasure channel in which input $X \in \{0, 1\}$, output $Y \in \{0, 1, \textsf{erasure}\}$, and $Y = X$ with probability $1-p$ and $Y = \textsf{erasure}$ with probability p where $p \in [0, 1]$. An information theorist claims that the capacity of the erasure channel is a sole function of p, regardless of transmission and reception strategies. Is the claim true? Also explain why.

Prob 1.5 (Jensen's inequality) Suppose that a function f is concave and X is a discrete random variable. Show that

$$\mathbb{E}[f(X)] \leq f(\mathbb{E}[X]).$$

Also identify conditions under which the equality holds.

Prob 1.6 (An empirical estimate of entropy) The table below shows the frequency of letter usage in a particular sample of an English text. Suppose that the text sample is sufficiently large such that the frequencies are precise enough. Then, one natural way to estimate the probability mass function (pmf) of English letter is to employ such frequencies:

$$\mathbb{P}(\text{English letter} = \text{A}) \approx \frac{\#\text{ of A's in the text}}{\text{total }\#\text{ of letters in the text}} \approx 0.0817.$$

Please see below for the estimates of other letters. Using Python, compute the entropy of English letter with these estimates.

A	8.17 %	H	6.09%	O	7.51 %	V	0.98%
B	1.49 %	I	6.97%	P	1.93 %	W	2.36%
C	2.78 %	J	0.15%	Q	0.10 %	X	0.15%
D	4.25 %	K	0.77%	R	5.99 %	Y	1.97%
E	12.7 %	L	4.03%	S	6.33 %	Z	0.07%
F	2.23 %	M	2.41%	T	9.06 %		
G	2.02 %	N	6.75%	U	2.76 %		

Prob 1.7 (Joint entropy) Suppose X and Y are binary random variables with $\mathbb{P}(X = 0) = 0.2$ and $\mathbb{P}(Y = 0) = 0.4$. A student claims that the joint distribution that maximizes joint entropy $H(X, Y)$ is:

$$\mathbb{P}(X = 0, Y = 0) = 0.08;$$
$$\mathbb{P}(X = 0, Y = 1) = 0.12;$$
$$\mathbb{P}(X = 1, Y = 0) = 0.32;$$
$$\mathbb{P}(X = 1, Y = 1) = 0.48.$$

Prove or disprove it.

Prob 1.8 (Independence) Suppose that two binary random variables X_1 and X_2 satisfy:

$$\mathbb{P}(X_1 = i_1, X_2 = i_2) = \frac{1}{4}$$

for all possible sequence patterns $(i_1, i_2) \in \{(0,0),(0,1),(1,0),(1,1)\}$. Show that X_1 and X_2 are independent and identically distributed (i.i.d.), each taking 0 or 1 with probability $\frac{1}{2}$. Also compute $H(X_1, X_2)$ and $H(X_1) + H(X_2)$.

Prob 1.9 (Conditional entropy) Let X and Y be discrete random variables.

(*a*) Show that

$$H(Y) \geq H(Y|X).$$

Do you think that the above inequality make an intuitive sense? If so, explain why.

(*b*) A curious student claims that for any $x \in \mathcal{X}$

$$H(Y) \geq H(Y|X = x).$$

Either prove or disprove it.

Prob 1.10 (Chain rule & conditional entropy) Let $\{X_i\}$ be a discrete random process with a joint distribution $p(x_1, x_2, \ldots, x_n)$. In view of the entropy (defined w.r.t. a single random variable) and the joint entropy (defined w.r.t. two random variables), a natural way to define the entropy for a *random process* is as follows:

$$H(X_1, X_2, \ldots, X_n) := \mathbb{E}\left[\log \frac{1}{p(X_1, X_2, \ldots, X_n)}\right].$$

(*a*) Derive the chain rule for the random process:

$$\begin{aligned} &H(X_1, X_2, \ldots, X_n) \\ &\quad = H(X_1) + H(X_2|X_1) + \cdots + H(X_n|X_1, X_2, \ldots, X_{n-1}). \end{aligned}$$

(*b*) Show that when X_1 and X_2 are independent,

$$H(X_2|X_1) = H(X_2).$$

Considering the interpretation for conditional entropy that we learned in Section 1.2, this result makes a perfect sense. When X_1 has nothing to do with X_2 (statistically speaking, being independent), the uncertainty of X_2 remains the same whether X_1 is revealed.

(*c*) Let $H(X_2|X_1 = x_1) := \mathbb{E}_{p(x_2|x_1)}\left[\log \frac{1}{p(X_2|X_1=x_1)}\right]$. Show that

$$H(X_2|X_1) = \sum_{x_1 \in \mathcal{X}_1} p(x_1) H(X_2|X_1 = x_1)$$

where $\mathcal{X}_1$ indicates the range of X_1.

Prob 1.11 (Mutual information) Recall the mutual information that we defined in Section 1.3:

$$I(X;Y) := H(Y) - H(Y|X)$$

where X and Y denote discrete random variables. One interpretation that we made is that mutual information captures *common* information between the two random variables involved, as it represents the overlapped area in the Venn diagram. Consider another discrete random variable Z.

(*a*) A curious student claims that if X and Y are independent, so is it even when Z is given:

$$I(X;Y) = 0 \Longrightarrow I(X;Y|Z) = 0$$

where $I(X;Y|Z) := H(Y|Z) - H(Y|Z,X)$. Either prove or disprove it.

(*b*) A creative student wishes to capture common information across *three* random variables. To this end, the student defines triple mutual information as follows:

$$I(X;Y;Z) := I(X;Y) - I(X;Y|Z).$$

And then she/he interprets $I(X;Y|Z)$ as the overlapped area between two parts, each being reflected in $H(X|Z)$ and $H(Y|Z)$ respectively. With this interpretation, the student feels happy about her/his definition because in that way $I(X;Y;Z)$ indeed indicates the overlapped area between three circles (each reflected in $H(X)$, $H(Y)$, $H(Z)$). With the faith that the area must be non-negative, the student claims:

$$I(X;Y;Z) \geq 0,$$

as in the conventional case $I(X;Y) \geq 0$. Either prove or disprove it.

Prob 1.12 (Kullback-Leibler divergence) Let p and q be two distributions on $\mathcal{X}$. Let

$$\mathsf{KL}(p\|q) := \mathbb{E}_p\left[\log\frac{p(X)}{q(X)}\right].$$

(*a*) Either prove or disprove that $\mathsf{KL}(p\|q) = \mathsf{KL}(q\|p)$.

(*b*) Show that $\mathsf{KL}(p\|q) \geq 0$.

(*c*) Show that the equality in the above inequality in part (*b*) holds if and only if $p = q$.

Prob 1.13 (Mutual information vs. KL divergence) In Section 1.3, we learned about one specific yet insightful expression that connects mutual information to the KL divergence:

$$I(X;Y) = \mathsf{KL}(\mathbb{P}_{X,Y} \| \mathbb{P}_X \mathbb{P}_Y)$$

where $\mathbb{P}_X(x)$ and $\mathbb{P}_Y(y)$ indicate probability distributions of discrete random variables $X \in \mathcal{X}$ and $Y \in \mathcal{Y}$, respectively; and $\mathbb{P}_{X,Y}(x,y)$ denotes the joint distribution. There is another insightful expression that relates mutual information to the KL divergence. In this problem, you are asked to establish the expression. The expression gives insights into GANs (Goodfellow *et al.*, 2014) and fair classifiers that we will study in Part III.

(*a*) Let $\mathbb{P}_{Y|X=x}(y)$ be the conditional distribution of Y given $X = x$. Show that

$$I(X;Y) = \sum_{x \in \mathcal{X}} \mathbb{P}_X(x) \mathsf{KL}(\mathbb{P}_{Y|X=x} \| \mathbb{P}_Y).$$

(*b*) Suppose $X \sim \mathsf{Bern}(\frac{1}{2})$ and

$$Y = \begin{cases} Y_{\mathsf{real}}, & \text{if } X = 1; \\ Y_{\mathsf{fake}}, & \text{if } X = 0 \end{cases}$$

where $Y_{\mathsf{real}} \in \mathcal{Y}$ and $Y_{\mathsf{fake}} \in \mathcal{Y}$ denote other discrete random variables. Show that

$$I(X;Y) = \mathsf{JS}(\mathbb{P}_{Y_{\mathsf{real}}} \| \mathbb{P}_{Y_{\mathsf{fake}}})$$

where $\mathsf{JS}(\mathbb{P}_{Y_{\mathsf{real}}} \| \mathbb{P}_{Y_{\mathsf{fake}}})$ is the Jensen-Shannon divergence (another well-known divergence measure in information theory and statistics) defined as:

$$\begin{aligned} \mathsf{JS}(\mathbb{P}_{Y_{\mathsf{real}}} \| \mathbb{P}_{Y_{\mathsf{fake}}}) := & \frac{1}{2}\mathsf{KL}\left(\mathbb{P}_{Y_{\mathsf{real}}} \Big\| \frac{\mathbb{P}_{Y_{\mathsf{real}}} + \mathbb{P}_{Y_{\mathsf{fake}}}}{2}\right) \\ & + \frac{1}{2}\mathsf{KL}\left(\mathbb{P}_{Y_{\mathsf{fake}}} \Big\| \frac{\mathbb{P}_{Y_{\mathsf{real}}} + \mathbb{P}_{Y_{\mathsf{fake}}}}{2}\right). \end{aligned}$$

Remark: Those who are familiar with GANs may be able to see a connection between GANs and mutual information. Otherwise, don't worry. We will elaborate the connection in Part III.

Prob 1.14 (Mutual information expressed in terms of optimization) Suppose $X \sim \mathsf{Bern}(\frac{1}{2})$ and

$$Y = \begin{cases} Y_{\mathsf{real}}, & \text{if } X = 1; \\ Y_{\mathsf{fake}}, & \text{if } X = 0 \end{cases}$$

where $Y_{\text{real}} \in \mathcal{Y}$ and $Y_{\text{fake}} \in \mathcal{Y}$ denote discrete random variables with $\mathbb{P}_{Y_{\text{real}}}(\cdot)$ and $\mathbb{P}_{Y_{\text{fake}}}(\cdot)$, respectively. In Prob 1.13, it was shown that

$$\begin{aligned} I(X;Y) &= \mathsf{JS}(\mathbb{P}_{Y_{\text{real}}} \| \mathbb{P}_{Y_{\text{fake}}}) \\ &= \frac{1}{2}\mathsf{KL}\left(\mathbb{P}_{Y_{\text{real}}} \| \frac{\mathbb{P}_{Y_{\text{real}}} + \mathbb{P}_{Y_{\text{fake}}}}{2}\right) + \frac{1}{2}\mathsf{KL}\left(\mathbb{P}_{Y_{\text{fake}}} \| \frac{\mathbb{P}_{Y_{\text{real}}} + \mathbb{P}_{Y_{\text{fake}}}}{2}\right). \end{aligned}$$

Show that

$$\begin{aligned} I(X;Y) = \max_{D(\cdot)} \frac{1}{2} &\sum_{y_{\text{real}} \in \mathcal{Y}} \mathbb{P}_{Y_{\text{real}}}(y_{\text{real}}) \log D(y_{\text{real}}) \\ &+ \frac{1}{2} \sum_{y_{\text{fake}} \in \mathcal{Y}} \mathbb{P}_{Y_{\text{fake}}}(y_{\text{fake}}) \log(1 - D(y_{\text{fake}})) + H(X). \end{aligned} \tag{1.18}$$

Remark: Those who are familiar with GANs may be able to see a *closer* connection between GANs and mutual information. Don't you see that connection yet? Don't worry. You will see more details later in Part III.

Prob 1.15 (Conditional entropy) Let X and Y be random variables that take values in finite sets $\mathcal{X}$ and $\mathcal{Y}$ respectively. You are given that $H(X) = 11$ and $H(Y|X) = H(X|Y)$. A student claims that $|\mathcal{Y}| \geq 3$. Either prove or disprove the claim.

Prob 1.16 (Chain rule) Let $\{X_i\}$ be a discrete random process. A student claims that

$$H(X_1, \ldots, X_n) \leq \frac{1}{n-1} \sum_{i=1}^{n} H(X_1, \ldots, X_{i-1}, X_{i+1}, \ldots, X_n).$$

Prove or disprove this statement.

Prob 1.17 (A lower bound) Suppose that random variables $X, Y, \bar{X}, \bar{Y}$ take values on the same set, and $\bar{X}$ and $\bar{Y}$ are independent. Let $E = \mathbf{1}\{X \neq Y\}$ and $\bar{E} = \mathbf{1}\{\bar{X} \neq \bar{Y}\}$ where $\mathbf{1}\{\cdot\}$ denotes an indicator function that returns 1 when the event is true; 0 otherwise. Show that

$$I(X;Y) \geq \mathsf{KL}\left(\mathbb{P}_E \| \mathbb{P}_{\bar{E}}\right) - \mathsf{KL}(\mathbb{P}_X \| \mathbb{P}_{\bar{X}}) - \mathsf{KL}(\mathbb{P}_Y \| \mathbb{P}_{\bar{Y}}).$$

Prob 1.18 (Mutual information in the play-off game) The play-off is a five-game series that terminates as soon as either team wins three games. Let X be the random variable that represents the outcome of a play-off between teams A and B; possible values of X are *AAA*, *BABAB* and *BBAAA*. Let Y be the number of games played, which ranges from 3 to 5. Assuming that A and B are equally matched and that the games are independent, calculate $I(X;Y)$.

Prob 1.19 (True or False?)

(*a*) Let X be a discrete random variable taking values in an alphabet $\mathcal{X}$ where $|\mathcal{X}| \geq 8$. The value of X is revealed to Alice, but not to Bob. Bob wishes to figure out the value of X by asking to Alice questions of the following type: if $X = i$, say 0; if $X = j$, say 1; if $X = k$, say 2; otherwise, say 3. Then, the minimum number of questions (on average) required to uncover the value of X is in between $2H(X)$ and $2H(X) + 1$.

(*b*) Consider weathers in Daejeon and Seoul. Assume that Daejeon's weather is sunny w.p. 0.3 and cloudy w.p. 0.7; Seoul's weather is the same as Daejeon's w.p. 0.3, and different w.p. 0.7. Conditioned on Daejeon's weather being cloudy, Seoul's weather is more predictable than that without any information on Daejeon's weather.

(*c*) In Section 1.3, we learned that conditioning reduces entropy: $H(Y|X) \leq H(Y)$ for discrete random variables (X,Y). The same argument holds with regard to mutual information, i.e., conditioning reduces mutual information: $H(Y|Z) - H(Y|Z,X) =: I(X;Y|Z) \leq I(X;Y)$ for any discrete random variable Z.

(*d*) If X and Y are independent random variables, then $I(X;Y) = 0$. However, the converse does not always hold.

(*e*) Consider triplet mutual information defined w.r.t. three random variables: $I(X;Y;Z) := I(X;Y) - I(X;Y|Z)$. As in the conventional case $I(X;Y) = I(Y;X)$, the symmetry holds:

$$I(X;Y;Z) = I(X;Z;Y) = \cdots = I(Z;Y;X).$$

(*f*) Let p and q be distributions. Define:

$$H(p,q) := H(p) + \mathsf{KL}(p\|q)$$

where $H(p)$ indicates the entropy of a random variable having the distribution of p. Then, $H(p,q)$ is convex in q.

(*g*) For any discrete random variables (X,Y,Z),

$$H(X,Y,Z) + H(Y) \leq H(X,Y) + H(Y,Z).$$

(*h*) Suppose two random variables $X \in \mathcal{X}$ and $Y \in \mathcal{Y}$ are independent, $|\mathcal{X}| = |\mathcal{Y}| = 2$ and $\mathcal{X} \cap \mathcal{Y} = \emptyset$. Then,

$$H(X+Y) = H(X) + H(Y).$$

(*i*) Let X be a discrete random variable taking values in an alphabet $\mathcal{X}$ where $|\mathcal{X}| \geq 6$. The value of X is revealed to Alice, but not to Bob. Bob wishes to figure out the value of X by asking to Alice questions of the following type: if $X = i$, say 0; if $X = j$, say 1; otherwise, say 2. Then, the minimum number of questions (on average) required to determine the value of X cannot exceed the entropy $H(X)$.

(*j*) Suppose X and Y are binary random variables with $\mathbb{P}(X = 0) = 0.2$ and $\mathbb{P}(Y = 0) = 0.4$. The joint distribution that maximizes the joint entropy $H(X, Y)$ is:

$$\mathbb{P}(X = 0, Y = 0) = 0.08;$$

$$\mathbb{P}(X = 0, Y = 1) = 0.12;$$

$$\mathbb{P}(X = 1, Y = 0) = 0.32;$$

$$\mathbb{P}(X = 1, Y = 1) = 0.48.$$

(*k*) Let X be a discrete random variable taking values in $\mathcal{X}$. For an one-to-one mapping function f and an arbitrary function g defined on $\mathcal{X}$,

$$H(f(X)|g(X)) = H(X|g(X)).$$

(ℓ) Suppose that X and Y are independent binary random variables. Then, there exists a distribution of $p(x, y)$ such that

$$H(X + Y) = \log 3.$$

1.4 Source Coding Theorem for i.i.d. Sources (1/3)

Recap In the previous sections, we delved into the fundamental concepts of information theory, including Shannon's two-stage architecture that involves splitting the encoder into a source encoder and a channel encoder. The purpose of this architecture is to convert information from different sources into a common currency: bits. Shannon's work resulted in two theorems that describe the efficiency of the two encoder blocks and, in turn, limit the amount of information that can be transmitted over a channel. These theorems are known as the source coding theorem and the channel coding theorem. Having established these key concepts, we are now prepared to prove the theorems. Over the next five sections, including this one, we will focus on proving the source coding theorem.

Outline The source coding theorem quantifies the minimum number of bits needed to represent an information source without losing any information. The information source can be composed of various elements, such as dots and lines (as in Morse code), English text, speech signals, video signals, or image pixels. Therefore, it consists of multiple components. For instance, a text is made up of multiple English letters, and speech signals contain multiple points, each indicating the signal's magnitude at a specific time instant. From the perspective of a receiver who is unaware of the signal, it can be considered a random signal. As a result, the source is modeled as a *random process* comprising random variables. Let $\{X_i\}$ denote the random process, where X_i represents a "symbol" in the source coding literature. To simplify matters, let us begin with a simple scenario in which X_i's are independent and identically distributed (i.i.d.). We denote by X a generic random variable that represents each individual instance of X_i. The source coding theorem in the i.i.d. case is stated as follows:

The minimum number of bits required to represent
the i.i.d. source X_i per symbol is $H(X)$.

In the present section, we will make an effort to prove this theorem. After completing the i.i.d. case, we will extend it to a more realistic non-i.i.d. distribution that the source may follow.

Symbol-by-symbol source encoder Since an information source consists of multiple components, an input to source encoder contains multiple symbols. So the encoder acts on multiple symbols in general. To understand what it means, consider a concrete example where source encoder acts on three consecutive symbols. What it means by acting on multiple symbols is that an output is a function of the three consecutive symbols. But for simplicity, we are going to consider a much simpler case for the time being in which the encoder *acts on each individual symbol, being independent of other symbols*. It means that the encoder produces bits in a *symbol-by-symbol basis*: a symbol X_1 yields a corresponding binary string, and independently another binary string w.r.t. the next symbol X_2 follows, and this goes on for other follow-up symbols. The reason that we consider this simple yet restrictive setting is that this case provides enough insights into a general case. It contains every key insight needed for generalization. Building upon the insights that we will obtain from this simple case, we will later address the general case.

The simple case allows us to simplify notations. First it suffices to focus on one symbol, say X. The encoder is nothing but a function of X. Let us denote the function by C. Please do not be confused with the same notation that we used to indicate channel capacity. The reason that we employ the same notation is that the output $C(X)$ is called codeword. Let $\ell(X)$ be the length of codeword $C(X)$. For example, consider $X \in \{a, b, c, d\}$ in which $C(a) = 0, C(b) = 10, C(c) = 110, C(d) = 111$. In this case, $\ell(a) = 1, \ell(b) = 2, \ell(c) = 3, \ell(d) = 4$. Note that $\ell(X)$ is a function of a random variable X, hence it is also a random variable. So we are interested in a representative quantity of such varying quantity, which is the expected codeword length:

$$\mathbb{E}[\ell(X)] = \sum_{x\in\mathcal{X}} p(x)\ell(x).$$

An optimization problem The efficiency of source encoder is well reflected in the expected codeword length. Hence, we wish to minimize the expected codeword length. The optimization problem of our interest is then:

$$\min_{\ell(x)} \sum_{x\in\mathcal{X}} p(x)\ell(x). \tag{1.19}$$

There are many ways to estimate the distribution $p(x)$ of an information source. See Prob 1.6 for one way. Assume that $p(x)$ is given. In this case, $\ell(x)$'s are only variables that we can optimize over.

Next, consider constraints that the optimization variables $\ell(x)$'s are subject to. The obvious constraints are: $\ell(x) \geq 1$ and $\ell(x) \in \mathbb{N}$. Are these constraints enough? No. If they were enough, the solution to this problem would become trivial. It would be 1. One can set $\ell(x) = 1$ for all x's to obtain 1. But this is too good to be true. In fact, there is another constraint on $\ell(x)$, concerning the condition that a valid code should satisfy.

A naive condition: Non-singularity For the validity of a code, the encoder function must be one-to-one. The reason is that if it is not one-to-one, there is no way to reconstruct the input from the output. In the source coding literature, a code is said to be *non-singular* if it is one-to-one mapping. Mathematically, the non-singularity condition reads:

$$C(x) = C(x') \implies x = x'.$$

Here is an example that respects this condition:

$$C(a) = 0; \quad C(b) = 010; \quad C(c) = 01; \quad C(d) = 10. \tag{1.20}$$

Note that every codeword is distinct, ensuring one-to-one mapping.

Is this non-singularity condition enough to ensure the validity of a code? Unfortunately, no. What we care about is a *sequence* of multiple symbols. What we get in the output is the sequence of binary strings which corresponds to a concatenation of such multiple symbols: $X_1X_2 \cdots X_n \implies C(X_1)C(X_2)\cdots C(X_n)$. Remember we assume the symbol-by-symbol encoder; hence we get $C(X_1)C(X_2)\cdots C(X_n)$ instead of $C(X_1X_2\cdots X_n)$. By non-singularity of the extended code, one should be able to reconstruct the sequence $X_1X_2\cdots X_n$ of input symbols from that output $C(X_1)C(X_2)\cdots C(X_n)$. But in the above example (1.20), there is ambiguity in decoding the sequence of input symbols. Here is a concrete example where one can see the ambiguity. Suppose that the output sequence reads:

output sequence: 010

Then, what are the corresponding input sequence? One possible input would be "*b*" ($C(b) = 010$). But there are some other patterns that yield the same output: "*ca*" ($C(c)C(a) = 010$) and "*ad*" ($C(a)C(d) = 010$). We have multiple candidates that agree upon the same output. This is problematic because we cannot tell which input sequence is fed into. In other words, we cannot uniquely figure out the input sequence.

A stronger condition: Unique decodability What additional condition do we need to satisfy in order to make a code valid? What we need is that for any encoded bit sequence, there must be no decoding ambiguity, in other words, there must be only one matching input sequence. This property is called *unique decodability*. This is equivalent to the one-to-one mapping constraint w.r.t. the *sequence* of source symbols with an arbitrary length. Here is a mathematical expression for unique decodability: for any n and m,

$$C(x_1)C(x_2)\cdots C(x_n) = C(x_1')C(x_2')\cdots C(x_m')$$
$$\implies x_1x_2\cdots x_n = x_1'x_2'\cdots x_m'.$$

A uniquely decodable example Let us give you an example where the unique decodability condition holds:

$$C(a) = 10; \quad C(b) = 00; \quad C(c) = 11; \quad C(d) = 110. \tag{1.21}$$

To verify unique decodability, we can follow these steps. Let's consider an output sequence:

output sequence: $10110101111\cdots$

First we read a binary string until we find a *matching codeword* or a codeword which includes the string in part. In this example, the first read must be 10 because there is only one corresponding codeword: $C(a)$. The corresponding input is "a". What about the next read? An ambiguity arises in the next two bits: 11. We have two possible candidates: (i) a matching codeword $C(c) = 11$; and (ii) another codeword $C(d) = 110$ which includes the string 11 in part. Here the "11" is either from "c" or from "d". It looks like this code is not uniquley decodable. But it is actually uniquely decodable – we can tell which is the correct one. The way to check is to look at the future string. What does this mean? Suppose we see one more bit after "11", i.e., we read 110. Still there is no way to figure out which is correct one. However, suppose we see two more bits after "11", i.e., we read 1101. We can then tell which symbol is fed into. That is, "d". Why? Another possibility "cb" ($C(c)C(b) = 1100$) does not agree upon 1101. So it is eliminated. We repeat this. If the input sequence can be decoded in a unique way using this method, then the code is considered to be uniquely decodable. In fact, one can verify that the above mapping (1.21) ensures unique decodability.[7]

Constraints on $\ell(x)$ induced by uniquely decodable property? Recall our goal: finding constraints on $\ell(x)$ in the optimization problem (1.19). How to translate the unique decodability property into a mathematical constraint in terms of $\ell(x)$'s? The translation is a bit difficult.

Fortunately, we have some positive news to share. The good news is that there exists a simpler and indirect method to determine the constraint. This approach is based on a type of uniquely decodable codes known as *prefix-free codes.* The prefix-free code that we will soon discuss has two key features: firstly, it imposes the same constraint as the one required for uniquely decodable codes (i.e., any uniquely decodable code must adhere to the prefix-free code constraint); secondly, it provides an easier method for identifying the constraint that a valid code must satisfy.

7. There is a rigorous way of checking unique decodability, proposed by Sardinas and Patterson. Please check Problem 5.27 in (Cover, 1999) for details.

The first feature implies that the prefix-free code constraint is both *necessary and sufficient* for a valid uniquely-decodable code. Therefore, it is enough to consider the prefix-free code when determining the constraint imposed by a valid code. Proving the first feature is not a straightforward task, but in Prob 2.3, we provide multiple subproblems that will assist you in proving it with relative ease. Nevertheless, the proof itself is non-trivial.

To comprehend the second feature, we need to understand the prefix-free code in more detail. We will first explain what the code is and also mention a significant advantage it has over non prefix-free codes that are also uniquely decodable.

Prefix-free codes Let us revisit the example (1.21) discussed earlier, which is a case of a uniquely decodable code. However, this example highlights a concern related to decoding complexity, which in turn motivates the use of prefix codes. The issue is that, as we observed in the previous example, decoding the second input symbol necessitates examining a *future* string, which implies that decoding is not immediate. This issue can be further exacerbated, particularly in the worst-case scenario, where the output sequence is as follows:

$$1100000000000000000000000000000000000001.$$

In this case, in order to decode even the first symbol, we have to take a look at many future strings.

In the prefix-free code that we will define soon, there is no such complexity issue. Here is an example of the prefix-free code:

$$C(a) = 0; \quad C(b) = 10; \quad C(c) = 110; \quad C(d) = 111. \tag{1.22}$$

One key property of this code is that *no codeword is a prefix of any other codeword.* This is why the code is named the *prefix-free code.* It is evident that the code in the previous example (1.21) is not prefix-free, despite being uniquely-decodable. One of its codewords serves as a prefix of another codeword, resulting in the need to examine future strings while decoding. In the worst-case scenario, the entire string must be examined to decode even the first input symbol. On the other hand, prefix-free codes like (1.22) do not have any codewords that act as prefixes of other codewords. This eliminates any ambiguity in decoding and avoids any decoding complexity issues, as there is no need to examine future strings to decode an input. The code can be decoded instantaneously, which is why prefix-free codes are also known as instantaneous codes.

Look ahead Remember the optimization problem (1.19) that we discussed earlier. The positive news is that: (i) the restriction on $\ell(x)$ that applies to the prefix-free code is the same as that imposed by the uniquely-decodable code, and (ii) it is easy to recognize the constraint caused by the prefix-free code. In the next section, we will determine the restriction that the prefix-free code property must satisfy. After that, we will approach the optimization problem.

1.5 Source Coding Theorem for i.i.d. Sources (2/3)

Recap In the preceding section, we made an attempt to prove the source coding theorem for i.i.d. sources. To begin with, we concentrated on a basic symbol-by-symbol encoder, where the code operates independently on each symbol, without any regard for the other symbols. Our aim was to create a code C that minimizes the expected codeword length $\mathbb{E}[\ell(X)]$. In order to accomplish this, we framed an optimization problem:

$$\begin{aligned} &\min_{\ell(x)} \sum_{x \in \mathcal{X}} p(x)\ell(x) \\ &\text{subject to some constraints on } \ell(x). \end{aligned} \tag{1.23}$$

The key to solving the problem is to come up with mathematical constraints on $\ell(x)$ that a valid code (fully specified by the unique-decodability property) should respect.

We acknowledged that deriving the constraints on $\ell(x)$ for the optimization problem can be challenging. Therefore, we decided to take a different approach based on the following facts: (1) the constraints on $\ell(x)$ that prefix-free codes (which are a subset of uniquely-decodable codes) satisfy are equivalent to those of uniquely-decodable codes; and (2) obtaining the mathematical constraints on $\ell(x)$ induced by the prefix-free code property is relatively straightforward. We postponed the proof of the first fact to Prob 2.3.

Outline In this section, we are going to derive the constraint due to the prefix-free code property, and will attack the optimization problem (1.23) accordingly.

Review of prefix-free codes We start by reviewing the prefix-free code example introduced in Section 1.4:

$$C(a) = 0; \quad C(b) = 10; \quad C(c) = 110; \quad C(d) = 111. \tag{1.24}$$

No codeword is a prefix of any other codeword. So it is indeed prefix-free.

From codeword to a binary code tree We present a visual depiction of the code that can assist us in determining the mathematical constraints on $\ell(x)$ more easily. This representation is based on the binary code tree, which was introduced earlier. In a binary code tree, each node (either the root or an internal node) has two branches. A one-to-one correspondence exists between a code mapping rule and the representation of a binary code tree.

To draw a binary code tree from a code mapping rule, we begin with the root node and draw two branches that originate from it. We then label the top branch

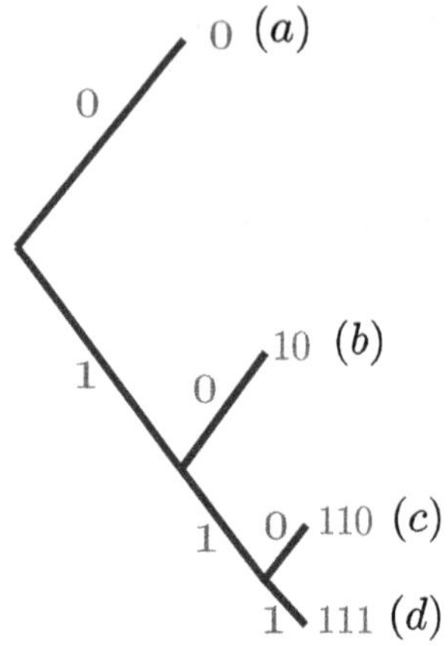

Figure 1.10. The codeword representation via a binary code tree.

with 0 and the bottom branch with 1. We may want to take the other way around: 1 for the top and 0 for the bottom. This is our own choice. We then have two nodes. Following that, we assign a binary label sequence to each node that represents the path from the root to that particular node. Specifically, we assign a label of 0 to the top branch and a label of 1 to the bottom branch. After that, we check if any codeword matches either of the two binary sequences 0 and 1. Since the codeword $C(a)$ matches the sequence 0 assigned to the top node, we label the top node with "*a*". A visual representation of this process can be found in Fig. 1.10.

We follow a similar process for the bottom node. However, since there is no matching codeword, we generate two additional branches from the node and assign a label of 0 on the top and 1 on the bottom. We assign the sequence of binary labels "10" to the top node and "11" to the bottom node. Since the codeword $C(b)$ is identical to "10", we assign "*b*" to the top node. There is no matching codeword for "11", so we split the bottom node into two, assigning "110" to the top and "111" to the bottom. Finally, we assign "*c*" to the top and "*d*" to the bottom.

Representing the code using a binary code tree helps in identifying a mathematical constraint on $\ell(x)$ that a prefix-free code should adhere to. The following two observations are helpful:

Observation #1 The first observation is regarding the location of nodes to which symbols are assigned. A tree consists of two types of nodes. One is an ending node (terminal) which has no further branch. We call that ending node a leaf. The second is a node from which another branch generates. We call it an internal node. Keeping these in our mind, let us take a look at the earlier binary code tree illustrated in Fig. 1.10. Notice that all codewords are assigned to leaves only. In other words, there is no codeword that is assigned to an internal node. Can you see why that is the case? If there were a codeword that is assigned to an internal code, we would violate the prefix-free code property because that codeword is a prefix of some other codeword which lives in the associated leaf. So the first observation that one can make from the prefix-free code is:

Observation #1: Codeword must be a leaf in a binary code tree.

Observation #2 Let us move on to the second observation that serves to relate the prefix-free code property to the mathematical constraint on $\ell(x)$. That is,

Observation #2: Codeword can be mapped to a subinterval in $[0, 1]$.

What does this mean? We can illustrate this by adding a new diagram to the binary code tree shown in Fig. 1.10. In this diagram, we associate an interval $[0, 1]$ with the set of all codewords. A line is drawn through the root of the tree, and the midpoint (0.5) of the associated interval is assigned to the point where the line intersects with the $[0, 1]$ interval. We then assign a codeword to the subinterval $[0, 0.5]$, since there is only one codeword above the root level. However, below the root level, there are multiple codewords. To address this, we draw another line on the central interior node in the bottom level, and assign the midpoint (0.75) of the $[0.5, 1]$ interval to the point where the line intersects with the $[0.5, 1]$ interval. We then assign the codeword $C(b)$ to the subinterval $[0.5, 0.75]$. We continue this process until all codewords are assigned to subintervals, resulting in the diagram shown in Fig. 1.11. As a result, each codeword is mapped to a unique subinterval of $[0, 1]$:

$$
\begin{aligned}
C(a) &\leftrightarrow [0, 0.5];\\
C(b) &\leftrightarrow [0.5, 0.75];\\
C(c) &\leftrightarrow [0.75, 0.875];\\
C(d) &\leftrightarrow [0.875, 1].
\end{aligned}
$$

This naturally leads to the following two facts: (1) subinterval size $= 2^{-\ell(x)}$; and (2) there is no overlap between subintervals. The second fact comes from the first observation: an interior node cannot be a codeword. This yields the following constraint:

$$\sum_{x \in \mathcal{X}} 2^{-\ell(x)} \leq 1. \tag{1.25}$$

This is called *Kraft's inequality.*

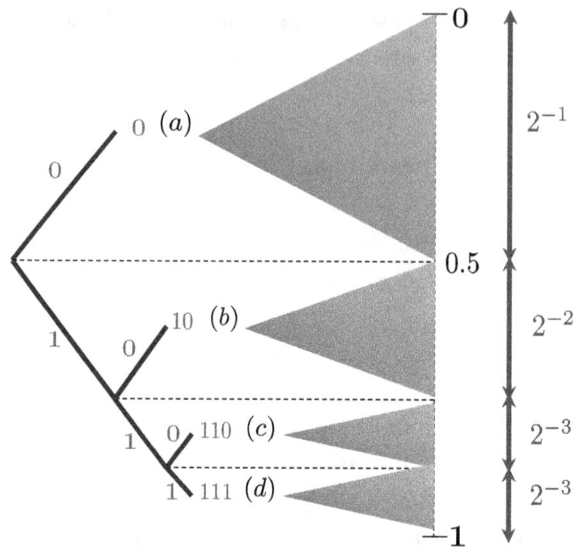

Figure 1.11. Observation #2: Any codeword can be mapped to a subinterval in [0, 1].

An optimization problem Using Kraft's inequality, we can formulate the optimization problem as:

$$\min_{\ell(x)} \sum_{x \in \mathcal{X}} p(x)\ell(x)$$
$$\text{subject to} \sum_{x \in \mathcal{X}} 2^{-\ell(x)} \leq 1, \quad \ell(x) \in \mathbb{N}, \quad \ell(x) \geq 1.$$

One can ignore the constraint $\ell(x) \geq 1$. Why? Otherwise, the Kraft's inequality $\sum_{x \in \mathcal{X}} 2^{-\ell(x)} \leq 1$ is violated. So the simplified problem reads:

$$\begin{aligned} &\min_{\ell(x)} \sum_{x \in \mathcal{X}} p(x)\ell(x) \\ &\text{subject to} \sum_{x \in \mathcal{X}} 2^{-\ell(x)} \leq 1, \quad \ell(x) \in \mathbb{N}. \end{aligned} \tag{1.26}$$

Non-convex optimization The optimization problem (1.26) is widely known for its difficulty in finding a solution. It falls under a category of optimization problems that are generally considered challenging. To see this, remember one definition that we introduced in Section 1.2. That is, concave functions. We say that a function is concave if $\forall x_1, x_2$ and $\lambda \in [0, 1]$, $\lambda f(x_1) + (1 - \lambda)f(x_2) \leq f(\lambda x_1 + (1 - \lambda)x_2)$. There is another type of functions which are defined in a similar yet opposite manner: *convex* functions. We say that a function f is convex if $-f$

is concave, i.e., $\forall x_1, x_2$ and $\lambda \in [0, 1]$,

$$\lambda f(x_1) + (1-\lambda) f(x_2) \geq f(\lambda x_1 + (1-\lambda) x_2). \tag{1.27}$$

The inequality has an opposite direction compared to the one used in defining concave functions.

If we consider the optimization problem (1.26) with the concept of convex functions, we can observe that the objective function is convex in $\ell(x)$. Additionally, the left-hand side of the inequality constraint (which can be transformed so that the right-hand side is 0) is $\sum_{x \in \mathcal{X}} 2^{-\ell(x)} - 1$, which is also convex in $\ell(x)$. An optimization problem is deemed convex when both the objective function and the left-hand sides of the constraints are convex in the variables (Boyd and Vandenberghe, 2004). Conversely, if an optimization problem contains any non-convex objective function and/or non-convex functions in the inequalities, it is considered to be non-convex.

In (1.26), the objective function and the function in the inequality constraint are both convex. However, when considering the integer constraint $\ell(x) \in \mathbb{N}$, we must take into account the definition of convexity with respect to a set. A set is considered convex if any linear combination of two points in the set is also in the set. If not, the set is non-convex. In this case, $\mathbb{N}$ is a non-convex set. To see this, consider two integer points, 1 and 2, and one linear combination of them, 1.5. Clearly, 1.5 is not an integer. As a result, the integer constraint $\ell(x) \in \mathbb{N}$ is non-convex, making the optimization problem non-convex as well. Problems with non-convex constraints, particularly integer constraints, are notoriously difficult to solve. As a result, the optimization problem at hand is extremely challenging and remains unsolved to date.

Approximate! To tackle this challenge, Shannon took a different approach. Recognizing the difficulty of the problem, he aimed to provide insight into the solution rather than solving it exactly. He proposed that while the problem is difficult, it may be possible to find an approximate solution that is similar to the exact solution. Therefore, his objective was to create upper and lower bounds on the solution that are sufficiently close to each other. In other words, Shannon attempted to approximate the solution of the problem rather than solving it exactly.

A lower bound The goal of the optimization problem presented in (1.26) is to minimize the objective function. Therefore, one would expect that a larger search space for $\ell(x)$ would lead to a smaller or equal solution compared to the exact solution in the original problem. Shannon was motivated by this insight to expand the search space in order to obtain a lower bound. One way to achieve this is by removing a constraint. Unsurprisingly, Shannon chose to remove the integer constraint, which includes a non-convex set, and is the main reason that makes the problem challenging. Here is the relaxed version of the problem that Shannon formulated:

$$\begin{aligned} \underline{\mathcal{L}} := \min_{\ell(x)} & \sum_{x\in\mathcal{X}} p(x)\ell(x) \\ \text{subject to} & \sum_{x\in\mathcal{X}} 2^{-\ell(x)} \le 1. \end{aligned} \tag{1.28}$$

Look ahead The optimization problem we have presented aims to minimize the expected length of codewords, taking into account Kraft's inequality and the integer constraint on $\ell(x)$ in the original formulation. However, we relaxed the integer constraint, which allowed us to convert the non-convex problem into a manageable convex optimization problem. In the following section, we will solve the convex optimization problem to obtain a lower bound. Subsequently, we will derive an upper bound and use both bounds to prove the source coding theorem.

1.6 Source Coding Theorem for i.i.d. Sources (3/3)

Recap In the previous section, we formulated an optimization problem that aims to minimize the expected codeword length while satisfying both Kraft's inequality and the integer constraint on $\ell(x)$. However, we encountered a challenge as the problem is non-convex and generally intractable. To make progress, we followed Shannon's approach of approximation by deriving lower and upper bounds that are as close as possible. To obtain a lower bound, we employed a trick of relaxing constraints and expanding the search space, which involved removing the non-convex integer constraint. This allowed us to convert the problem into a tractable convex optimization problem:

$$\begin{aligned} \underline{\mathcal{L}} := \min_{\ell} & \sum_{x\in\mathcal{X}} p(x)\ell(x) \\ \text{subject to} & \sum_{x\in\mathcal{X}} 2^{-\ell(x)} \leq 1. \end{aligned} \tag{1.29}$$

Outline In this section, we will derive the lower bound. We will also derive an upper bound by introducing another trick. Based on the lower and upper bounds, we will finally complete the proof of the source coding theorem.

A lower bound Convex optimization problems, where both the objective function and constraint functions are convex, have been extensively studied and there exist numerous methods to solve them. One such method is the *Lagrange multiplier method*, which is commonly taught in Calculus courses (Stewart, 2015; Boyd and Vandenberghe, 2004). The idea behind this method is to introduce a new variable called the Lagrange multiplier, denoted by λ, and define a Lagrange function that involves both the optimization variables $\ell(x)$'s and the Lagrange multiplier:

$$\mathcal{L}(\ell(x),\lambda) = \sum_{x\in\mathcal{X}} p(x)\ell(x) + \lambda\left(\sum_{x\in\mathcal{X}} 2^{-\ell(x)} - 1\right).$$

In the canonical form, the number of Lagrange multipliers equals the number of constraints. In this case, we only have one constraint, so we have one Lagrange multiplier. The Lagrange function consists of two parts: (i) the objective function, and (ii) the product of the Lagrange multiplier and the left-hand side of the inequality constraint.[8]

8. In the canonical form of inequality constraints, the right-hand-side reads 0 and the inequality direction is $\leq$.

How does the Lagrange multiplier method work? We take a derivative of the Lagrange function w.r.t. optimization variables $\ell(x)$'s. We also take a derivative w.r.t. the Lagrange multiplier λ. Setting these derivatives to zero, we get:

$$\frac{\mathcal{L}(\ell(x), \lambda)}{d\ell(x)} = 0;$$
$$\frac{\mathcal{L}(\ell(x), \lambda)}{d\lambda} = 0.$$

It has been found that the solution can be obtained by solving these equations[9] under the constraint of $\lambda \geq 0$. However, we will not use this method for two reasons. Firstly, this method is somewhat complicated and messy. Secondly, it is not quite intuitive as to why it should work, as there is a deep underlying theorem called the strong duality theorem (Boyd and Vandenberghe, 2004) which proves that this approach leads to the optimal solution under convexity constraints. Since we will not deal with the proof of this theorem, it is reasonable not to take an approach that relies on it. In Prob 2.4, you will have an opportunity to use the Lagrange multiplier method to solve the problem.

Rather than following the conventional approach, we will adopt a much simpler and intuitive alternative. Before delving into the specifics of the method, let us streamline the optimization problem even further. It is worth noting that we can disregard situations where the strict inequality is satisfied: $\sum_{x \in \mathcal{X}} 2^{-\ell(x)} < 1$. Suppose there exists an optimal solution, say $\ell^*(x)$, such that the strict inequality holds: $\sum_{x \in \mathcal{X}} 2^{-\ell^*(x)} < 1$. Then, one can always come up with a better solution, say $\ell'(x)$, such that: for some x_0,

$$\ell'(x_0) < \ell^*(x_0);$$
$$\ell'(x) = \ell^*(x) \quad \forall x \neq x_0;$$
$$\sum_{x \in \mathcal{X}} 2^{-\ell'(x)} = 1.$$

We reduced $\ell^*(x_0)$ a bit for one particular symbol x_0, in an effort to increase $2^{-\ell^*(x_0)}$ so that we achieve $\sum 2^{-\ell'(x)} = 1$. This is indeed a better solution as it yields a smaller objective solution due to $\ell'(x_0) < \ell^*(x_0)$. This is contradiction, implying that an optimal solution occurs only when the equality constraint holds. Hence, it suffices to consider the equality constraint. The optimization can then be

9. These are called the KKT conditions in the optimization literature (Karush, 1939; Kuhn and Tucker, 2014).

simplified as:

$$\begin{aligned} \underline{\mathcal{L}} := \min_{\ell(x)} & \sum_{x\in\mathcal{X}} p(x)\ell(x) \\ \text{subject to } & \sum_{x\in\mathcal{X}} 2^{-\ell(x)} = 1. \end{aligned} \tag{1.30}$$

The approach that we will take is based on a method called "change of variables". Let $q(x) = 2^{-\ell(x)}$. Then, the equality constraint becomes $\sum_{x\in\mathcal{X}} q(x) = 1$, and $\ell(x)$ in the objective function should be replaced with $\log \frac{1}{q(x)}$:

$$\begin{aligned} \underline{\mathcal{L}} = \min_{q(x)} & \sum_{x\in\mathcal{X}} p(x)\log\frac{1}{q(x)} \\ \text{subject to } & \sum_{x\in\mathcal{X}} q(x) = 1, \quad q(x) \geq 0. \end{aligned} \tag{1.31}$$

Observe that the constraint $q(x) \geq 0$ is introduced as $q(x) = 2^{-\ell(x)}$. When implementing a "change of variable", we need to be mindful of any inherent restrictions on the novel variables that were not present in the initial optimization problem. Now take note that $\sum_{x\in\mathcal{X}} q(x) = 1$. What does this trigger in your memory? A probability mass function! Remember the axiom that the pmf must satisfy.

The objective function bears a resemblance to the one presented earlier, namely entropy $H(X)$. We now assert that the solution to the optimization problem is $H(X)$, and the minimizing function $q^*(x)$ (that minimizes the objective function) is equal to $p(x)$. Here is the reasoning behind this assertion.

If we subtract the objective function from $H(X)$, we obtain:

$$\begin{aligned} & \sum_{x\in\mathcal{X}} p(x)\log\frac{1}{q(x)} - \sum_{x\in\mathcal{X}} p(x)\log\frac{1}{p(x)} \\ & = \sum_{x\in\mathcal{X}} p(x)\log\frac{p(x)}{q(x)} \\ & = \mathbb{E}_p\left[\log\frac{p(X)}{q(X)}\right] \end{aligned}$$

What does the final term bring to mind? That is, the Kullback-Leibler (KL) divergence that we covered in Section 1.3. By leveraging a crucial fact about the KL divergence, namely that it is non-negative (verified in Prob 1.12), we can readily observe that the objective function is minimized when $\underline{\mathcal{L}} = H(X)$, and the minimizer is:

$$q^*(x) = p(x), \quad \text{i.e., } \ell^*(x) = \log \frac{1}{p(x)}. \tag{1.32}$$

An upper bound Let's now shift our focus to the issue of an upper bound. To generate a lower bound, we expanded the search space. Conversely, what is a natural approach to generating an upper bound? The answer is to narrow down the search space. We will adopt a basic method for narrowing the search space: selecting a specific choice for the optimization variables $\ell(x)$'s.

What is the particular choice we want to make for $\ell(x)$? Choosing $\ell(x)$ randomly could result in a potentially weak upper bound. Therefore, we need to be cautious when making our selection. One might speculate that a good choice would be similar to the optimal solution in the relaxed optimization problem (without the integer constraint). In this respect, the minimizing function $q^*(x)$ for the relaxed optimization problem can provide some guidance. Recall the minimizer in the instance:

$$q^*(x) = p(x).$$

Since $q(x) := 2^{-\ell(x)}$, in terms of $\ell(x)$, it would be:

$$\ell^*(x) = \log \frac{1}{p(x)}.$$

If $\ell^*(x)$'s were integers, we are happy as we can obtain the exact solution to the original problem. In general, however, $\ell^*(x)$'s are not necessarily integers. One natural choice for $\ell(x)$ is then to take an integer which is as close to $\ell^*(x)$ as possible. So one can think of two options: (i) $\lfloor \log \frac{1}{p(x)} \rfloor$; and (ii) $\lceil \log \frac{1}{p(x)} \rceil$. Which of the two options would you like to choose? In reality, the first option is invalid. Why? Consider Kraft's inequality. Therefore, the appropriate selection is the second one. With the second choice, we obtain:

$$\begin{aligned} \mathcal{L}^* &\le \sum_{x \in \mathcal{X}} p(x) \left\lceil \log \frac{1}{p(x)} \right\rceil \\ &\le \sum_{x \in \mathcal{X}} p(x) \left(\log \frac{1}{p(x)} + 1 \right) \\ &= H(X) + 1. \end{aligned}$$

Are the bounds tight? In summary, what we can say for $\mathcal{L}^*$ is :

$$H(X) \le \mathcal{L}^* \le H(X) + 1.$$

First take a look at the lower bound. The lower bound is tight when $\log \frac{1}{p(x)}$'s are integers (i.e., $p(x)$'s are integer powers of 2) and therefore $\ell^*(x)$ can be chosen as $\log \frac{1}{p(x)}$ without violating the integer constraint. However, this is a particular case because in general $p(x)$ is not limited to that particular type. As for the upper bound $H(X)+1$, if $H(X)$ is large enough, the gap of 1 would be negligible. However, if $H(X)$ is comparable to (or much smaller than) 1, the bounds are loose. For instance, consider a case in which X is a binary random variable with $\mathbb{P}(X=0)=p$ where $p \ll 1$. In this case, $H(X)$ is close to 0; hence, the bounds play almost no role in the case.

General source encoder Despite our significant efforts to approximate $\mathcal{L}^*$, we discovered that the bounds are generally not tight, rendering our efforts useless. However, the methods employed to derive these bounds play a crucial role in demonstrating the source coding theorem. Here's why.

Our analysis has been limited to a particular scenario in source encoding. We have focused on a symbol-by-symbol encoder that processes each symbol independently. However, the encoder can handle an arbitrary length of input sequence to produce an output. As a result, it can operate on multiple symbols. For instance, one could take an n-length sequence of $Z_n := (X_1, X_2, \ldots, X_n)$, which is called a super symbol, and generate an output like $C(X_1, X_2, \ldots, X_n)$. The sequence length n is a design parameter that can be chosen as desired.

Interestingly, if we apply the bounding methods we have learned to this general situation, we can readily demonstrate the source coding theorem. Let $\mathcal{L}_n^*$ be the minimum expected codeword length concerning the super symbol Z_n:

$$\mathcal{L}_n^* = \min_{\ell(z)} \sum_{z \in \mathcal{Z}} p_{Z_n}(z)\ell(z).$$

Applying the same bounding techniques to $\mathcal{L}_n^*$, we get:

$$H(Z_n) \le \mathcal{L}_n^* \le H(Z_n) + 1.$$

Since the codeword length *per symbol* is of our interest, what we care about is the one divided by n:

$$\frac{H(Z_n)}{n} \le \frac{\mathcal{L}_n^*}{n} \le \frac{H(Z_n)+1}{n}. \tag{1.33}$$

Remember that the length n of the super symbol is of our design choice and hence we can choose it so that it is an arbitrary integer. In an extreme case, we can make it arbitrarily large. This is the way that we achieve the limit. Note that the lower and upper bounds coincide with $\frac{H(Z_n)}{n}$ as n tends to infinity.

We are almost done with the proof. What remains is to calculate such matching quantity. Using the chain rule (a generalized version of the chain rule – check in Prob 1.10), we get:

$$\begin{aligned}
&\frac{H(X_1, X_2, \ldots, X_n)}{n} \\
&\quad = \frac{H(X_1) + H(X_2|X_1) + \cdots + H(X_n|X_1, \cdots, X_{n-1})}{n} \\
&\quad \overset{(a)}{=} \frac{H(X_1) + H(X_2) + \cdots + H(X_n)}{n} \\
&\quad \overset{(b)}{=} H(X)
\end{aligned}$$

where (a) follows from the independence of $(X_1, X_2, \ldots, X_n)$ and the fact that $H(X_2|X_1) = H(X_2)$ when X_1 and X_2 are independent (check in Prob 1.10(b)); and (b) is due to the fact that $(X_1, X_2, \ldots, X_n)$ are identically distributed.

Applying this to (1.33) together with the sandwidth theorem, we obtain:

$$\lim_{n \to \infty} \frac{\mathcal{L}_n^*}{n} = H(X).$$

This proves the source coding theorem for i.i.d. sources: the maximum compression rate of an information source per symbol is $H(X)$.

Look ahead The source coding theorem has been proven, showing that optimal codes exist that can achieve the limit. The optimization problem was formulated and it was shown that the solution to the problem is entropy. However, we did not discuss the explicit sequence pattern of the optimal codes, which means that we did not address how to design the optimal codes. In the next section, we will delve into this issue more deeply.

Problem Set 2

Prob 2.1 (Unique decodability) Consider a binary code tree in Fig. 1.12 for a symbol $X \in \{a, b, c, d\}$. Is the code uniquely decodable? Also explain why. You don't need to prove your answer rigorously. A non-rigorous yet simple explanation based on what we learned in Section 1.4 suffices.

Prob 2.2 (Prefix-free codes vs. Kraft's inequality) Consider prefix-free codes for $X \in \{1, 2, \ldots, M\}$ where M is an arbitrary positive integer. Let $\ell(x)$ be the codeword length for $X = x$. In Section 1.5, we showed that such prefix-free codes satisfy Kraft's inequality:

$$\sum_{x \in \mathcal{X}} 2^{-\ell(x)} \leq 1.$$

In this problem, you are asked to prove the *converse*: showing that if Kraft's inequality holds, then there exists a prefix-free code with such $\ell(x)$'s.

Prob 2.3 (Proof of Kraft's inequality for uniquely decodable codes) In Section 1.5, it was claimed that a direct way to come up with a mathematical constraint on $\ell(x)$'s that uniquely decodable codes should satisfy is difficult. In this problem, you are asked to develop the constraint. Consider a symbol-by-symbol source code in which the code acts on each individual symbol independently, i.e., an input sequence of $X_1 X_2 \cdots X_n$ yields the output $C(X_1)C(X_2) \cdots C(X_n)$

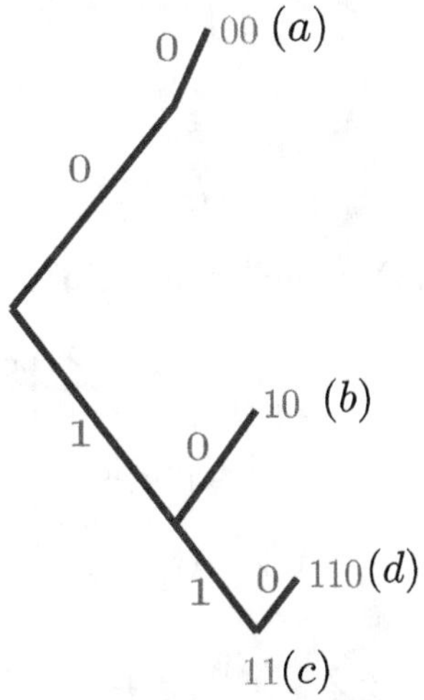

Figure 1.12. Illustration of a binary code tree with four codewords.

where $C(X)$ indicates a codeword for a symbol X. Each symbol X takes on one of the following values $1, 2, \ldots, M$. Let $\ell(i)$ be the length of codeword $C(i)$ where $i \in \{1, 2, \ldots, M\}$. Suppose that the source code is uniquely decodable. Through the following subproblems, you are asked to show that Kraft's inequality holds for the uniquely decodable code:

$$\sum_{i=1}^{M} 2^{-\ell(i)} \leq 1.$$

Remark: This implies that Kraft's inequality is also necessary, thus proving that Kraft's inequality (induced by prefix-free codes) is indeed a necessary and sufficient condition that uniquely decodable codes should satisfy.

(*a*) For a positive integer n, show that

$$\left[\sum_{i=1}^{M} 2^{-\ell(i)}\right]^n = \sum_{i_1=1}^{M} \sum_{i_2=1}^{M} \cdots \sum_{i_n=1}^{M} 2^{-(\ell(i_1)+\ell(i_2)+\cdots+\ell(i_n))}.$$

(*b*) Consider a concatenation of n symbols $X_1 X_2 \cdots X_n = i_1 i_2 \cdots i_n$. What is the codeword corresponding to such a concatenation? Also what is the length of the corresponding codeword?

(*c*) Let $\ell_{\max} := \max_{1 \leq i \leq M} \ell(i)$ and $\ell_{\min} := \min_{1 \leq i \leq M} \ell(i)$. Let N_ℓ be the number of n-fold concatenated sequences which yield codewords of length ℓ. Using parts (*a*) and (*b*), show that

$$\left[\sum_{i=1}^{M} 2^{-\ell(i)}\right]^n = \sum_{\ell=n\ell_{\min}}^{n\ell_{\max}} N_\ell 2^{-\ell}.$$

(*d*) Show that

$$N_\ell \leq 2^\ell.$$

(*e*) Using parts (*c*) and (*d*), complete the proof of Kraft's inequality.

Prob 2.4 (Convex optimization) Let $X \in \mathcal{X}$ be a discrete random variable with pmf $p(x)$. In Section 1.6, in the process of proving the source coding theorem, we considered the following optimization problem:

$$\min_{\ell(x)} \sum_{x \in \mathcal{X}} p(x)\ell(x) :$$

$$\sum_{x \in \mathcal{X}} 2^{-\ell(x)} \leq 1.$$

(*a*) State the definition of a *convex set*.
(*b*) Consider a set $\mathcal{A} := \{\ell(x) : \sum_{x\in\mathcal{X}} 2^{-\ell(x)} \leq 1\}$. Prove that the set $\mathcal{A}$ is convex.
(*c*) State the definition of *convex optimization*.
(*d*) Is the above optimization problem convex? Also explain why.
(*e*) Using the Lagrange multiplier method that we discussed in Section 1.6, derive the solution to the optimization problem as well as the minimizer $\ell^*(x)$.

Prob 2.5 (True or False?)

(*a*) Consider a discrete random variable $X \in \mathcal{X}$. Suppose a source code w.r.t. X satisfies:

$$\sum_{x\in\mathcal{X}} 2^{-\ell(x)} \leq 1$$

where $\ell(x)$ denotes the codeword length w.r.t. x. Then, there always exists a prefix-free code that satisfies the above.

(*b*) Any codeword of a uniquely decodable code cannot be mapped to an internal node in the corresponding binary code tree.

(*c*) Consider a source symbol $X \in \{a, b, c, d\}$. Consider a source code:

$$C(a) = 0; C(b) = 11; C(c) = 111; C(d) = 010.$$

This code is uniquely decodable.

(*d*) If the most probable letter in an alphabet has probability less than $\frac{1}{3}$, any prefix-free code will assign a codeword length of at least 2 to that letter.

(*e*) In Section 1.5, we learned that the integer set $\mathbb{N} = \{\ldots, -2, -1, 0, 1, 2, \ldots\}$ is *non-convex*. On the other hand, the real set $\mathbb{R}$ is *convex*.

1.7 Source Code Design

Recap In the preceding sections, we established the source coding theorem for the i.i.d. source scenario. To gain insights, we first examined a simple but restricted context - the symbol-by-symbol encoder, where the source encoder operates on each symbol independently. Subsequently, we formulated an optimization problem that aims to minimize the expected codeword length in this scenario. However, due to the problem's intractability, we attempted to approximate the solution by deriving reasonably tight lower and upper bounds. By applying simple yet powerful bounding techniques, we were able to derive lower and upper bounds that differ by 1.

Furthermore, we extended these bounding techniques to a general setting in which the source encoder operates on multiple symbols of possibly varying lengths, resulting in the following:

$$\frac{H(X_1,\ldots,X_n)}{n} \leq \frac{\mathcal{L}_n^*}{n} \leq \frac{H(X_1,\ldots,X_n)+1}{n}$$

where $\mathcal{L}_n^*$ indicates the minimum expected codeword length w.r.t. a super symbol $Z_n := (X_1,\ldots,X_n)$. In the limit of n, this gives:

$$\frac{\mathcal{L}_n^*}{n} \longrightarrow H(X). \tag{1.34}$$

Outline The implication of (1.34) is that there are optimal codes that can reach the limit, but we haven't discussed the specific sequence patterns of these optimal codes. In other words, we haven't explained how to design these optimal codes. This issue will be thoroughly investigated in this section.

Regimes in which one can achieve the limit Recall the setting where we can achieve the limit, i.e., the setting in which a super symbol is taken while its length tends to infinity. In order to design codes, we need to specify the following two: the *length* and *pattern* of codewords. The minimizer of the relaxed optimization problem (see (1.32)) gives insights into the length of optimal codewords. Recall:

$$\ell^*(X^n) = \log \frac{1}{p(X^n)}$$

where X^n denotes the shorthand notation of the sequence: $X^n := (X_1,\ldots,X_n)$. Hence, in order to figure out how the optimal length $\ell^*(X^n)$ looks like in the interested regime of large n, we should take a look at the behavior of $p(X^n)$ for the regime.

Behavior of $p(X^n)$ for a large value of n We focus on the binary alphabet case in which $X_i \in \{a, b\}$ and $\mathbb{P}(X_i = a) = p$. Later we will consider the general alphabet case. The sequence consists of a certain combination of two symbols: *a*'s and *b*'s. For very large n, an interesting behavior occurs on two quantities regarding the sequence. One is the faction of symbol "*a*", represented as the number of *a*'s divided by n. This is an empirical mean of occurrences of symbol "*a*". The second is the symbol "*b*" counterpart, the fraction of symbol *b*.

To see this, we compute the probability of observing X^n. Since X^n is i.i.d., $p(X^n)$ is simply the product of individual probabilities:

$$p(X^n) = p(X_1)p(X_2)\cdots p(X_n).$$

Each individual probability is p or $1-p$ depending on the value of X_i. So the result would be of the following form:

$$p(X^n) = p^{\{\#\text{ of } a\text{'s}\}}(1-p)^{\{\#\text{ of } b\text{'s}\}}.$$

Consider $\log \frac{1}{p(X^n)} (= \ell^*(X^n))$ that we are interested in:

$$\log \frac{1}{p(X^n)} = \{\#\text{ of } a\text{'s}\} \cdot \log \frac{1}{p} + \{\#\text{ of } b\text{'s}\} \cdot \log \frac{1}{1-p}.$$

Dividing by n on both sides, we obtain:

$$\frac{1}{n} \log \frac{1}{p(X^n)} = \frac{\{\#\text{ of } a\text{'s}\}}{n} \cdot \log \frac{1}{p} + \frac{\{\#\text{ of } b\text{'s}\}}{n} \cdot \log \frac{1}{1-p}. \tag{1.35}$$

What can we say about this in the limit of n? Your intuition says: $\frac{\{\#\text{ of } a\text{'s}\}}{n} \to p$ as $n \to \infty$. One can naturally expect that as $n \to \infty$, the fraction would be concentrated around $\mathbb{P}(X = a)$. This is indeed the case. Relying on a well-known theorem, called the *Weak Law of Large Numbers* (WLLN[10]), one can prove this. Let $Y_i = \mathbf{1}\{X_i = a\}$ where $\mathbf{1}\{\cdot\}$ is an indicator function that returns 1 when the event $(\cdot)$ is true; 0 otherwise. Consider:

$$S_n := \frac{Y_1 + Y_2 + \cdots + Y_n}{n}$$

where S_n indicates the fraction of *a*'s. Obviously it is a sequence of n.

Now consider the convergence of the sequence. From Calculus, you learned about the convergence of sequences, which are *deterministic*. But the situation is a bit different here. The reason is that S_n is a *random variable* (not deterministic). We need to consider the convergence w.r.t. a random process. There are multiple types of convergence w.r.t. random processes. One type of the convergence that is

10. This is the law (discovered by Jacob Bernoulli) mentioned in Section 1.3

needed to be explored in our problem context is the *convergence in probability*. In fact, what the WLLN that we mentioned above says is that

WLLN: S_n *converges to* $\mathbb{E}[Y] = p$ *in probability.*

Simply speaking, it means that S_n converges to p *with high probability (w.h.p.)*. But in mathematics, the meaning should be rigorously stated. What it means by this in mathematics is that for any $\epsilon > 0$,

$$\mathbb{P}(|S_n - p| \leq \epsilon) \rightarrow 1 \qquad \text{as } n \rightarrow \infty.$$

In other words, S_n is within $p \pm \epsilon$ w.h.p.

Applying the WLLN to $\frac{\{\#\text{ of } a\text{'s}\}}{n}$ and $\frac{\{\#\text{ of } b\text{'s}\}}{n}$, we get: for any $\epsilon_1, \epsilon_2 > 0$,

$$\frac{\{\#\text{ of } a\text{'s}\}}{n} \text{ is within } p \pm \epsilon_1,$$
$$\frac{\{\#\text{ of } b\text{'s}\}}{n} \text{ is within } 1 - p \pm \epsilon_2,$$

as $n \rightarrow \infty$. Applying this to (1.35), we can say that in the limit of n,

$$\begin{aligned}\frac{1}{n} \log \frac{1}{p(X^n)} &= (p \pm \epsilon_1) \log \frac{1}{p} + (1 - p \pm \epsilon_2) \log \frac{1}{1-p} \\ &= H(X) \pm \epsilon\end{aligned}$$

where $\epsilon := \epsilon_1 \log \frac{1}{p} + \epsilon_2 \log \frac{1}{1-p}$.

Manipulating the above, we get:

$$p(X^n) = 2^{-n(H(X) \pm \epsilon)} \text{ holds w.h.p.} \tag{1.36}$$

The key observation is that for a very large value of n, ϵ can be made arbitrarily close to 0 and thus $p(X^n)$ is almost the same for all such X^n. We call such sequences *typical sequences*. The set that contains typical sequences is said to be a *typical* set, formally defined as below.

$$A_\epsilon^{(n)} := \{x^n : 2^{-n(H(p)+\epsilon)} \leq p(X^n) \leq 2^{-(H(p)-\epsilon)}\}.$$

Notice that $p(X^n) \approx 2^{-nH(X)}$ is almost uniformly distributed. This property plays a crucial role to design a optimal code. In the sequel, we will use this property to design the code.

Prefix-free code design What (1.36) suggests is that any arbitrary sequence is asymptotically *equiprobable*. Remember in Section 1.1 that a good source code assigns a short-length codeword to a frequent symbol, while assigning a long-length codeword to a less frequent symbol. For a very large value of n, any sequence is almost equally probable. This implies that the optimal length of a codeword assigned for any arbitrary sequence would be roughly the same as:

$$\ell^*(X^n) = \log \frac{1}{p(X^n)} \approx nH(X).$$

A binary code tree can be constructed where the depth of the tree is approximately $nH(X)$, and the codewords are allocated to the leaves. It is worth noting that in a prefix-free code, the codewords are located only in the leaves.

Let us check if we can map all the possible sequence patterns of X^n into leaves. To this end, we need to check two values: (1) the total number of leaves; and (2) the total number of possible input sequences. First of all, the total number of leaves is roughly $2^{nH(X)}$. What about the second value? The total number of input sequence patterns is 2^n because each symbol can take on one of the two values ("a" and "b") and we have n of them. But in the limit of n, the sequence X^n behaves in a particular manner, more specifically, in a manner that $p(x^n) \approx 2^{-nH(X)}$; hence, the number of such sequences is not the maximum possible value of 2^n. Then, how many sequences such that $p(x^n) \approx 2^{-nH(X)}$? To see this, consider:

$$\sum_{x^n : p(x^n) \approx 2^{-nH(X)}} p(x^n) \approx |\{x^n : p(x^n) \approx 2^{-nH(X)}\}| \times 2^{-nH(X)}.$$

The aggregation of all the probabilities of such sequences cannot exceed 1; hence,

$$|\{x^n : p(x^n) \approx 2^{-nH(X)}\}| \lesssim 2^{nH(X)}.$$

Note that the cardinality of the set does not exceed the total number of leaves $\approx 2^{nH(X)}$. So by using parts of leaves, we can map all of such sequences. This completes the design of an optimal prefix-free code. See Fig. 1.13.

Finally let us check if this code indeed achieves the fundamental limit $H(X)$. Notice that the length of every codeword is $\approx nH(X)$. So the expected codeword length per symbol would be $\approx \frac{nH(X)}{n} = H(X)$.

Remark: The argument for the source code design is based on approximation. You will have a chance to do this rigorously in Prob 3.7.

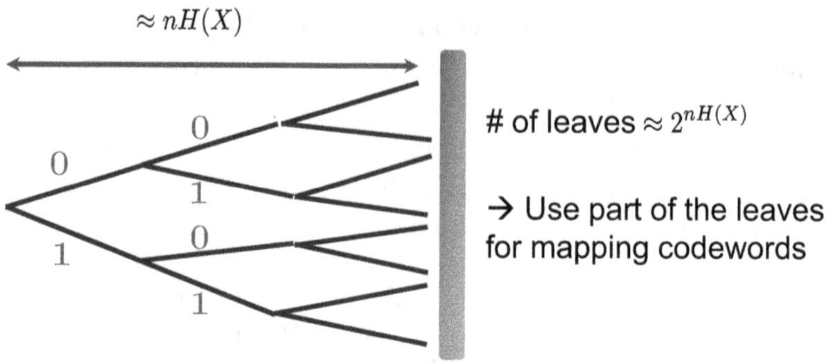

Figure 1.13. Design of optimal prefix-free codes.

Extension to non-binary sources We have considered the binary alphabet case. How about for non-binary sources? Using the WLLN, one can prove that:

$$\frac{1}{n}\log\frac{1}{p(X^n)} \longrightarrow H(X) \quad \text{in prob.}$$

although X is an arbitrary random variable, not limited to the binary one. Please check this is indeed the case in Prob 3.4. Roughly speaking, this implies that $p(X^n) \approx 2^{-nH(X)}$ w.h.p. – any arbitrary sequence is asymptotically equally probable. Using this and applying the code design rule based on a binary code tree, we can easily construct an optimal source code. Every input sequence is mapped to a leaf in a binary code tree with depth $\approx nH(X)$ and hence, the expected codeword length per symbol is $H(X)$. The sequence pattern of a codeword is determined by which leaf the codeword is assigned to.

Look ahead We have focused on i.i.d. sources so far. However, in real-world scenarios, many information sources exhibit non-i.i.d. behavior. Therefore, it is crucial to investigate the source coding theorem and optimal code design for non-i.i.d. sources. In the next section, we will delve into these practically-relevant scenarios.

1.8 Source Coding Theorem for General Sources

Recap In the previous section, we examined the method for creating an optimal code that can achieve the entropy promised by the source coding theorem for the i.i.d. source case. The approach utilized a super symbol-based technique, where the source encoder operates on a sequence of multiple input symbols (n symbols), and the value of n is chosen by the designer. The construction aimed to increase the size of the super symbol to a sufficiently large value. As we analyzed the technique for large n, we discovered using the WLLN that:

$$\frac{1}{n} \log \frac{1}{p(X^n)} \xrightarrow{\text{in prob.}} H(X) \tag{1.37}$$

i.e., $\frac{1}{n} \log \frac{1}{p(X^n)}$ lies in between $H(X) - \epsilon$ and $H(X) + \epsilon$ for any $\epsilon > 0$, as n tends to infinity. Inspired by the fact that the codeword length solution for the lower bound in the interested optimization problem is $\ell^*(x^n) = \log \frac{1}{p(x^n)}$, we focused on the quantity of $\frac{1}{p(x^n)}$. From (1.37), we observed that in the limit of n, the quantity becomes:

$$\log \frac{1}{p(X^n)} \approx nH(X).$$

As a result, we were prompted to investigate a prefix-free code where a given input sequence x^n is assigned to a leaf located at the level with a tree depth of approximately $nH(X)$. By doing so, we guarantee that the expected codeword length per symbol is approximately $H(X)$, meeting the expected limit. Additionally, we verified that the number of possible input sequences with probability $p(x^n)$ approximately equal to $2^{-nH(X)}$ is lower than the total number of leaves. This guarantees a unique mapping for each of these input sequences.

Outline Our focus was solely on i.i.d. sources. However, when it comes to non-i.i.d. sources, one might wonder what the corresponding source coding theorem is, as well as how we can design optimal codes for such sources. In this section, we will explore these inquiries.

General non-i.i.d. sources In real-world scenarios, most information sources deviate significantly from the i.i.d. assumption. One prime example of such a source is an English text. To illustrate this, let's take the example where the first and second letters are "t" and "h" respectively. In this case, it is reasonable to expect that the third letter would be "e" due to the high frequency of the word "the" in typical English texts. This example highlights the strong correlation between symbols in a text, indicating that the sequence is dependent and not i.i.d. Many other information sources exhibit similar characteristics, making it imperative to explore the source coding theorem and optimal code design for non-i.i.d. sources.

Source coding theorem for general sources By utilizing the bounding techniques we have learned, we can easily address the non-i.i.d. source case. If we utilize a super symbol-based source code with a super symbol size of n, and let $\mathcal{L}_n^* = \mathbb{E}[\ell(X_n)]$, we can apply the same lower and upper bound techniques to show that:

$$H(X_1, X_2, \ldots, X^n) \leq \mathcal{L}_n^* \leq H(X_1, X_2, \ldots, X_n) + 1.$$

Dividing the above by n, we get:

$$\frac{H(X_1, X_2, \ldots, X^n)}{n} \leq \frac{\mathcal{L}_n^*}{n} \leq \frac{H(X_1, X_2, \ldots, X_n) + 1}{n}.$$

The expected codeword length per symbol is related to the following quantity:

$$\lim_{n \to \infty} \frac{H(X_1, X_2, \ldots, X_n)}{n}.$$

Is the limit present? If so, we can conclude the task. However, it is not always the case, as there are certain artificial instances where the limit does not exist, as shown in page 75 of (Cover, 1999). Nonetheless, in numerous cases that are of practical importance, the limit does exist. Therefore, let us only focus on those cases where the limit exists, and we can express the source coding theorem as follows:

$$\textit{Minimum \# of bits that can represent a general source per symbol}$$
$$= \lim_{n \to \infty} \frac{H(X_1, X_2, \ldots, X_n)}{n}.$$

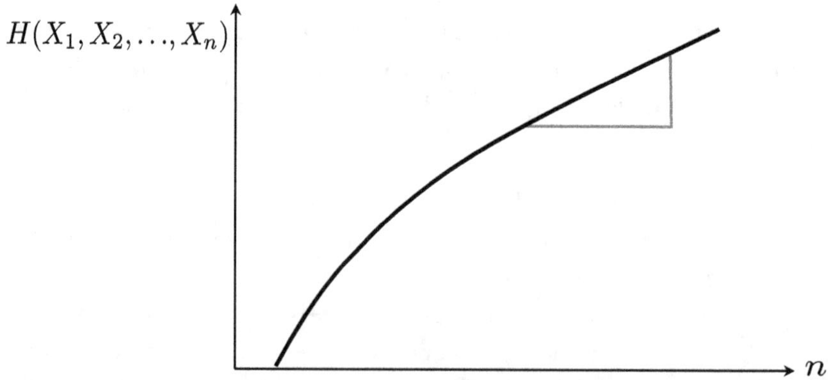

Figure 1.14. $\lim_{n \to \infty} \frac{H(X_1, X_2, \ldots, X_n)}{n}$ means the growth rate of the sequence uncertainty w.r.t. n, so it is called the entropy rate.

There is a term that refers to this limit. In order to comprehend why this term is used, take a look at a graph where the n and $H(X_1, X_2, \ldots, X_n)$ are represented on the x and y axes, respectively. This can be observed in Fig. 1.14. What the above limit means is the slope. In other words, it means the growth rate of the sequence uncertainty w.r.t. n. Hence, it is called the *entropy rate*.

Stationary process You might be wondering how to calculate the entropy rate. In many cases of practical significance, computing the entropy rate is a straightforward task. One such example is a stationary process. A random process is considered stationary if $\{X_i\}$ has the same statistical properties (such as joint distribution) as its shifted version $\{X_{i+\ell}\}$ for any non-negative integer ℓ. An example of this is English text. The statistics of a 10-year-old text would be almost identical to that of a present-day text; for instance, the frequency of the word "the" in an older text would be roughly the same as in a contemporary text.

When the limit is applied to a stationary process, it can be simplified even further. To illustrate this, let's consider:

$$\begin{aligned} &H(X_1, X_2, \ldots, X_n) \\ &\quad = H(X_1) + H(X_2|X_1) + \cdots + H(X_n|X_1, \ldots, X_{n-1}) \\ &\quad = \sum_{i=1}^{n} H(X_i|X_1, X_2, \ldots, X_{i-1}) \\ &\quad = \sum_{i=1}^{n} H(X_i|X^{i-1}) \end{aligned}$$

where the first equality is due to the chain rule and the last comes from the shorthand notation that we introduced earlier: $X^{i-1} := (X_1, \ldots, X_{i-1})$. Let $a_i = H(X_i|X^{i-1})$. We can see two properties of the deterministic sequence $\{a_i\}$. First it is non-negative. Second, it is non-increasing, i.e., $a_{i+1} \leq a_i$. To see this, consider:

$$\begin{aligned} a_{i+1} &= H(X_{i+1}|X_1, X_2, \ldots, X_i) \\ &\leq H(X_{i+1}| \quad X_2, \ldots, X_i) \\ &= H(X_i|X_1, \ldots, X_{i-1}) \\ &= a_i \end{aligned}$$

where the inequality follows from the fact that conditioning reduces entropy and the second last equality is due to the stationarity of the process. These two properties imply that the deterministic sequence $\{a_i\}$ has a limit. Why? Please check in

Prob 3.8. Now consider

$$\frac{H(X_1, \ldots, X_n)}{n} = \frac{1}{n} \sum_{i=1}^{n} a_i.$$

This quantity indicates the running average of the sequence $\{a_i\}$. Keeping in your mind that $\{a_i\}$ has a limit, what your intuition says is that the running average will converge to the limit because almost all the components in the running average will converge to the limit. This is indeed the case. The entropy rate of the stationary process is:

$$H(\mathcal{X}) = \lim_{i \to \infty} H(X_i | X^{i-1}). \tag{1.38}$$

Please see Prob 3.8 for the rigorous proof.

Next, how to design optimal codes for such a stationary process? We can apply the same methodology that we developed earlier. The first is to check that such a stationary sequence is also asymptotically equiprobable. In fact, one can resort to a generalized version of the WLLN to prove that this is indeed the case:

$$p(X^n) \approx 2^{-nH(\mathcal{X})}. \tag{1.39}$$

The proof of this is not that simple, requiring some non-trivial tricks. We will not deal with the proof here. If you are interested, you can try it via Prob 3.10. Recalling what we learned in the previous section. What (1.39) suggests is that the optimal code assigns the same codeword length for every sequence and the length should read roughly $nH(\mathcal{X})$. The sequence patterns will be determined by the binary code tree of depth $\approx nH(\mathcal{X})$ in which codewords are mapped to leaves.

From theory to practice Up to this point, we have discussed the limit (the entropy rate) in relation to the compression rate of an information source (which is almost always a stationary process in practical applications) and learned how to design optimal codes that achieve this limit. However, our focus has been on an idealistic scenario where the super symbol size n can be made infinitely large. In reality, n must be finite for two reasons. Firstly, the hardware used to implement the code cannot support an infinite n size. Secondly, the amount of available information from the source is limited. Therefore, the question arises: what are the optimal codes for a finite value of n? This was the question posed by Shannon and shared with one of the professors in MIT, Prof. Robert Fano. Prof. Fano then shared this question with students who took the information theory course that he held at that time.

As previously mentioned, solving the optimization problem for the finite n case is non-convex and extremely difficult. To date, a closed-form solution to the limit remains elusive.

However, surprisingly, one of Prof. Fano's students, David Huffman, devised a simple algorithm that leads to the optimal code. Although he did not provide an exact closed-form solution to the limit, he was able to give an explicit design rule in the form of an algorithm that generates the optimal code. This code is now known as the Huffman code, and it was developed as a term project in the class.

Look ahead In summary, we have demonstrated the source coding theorem for general information sources and discovered how to design codes that achieve the limit. Additionally, we provided a brief introduction to some practical codes such as the Huffman code. The next section will delve deeper into the Huffman code.

1.9 Huffman Code and Python Implementation

Recap We have demonstrated the source coding theorem for general information sources (stationary processes in practically-relevant scenarios). We also learned how to design optimal codes that attain the limit, which is the entropy rate. Nevertheless, our attention has been focused on an idealistic situation in which the super symbol size n is infinitely large, while in practice, n must be finite. The good news is that an optimal code construction was developed shortly after Shannon established the source coding theorem.

Outline In this section, we will delve into the study of the optimal code, known as the Huffman code, which was developed by David Huffman. The inspiration for the code came from "thinking outside the box". Huffman scrutinized an intuitive binary code tree and deduced several properties that an optimal binary code tree must satisfy. These properties led him to invent a natural algorithm that guarantees optimality. We will first examine the crucial properties that an optimal binary code tree should possess. Next, we will describe the functioning of the optimal algorithm. Finally, we will investigate the implementation of the algorithm using Python.

An optimization problem for $n = 1$ Let us start with the simplest case in which the super symbol size n is 1. Let a symbol X be an M-ary random variable with the pmf $p(x)$ in which $x \in \mathcal{X} = \{a_1, a_2, \ldots, a_M\}$. For notational simplicity, we will denote $p(a_i)$ by p_i for $i \in \{1, \ldots, M\}$. Let ℓ_i be the codeword length w.r.t. symbol a_i. Then, the optimization problem which aims to minimize the expected codeword length is:

$$\begin{aligned} &\min_{\ell_i} \sum_{i=1}^{M} p_i \ell_i \\ &\text{s.t.} \sum_{i=1}^{M} 2^{-\ell_i} \leq 1, \ \ell_i \in \mathbb{N}. \end{aligned} \tag{1.40}$$

As we learned in Section 1.5, this is a *non-convex* optimization problem. Especially when it involves an integer constraint, it is so called *integer programming* which is known to be notoriously difficult. In general, solving integer programming requires searching over all possible candidates for variables.

Since the intractability of the problem is well-known, in early days, only heuristics were suggested. One heuristic proposed by Shannon is to choose ℓ_i as (Shannon, 2001):

$$\ell_i = \left\lceil \log \frac{1}{p_i} \right\rceil.$$

People later called this *Shannon code*. When there is no ceiling, the minimizer is the solution to the relaxed optimization problem which ignores the integer constraint. Hence, it yields the exact solution when $\log \frac{1}{p_i}$'s are integers. In general, $\log \frac{1}{p_i}$'s are not necessarily integers, so the problem is not that simple. Actually even until now, the closed-form solution for the expected codeword length has been unknown.

Birth of the Huffman code Fano had developed another heuristic code, which was later named Shannon-Fano code (Salomon, 2004), but it was not generally optimal. He challenged the students in his information theory course at MIT to find a solution to the integer programming problem. Unexpectedly, a student named David Huffman came up with a simple algorithm that leads to the optimal code, which is now known as the Huffman code (Huffman, 1952).

Instead of attempting to solve the difficult problem directly, Huffman used a creative approach to gain insights from the properties that the optimal code must have. He studied the binary code tree and established three key properties that the optimal binary code tree should satisfy. These properties allowed him to develop an optimal algorithm.

Properties of optimal prefix-free codes Without loss of generality, assume that $p_1 \geq p_2 \geq \cdots \geq p_M$. Let ℓ_i^*'s be optimal codeword lengths. The first property is w.r.t. the relationship between p_i's and ℓ_i^*'s:

$$\textit{Property 1: } p_i > p_j \longrightarrow \ell_i^* \leq \ell_j^*. \tag{1.41}$$

This is very intuitive. Having a larger frequency (higher p_i), the corresponding optimal codeword length must be smaller (smaller ℓ_i^*). Here is a rigorous proof for this seemingly-trivial property. The idea is by contradiction. Suppose $\ell_1^* > \ell_2^*$ when $p_1 > p_2$. The optimal expected codeword length reads:

$$\mathcal{L}^* = p_1\ell_1^* + p_2\ell_2^* + \sum_{i=3}^{M} p_i\ell_i^*.$$

On the other hand, consider another expected codeword length, say $\hat{\mathcal{L}}$, w.r.t. the case in which we interchange the role of ℓ_1^* and ℓ_2^*:

$$\hat{\mathcal{L}} = p_1\ell_2^* + p_2\ell_1^* + \sum_{i=3}^{M} p_i\ell_i^*.$$

We then see that

$$\begin{aligned}\mathcal{L}^* - \hat{\mathcal{L}} &= p_1(\ell_1^* - \ell_2^*) - p_2(\ell_1^* - \ell_2^*)\\ &= (p_1 - p_2)(\ell_1^* - \ell_2^*) > 0.\end{aligned}$$

This contradicts with the hypothesis that $\mathcal{L}^*$ is the minimum expected codeword length.

The second property is regarding the relationship between the optimal lengths w.r.t. the last two least probable symbols:

$$\textit{Property 2: } \ell^*_{M-1} = \ell^*_M. \tag{1.42}$$

The immediacy of this statement arises from the fact that a binary code tree possesses only two branches. If we assume that $\ell^*_{M-1} \neq \ell^*_M$, there must exist a vacant leaf adjacent to the leaf designated for the least probable symbol a_M. This assumption is contradictory since a binary code tree with an unoccupied left leaf cannot produce the most efficient code. To further illustrate this point, let us examine a simple instance in which M equals 4, and the codewords are:

$$C(a_1) = 0, C(a_2) = 10, C(a_3) = 110, C(a_4) = 1110.$$

See Fig. 1.15. In this case, one can immediately find a better codeword that yields a shorter length: $C(a_4) = 111$.

The last third property is about the relationship between the last two least probable symbols:

$$\textit{Property 3: } (a_{M-1}, a_M) \textit{ can be assumed to be siblings,} \tag{1.43}$$

meaning that the codewords for a_{M-1} and a_M can always be mapped to the leaves which are neighbors with each other. Here is the proof. Suppose (a_{M-1}, a_M) are not siblings. Then, there must be another symbol, say a_{M-i}, $(i \geq 2)$, such that it is a sibling of a_M. Otherwise, one can immediately find a better code (Why?). Clearly $\ell^*_{M-i} = \ell^*_M$. This together with *Property 2* gives: $\ell^*_{M-1} = \ell^*_{M-i}$. Hence,

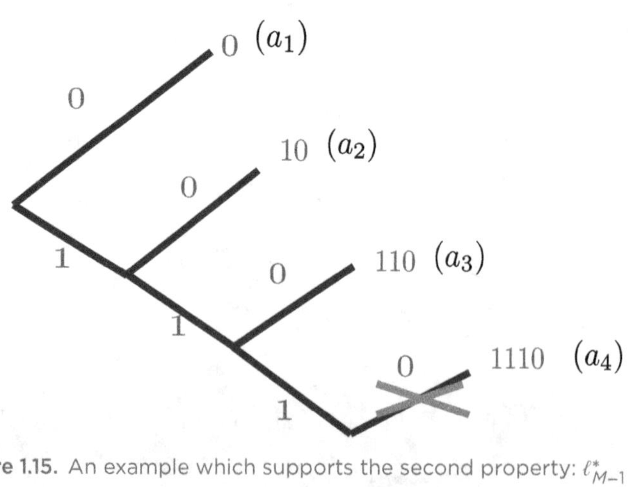

Figure 1.15. An example which supports the second property: $\ell^*_{M-1} = \ell^*_M$.

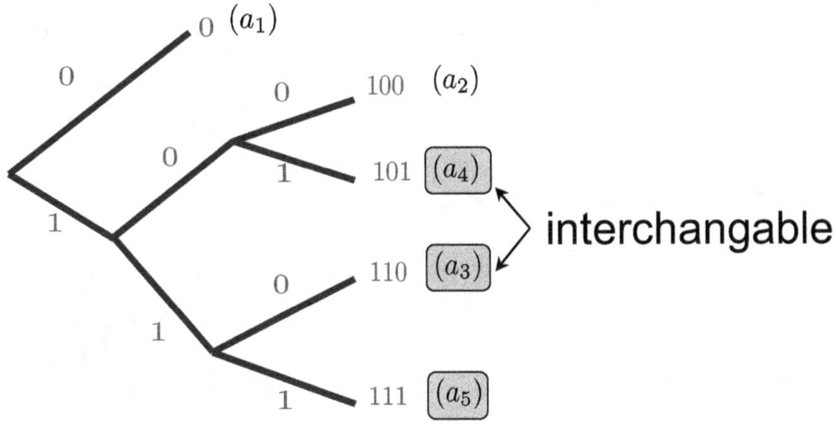

Figure 1.16. An example which supports the third property: two least probable symbols (a_4, a_5) can be assumed to be siblings.

one can interchange the codewords of a_{M-i} and a_{M-1} without loss of optimality (while keeping the expected codeword length). This ensures a_{M-1} to be a sibling of a_M. To see this clearly, consider an example in which $M = 5$ and codewords are:

$$C(a_1) = 0, C(a_2) = 100, C(a_3) = 110, C(a_4) = 110, C(a_5) = 111.$$

See Fig. 1.16. In this case, one can interchange the codewords of a_3 and a_4 so that a_4 is a sibling of a_5 while maintaining the expected codeword length.

An optimal algorithm The above three properties (1.41)–(1.43) enabled Huffman to come up with a simple and natural algorithm. Let us explore it by starting with the simplest case $M = 2$. In this case, the second property $\ell^*_{M-1} = \ell^*_M$ yields an obvious construction: $C(a_1) = 0$ and $C(a_2) = 1$.

Next consider $M = 3$. The second property gives $\ell^*_2 = \ell^*_3$. Obviously $\ell^*_2 = \ell^*_3 \geq 2$. Otherwise (i.e., $\ell^*_2 = \ell^*_3 = 1$), ℓ^*_1 must be greater than or equal to 2. But this violates the first property. Hence, $\ell^*_1 = 1$ and $\ell^*_2 = \ell^*_3 \geq 2$; and this yields a straightforward construction: Assigning 10 and 11 to two least probable symbols while mapping 0 to the most frequent symbol:

$$C(a_1) = 0, C(a_2) = 10, C(a_3) = 11. \tag{1.44}$$

Now consider $M = 4$. With the second and third properties, $\ell^*_3 = \ell^*_4$ and (a_3, a_4) can be assumed to be siblings. Consider the internal node associated with the two leaves w.r.t. (a_3, a_4). Let a'_3 be a virtual symbol which represents the two symbols (a_3, a_4). Map a'_3 to that internal node, and define

$$p'_3 := p(a'_3) = p_3 + p_4.$$

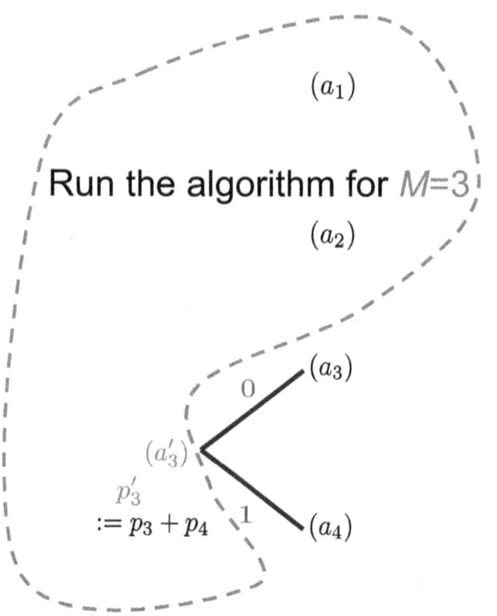

Figure 1.17. The Huffman algorithm for $M = 4$.

Let ℓ_3' be the codeword length w.r.t. a_3'. This then gives $\ell_3'^* + 1 = \ell_3^* = \ell_4^*$. Consider a new set of symbols (a_1, a_2, a_3') with (p_1, p_2, p_3'). The idea of the Huffman algorithm is to take a recursion: Running the $M = 3$ algorithm (described in (1.44)) for the new set of (a_1, a_2, a_3'). See Fig. 1.17.

The recursive approach outlined here results in the minimum expected length for codewords. While the verification process will be elaborated upon later, a comprehensive account of the algorithm for any given M is presented below:

1. If $M = 3$, we run the algorithm described in (1.44).
2. Otherwise, merge two least probable symbols (a_{M-1}, a_M) to generate a virtual symbol a_{M-1}' with $p_{M-1}' := p_{M-1} + p_M$. Run the same algorithm (performing procedures 1 and 2) for the new set of $M - 1$ symbols: $(a_1, a_2, \ldots, a_{M-2}, a_{M-1}')$.

We also provide an example for $M = 5$:

$$(p_1, p_2, p_3, p_4, p_5) = (0.4, 0.2, 0.15, 0.15, 0.1). \tag{1.45}$$

See Fig. 1.18. We first merge two least probable symbols (a_4, a_5) to generate a virtual symbol a_4' with $p_4' = 0.15 + 0.1 = 0.25$. Next we repeat the same now for the new set of four symbols with $(p_1, p_2, p_3, p_4') = (0.4, 0.2, 0.15, 0.25)$. We merge two least probable symbols (a_2, a_3) to generate a virtual symbol, say a_2', with

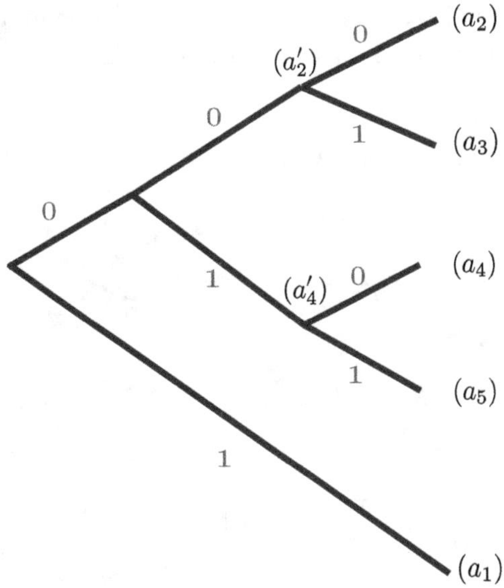

Figure 1.18. An example of how the Huffman algorithm runs.

$p_2' = 0.2 + 0.15 = 0.35$. This yields another set of symbols with $(p_1, p_2', p_4') = (0.4, 0.35, 0.25)$. Lastly we merge a_2' and a_4' to complete the algorithm.

Proof of the optimality Let us prove the optimality of the Huffman algorithm. Let

$$\mathcal{L} = \sum_{i=1}^{M} p_i \ell_i = p_{M-1}\ell_{M-1} + p_M \ell_M + \sum_{i=1}^{M-2} p_i \ell_i$$

Here (a_{M-1}, a_M) are the two least probable symbols and we merge them to generate a_{M-1}' with $p_{M-1}' = p_{M-1} + p_M$. Note that $\ell_{M-1}' + 1 = \ell_{M-1} = \ell_M$. Using this, we get:

$$\begin{aligned}\mathcal{L} &= p_{M-1}(\ell_{M-1}' + 1) + p_M(\ell_{M-1}' + 1) + \sum_{i=1}^{M-2} p_i \ell_i \\ &= p_{M-1} + p_M + \left[p_{M-1}'\ell_{M-1}' + \sum_{i=1}^{M-2} p_i \ell_i \right].\end{aligned} \tag{1.46}$$

Let $\mathcal{L}'$ be the expected codeword length w.r.t. the new set of symbols $(a_1, a_2, \ldots, a_{M-2}, a_{M-1}')$. Then, $\mathcal{L}' = p_{M-1}'\ell_{M-1}' + \sum_{i=1}^{M-2} p_i \ell_i$, which coincides with the second bracketed term in the second equality of (1.46). The key to observe is that minimizing $\mathcal{L}'$ yields the same solution as that aiming to minimize

$\mathcal{L}$, as the left-over term $p_{M-1} + p_M$ is irrelevant to the optimization variables ℓ_i's. Also Kraft's inequality for the new set of $(a_1, a_2, \ldots, a_{M-2}, a'_{M-1})$ holds:

$$
\begin{aligned}
2^{-\ell'_{M-1}} + \sum_{i=1}^{M-2} 2^{-\ell_i}
&= 2 \cdot 2^{-(\ell'_{M-1}+1)} + \sum_{i=1}^{M-2} 2^{-\ell_i} \\
&= 2^{-\ell_{M-1}} + 2^{-\ell_M} + \sum_{i=1}^{M-2} 2^{-\ell_i} \\
&= \sum_{i=1}^{M} 2^{-\ell_i} \leq 1.
\end{aligned}
$$

This proves the optimality of the Huffman algorithm.

Extension to an arbitrary super symbol size n To give you an idea, let us consider the case in which the size of the super symbol $n = 2$. Let $Z = (X_1, X_2)$. Then, the probability distribution w.r.t. Z reads:

$$p(a_1, a_1), p(a_1, a_2), \ldots, p(a_M, a_M). \tag{1.47}$$

We can then run the Huffman algorithm w.r.t. Z. The generalization to arbitrary n is straightforward. Let $Z = (X_1, X_2, \ldots, X_n)$. Then, the probability distribution is defined as in (1.47). Running the same algorithm, we obtain the optimal code.

Python implementation We explore how to implement the Huffman algorithm via Python. For illustrative purpose, consider a simple setup where $M = 5$ (say a, b, c, d, e) and the probability distribution is the same as in (1.45):

$$(p_1, p_2, p_3, p_4, p_5) = (0.4, 0.2, 0.15, 0.15, 0.1). \tag{1.48}$$

Since the algorithm is based on a binary tree, we start with constructing a binary tree class with the top and bottom branches.

```
class TreeNode(object):
    def __init__(self, top=None, bottom=None):
        self.top = top
        self.bottom = bottom

    def children(self):
        return self.top, self.bottom
```

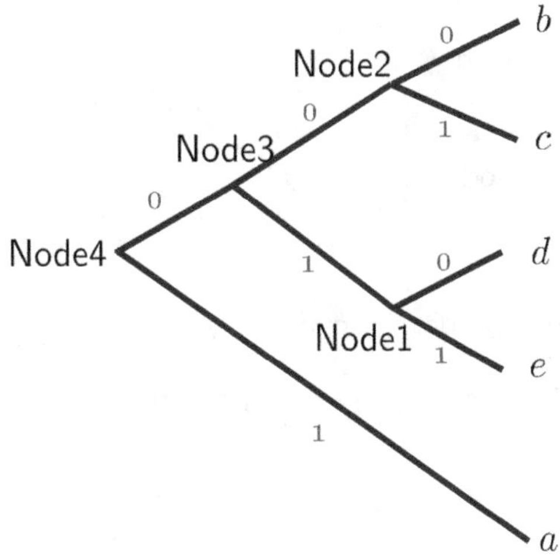

Figure 1.19. A binary code tree constructed by the Huffman algorithm.

This class is equipped with a method (named children) that returns the values assigned to the top and bottom branches.

Using this tree class, we wish to construct a binary code tree as illustrated in Fig. 1.19. We first create a tree (that we named Node 1) by merging the two least probable symbols (d, e). To Node 1, we then assign a new probability $p_4' = 0.15 + 0.1$. Next we repeat the same for the new set of four symbols (a, Node 1, b, c) with (p_1, p_4', p_2, p_3). We build another tree, Node 2 by merging the two least probable symbols (b, c) in the new set. For another set of symbols (a, Node 2, Node 1) with $(p_1, p_2', p_4') = (0.4, 0.35, 0.25)$, we create Node 3, taking Node 2 and Node 1 as the top and bottom children. Lastly we construct Node 4 (the root node) to complete the algorithm.

For code implementation, we represent the probability distribution via a dictionary where the keys and values are symbols (or TreeNode) and probabilities, respectively. To merge two least probable symbols, we sort the dictionary in a descending order, thus taking the last two. To this end, we employ sorted function. Here is code implementation.

```
# probability distribution
freq={"a":0.4, "b":0.2, "c":0.15, "d":0.15, "e":0.1}
# Sort in a descending order
freq=sorted(freq.items(),key=lambda x: x[1],reverse=True)

# Construct a binary code tree
while len(freq) > 1:
    # Retrieve the mimimum probability
    (key_b, c_b) = freq[-1]
```

```
    # Retrieve the 2nd minimum probability
    (key_t, c_t) = freq[-2]
    # Build a TreeNode taking minimum and 2nd minimum
    # as the bottom and the top child, respectively
    new_node = TreeNode(key_t,key_b)
    # Construct a new probability distribution with
    # remaining probabilities & that of the new NodeTree
    freq = freq[:-2]
    freq.append((new_node, c_t+c_b))
    # Sort in a descending order
    freq = sorted(freq, key=lambda x: x[1], reverse=True)
    print(freq)
```

```
[('a', 0.4),
(<__main__.TreeNode object at 0x000001344B434250>, 0.25),
('b',0.2), ('c', 0.15)]
[('a', 0.4),
(<__main__.TreeNode object at 0x000001344B434A30>, 0.35),
(<__main__.TreeNode object at 0x000001344B434250>, 0.25)]
[(<__main__.TreeNode object at 0x000001344B434580>, 0.6),
('a', 0.4)]
[(<__main__.TreeNode object at 0x000001344B4343A0>, 1.0)]
```

The tree construction stops when the newly updated dictionary contains only one element. In the above example, we have four steps in total. To see how each step works, we print out the updated dictionary. We also illustrate it via Fig. 1.20.

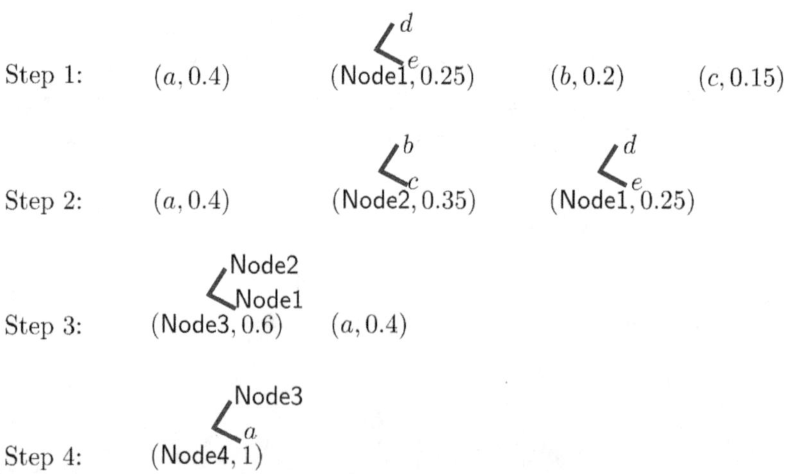

Figure 1.20. How each step works in the Huffman algorithm.

Next we construct an encoding function due to the Huffman algorithm. In the above example, we wish to generate a dictionary as below:

$$\{b:000, c:001, d:010, e:011, a:1\}.$$

To this end, we need to traverse all the leaves while producing a corresponding binary string w.r.t. each leaf. In the data structure literature, there are two major methods for traversing all the leaves in a binary tree: (i) Depth First Search (DFS); and (ii) Breath First Search (BFS). Here we take DFS and this can easiliy be implemented via recursion. Here is code implementation.

```
def huffman_code_tree(node, binString=''):
    # returns a dictionary where
    # (key, value) = (symbol, codeword)

    # if node is of string type, return (node,binString)
    if type(node) is str:
        return {node: binString}
    (top,bottom) = node.children()
    # initialize a dictionary for encoding
    d = dict()
    # top child: assign a label '0'
    d.update(huffman_code_tree(top, binString+'0'))
    # bottom child: assign a label '1'
    d.update(huffman_code_tree(bottom,binString+'1'))
    return d

enc_dict = huffman_code_tree(freq[0][0])
print(enc_dict)
```

```
{'b': '000', 'c': '001', 'd': '010', 'e': '011', 'a': '1'}
```

Notice that the resultant dictionary matches the desired codewords illustrated in Fig. 1.19.

Limitations of the Huffman code Although the Huffman algorithm provides an optimal code in practical scenarios with a finite number of source elements, its implementation has certain limitations. One limitation is the requirement for knowledge of source statistics to design the Huffman code, which means that the joint distribution of a super symbol, $p(x_1, x_2, \ldots, x_n)$, must be known. However, obtaining this statistical knowledge may not be straightforward in practice. Therefore, a pertinent question that arises in this practical scenario is: Is it possible to find optimal codes that do not rely on prior knowledge of source statistics?

Lempel-Ziv code (Ziv and Lempel, 1977, 1978) Lempel and Ziv are two individuals who could answer the question regarding whether there are optimal codes that do not rely on prior knowledge of source statistics. They developed a universal code called the Lempel-Ziv code, which can be applied to any type of information source without requiring statistical knowledge. The code has generated significant interest from system designers because it can approach the limit, i.e., the entropy rate. The Lempel-Ziv code has been implemented in various systems under different names, such as gzip, pkzip, and UNIX compression.

The basic idea behind the Lempel-Ziv code is straightforward and is something that people use in their daily lives. To illustrate the idea, let us consider the context of closing email phrases. People often use similar phrases such as "I look forward to your reply," "I look forward to seeing you," "I look forward to hearing from you," and "Your prompt reply would be appreciated." If we were to compress this English phrase as it is, it would require more bits than the number of alphabets in the phrase. However, the Lempel-Ziv code suggests that we can compress the phrase much more effectively using a dictionary. Since people use only a few closing phrases, we can create a dictionary that maps each phrase to an index like 1, 2, 3. This dictionary serves as the basis for the Lempel-Ziv code.

The Lempel-Ziv code operates in the following way: Firstly, a dictionary is created from a pilot sequence that forms part of the entire sequence. This dictionary is then shared between the source encoder and decoder. The source encoder only

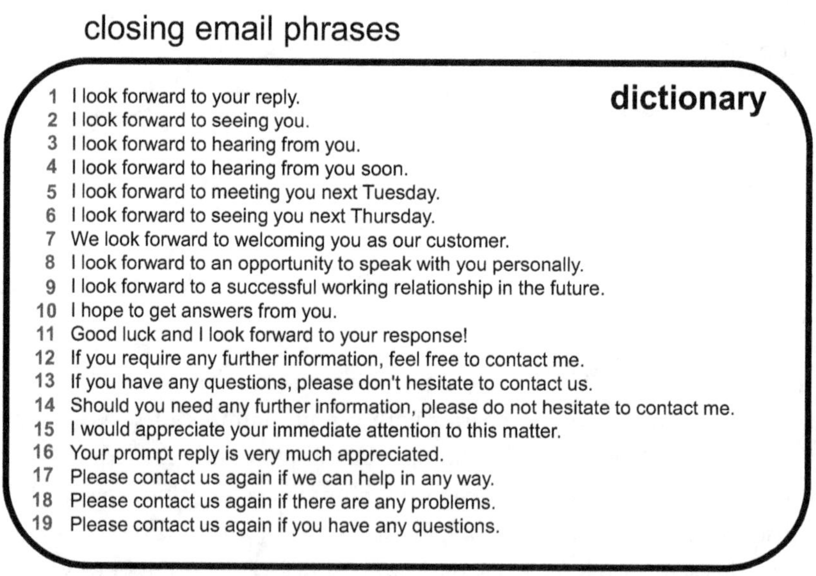

Figure 1.21. The idea of the Lempel-Ziv code. We construct a dictionary (e.g., assigning indices, marked in blue, for frequently used email phrases) using only part (pilot sequence) of the entire sequence. We then share the dictionary between source encoder and decoder. The decoder receives the indices and recovers the original sequence using the shared dictionary. This code is shown to achieve the limit (the entropy rate) as the dictionary size increases.

encodes the indices, while the decoder decodes the indices using the shared dictionary. By following this process, it is not necessary to have knowledge of the statistics of the information sources. Moreover, it has been demonstrated that this code can approach the limit, i.e., the entropy rate, as the dictionary size increases. However, since the focus of this book is not on this code, we will refrain from delving further into its details.

Look ahead Thus far, we have demonstrated the source coding theorem for general information sources and acquired knowledge about optimal code design. Additionally, we have gained insight into constructing an explicit optimal code for finite n, which is the Huffman code, along with its implementation using Python. These constitute the contents of Part I. In the ensuing section, we will commence with Part II.

Problem Set 3

Prob 3.1 (Markov's inequality) Consider a non-negative random variable X and a positive value $a > 0$.

(*a*) Show that $a \cdot \mathbf{1}\{X \geq a\} \leq X$ where $\mathbf{1}\{\cdot\}$ denotes an indicator function.
(*b*) Using part (*a*), prove Markov's inequality:

$$\mathbb{P}(X \geq a) \leq \frac{\mathbb{E}[X]}{a}.$$

Prob 3.2 (Chebychev inequality) Let X be a discrete random variable with mean μ and variance $\sigma^2 < \infty$. Show that for any $t > 0$,

$$\mathbb{P}(|X - \mu| \geq t) \leq \frac{\sigma^2}{t^2}.$$

Prob 3.3 (Weak Law of Large Numbers) Let $\{Y_i\}$ be an i.i.d. discrete random process with mean μ and variance $\sigma^2 < \infty$.

(*a*) State the Weak Law of Large Numbers (WLLN) w.r.t. $\{Y_i\}$.
(*b*) Prove the WLLN.

Prob 3.4 (Typical sequences) Let $\{X_i\}$ be an i.i.d. random process where $X_i \in \mathcal{X}$. Let $X^n := (X_1, X_2, \ldots, X_n)$. In Section 1.7, we claimed the following for an arbitrary alphabet case $|\mathcal{X}| = M$:

$$\frac{1}{n} \log \frac{1}{p(X^n)} \overset{\text{in prob.}}{\longrightarrow} H(X). \tag{1.49}$$

(*a*) Explain the meaning of the *convergence in prob.* in (1.49).
(*b*) Prove that (1.49) holds indeed.
(*c*) We say that a sequence x^n is ϵ-*typical* if it satisfies: for $\epsilon > 0$,

$$2^{-n(H(X)+\epsilon)} \leq p(x^n) \leq 2^{-n(H(X)-\epsilon)}.$$

Define a typical set that contains the typical sequences as elements:

$$A_\epsilon^{(n)} := \{x^n : 2^{-n(H(X)+\epsilon)} \leq p(x^n) \leq 2^{-n(H(X)-\epsilon)}\}. \tag{1.50}$$

Show that for any $\epsilon > 0$,

$$\mathbb{P}(A_\epsilon^{(n)}) := \mathbb{P}(X^n \in A_\epsilon^{(n)}) \longrightarrow 1 \text{ as } n \to \infty.$$

(*d*) Show that for any $\epsilon > 0$, $|A_\epsilon^{(n)}| \leq 2^{n(H(X)+\epsilon)}$.

Prob 3.5 (An example of ϵ-typical sequences) Let $\{X_i\}$ be an i.i.d. ternary random process:

$$X_i = \begin{cases} 1, & \text{w.p. } p; \\ 2, & \text{w.p. } q; \\ 3, & \text{w.p. } 1-p-q. \end{cases}$$

We say that a sequence x^n is *ϵ-typical* if it satisfies: for $\epsilon > 0$,

$$2^{-n(H(X)+\epsilon)} \leq p(x^n) \leq 2^{-n(H(X)-\epsilon)}. \tag{1.51}$$

Fix $\epsilon = 0.01$ and $(p,q) = (1/2, 1/4)$. Consider a sequence x^n such that

$$\frac{\#\text{ of 1's}}{n} = p + 0.05;$$
$$\frac{\#\text{ of 2's}}{n} = q - 0.03;$$
$$\frac{\#\text{ of 3's}}{n} = 1 - p - q - 0.02.$$

Is x^n ϵ-typical? Also explain why.

Prob 3.6 (A choice of codeword length) Let $\{X_i\}$ be an i.i.d. random process where $X_i \in \mathcal{X}$. Fix $\epsilon > 0$. In Section 1.7, we derived the following with the help of the WLLN:

$$n(H(X) - \epsilon) \leq \log \frac{1}{p(x^n)} \leq n(H(X) + \epsilon) \qquad \text{w.h.p.} \tag{1.52}$$

where w.h.p. stands for "with high probability". This motivated us to set the length of every codeword being equal to:

$$\ell(x^n) = \lceil n(H(X) + \epsilon) \rceil \quad \forall x^n \in A_\epsilon^{(n)} \tag{1.53}$$

where $A_\epsilon^{(n)}$ is a typical set defined as:

$$A_\epsilon^{(n)} := \{x^n : 2^{-n(H(X)+\epsilon)} \leq p(x^n) \leq 2^{-n(H(X)-\epsilon)}\}. \tag{1.54}$$

We also verified the validity of this choice by demonstrating that the number of leaves that codewords can be mapped in the design of a prefix-free code is greater than or equal to $|A_\epsilon^{(n)}|$.

On the other hand, one may suggest another choice:

$$\ell(x^n) = \lceil n(H(X) - \epsilon) \rceil \quad \forall x^n \in A_\epsilon^{(n)}. \tag{1.55}$$

In this problem, you are asked to prove that this choice is invalid.

(*a*) Fix $\epsilon > 0$. Show that for sufficiently large n,

$$\sum_{x \in A_\epsilon^{(n)}} p(x^n) \geq 1 - \epsilon. \tag{1.56}$$

(*b*) Fix $\epsilon > 0$. Show that for sufficiently large n,

$$|A_\epsilon^{(n)}| \geq (1 - \epsilon)2^{n(H(X)-\epsilon)}. \tag{1.57}$$

(*c*) Using part (*b*) or otherwise, show that the choice (1.55) on the codeword length is invalid in the limit of n.

Prob 3.7 (Source code design for an i.i.d. source) Let $\{X_i\}$ be an i.i.d. random process where $X_i \in \mathcal{X}$. Fix $\epsilon > 0$. In this problem, you are asked to construct a super symbol-based code of size n that achieves the expected codeword length (per symbol) of $H(X) + \epsilon$ in the limit of n. Notice that it can achieve the fundamental limit $H(X)$ (promised by the source coding theorem) with $\epsilon \to 0$.

(*a*) Let $A_\epsilon^{(n)}$ be a typical set defined as (1.54). Consider a prefix-free code C such that the first bit of $C(x^n)$ is assigned 0 if $x^n \in \mathcal{A}_\epsilon^{(n)}$; assigned 1 otherwise. The pattern of the remaining bits of $C(x^n)$ is specified by a binary code tree constructed as follows. To the internal node associated with the first upper branch (labeled 0), we attach a full subtree of depth $\lceil n(H(X) + \epsilon) \rceil$ so that the total number of leaves in the top subtree is $2^{\lceil n(H(X)+\epsilon)\rceil}$. If $x^n \in \mathcal{A}_\epsilon^{(n)}$, we map the typical sequence into one of the leaves in the top subtree. Different sequences are assigned distinct leaves.

To the other internal code associated with the first *lower* branch (labeled 1), we attach another full subtree of depth $\lceil n \log |\mathcal{X}| \rceil$ so that the total number of leaves in the bottom subtree is $2^{\lceil n \log |\mathcal{X}| \rceil}$. If $x^n \notin \mathcal{A}_\epsilon^{(n)}$, we map the non-typical sequence into one of the leaves in the bottom subtree. Show that this prefix-free code is valid, i.e., there are enough leaves for mapping all possible sequences x^n's.

(*b*) Suppose we employ the prefix-free code in part (*a*). Show that

$$\lim_{n \to \infty} \frac{\mathbb{E}[\ell(X^n)]}{n} = H(X) + \epsilon$$

where $\ell(X^n)$ indicates the length of $C(X^n)$.

Prob 3.8 (Convergence of a non-increasing & non-negative sequence) Consider a deterministic sequence $\{a_i\}$ satisfying the following two properties: (i) $a_i \geq 0$ and (ii) $a_i \geq a_{i+1}$.

(*a*) State the definition of the *existence of a limit.*
(*b*) Show that a_i converges to a limit: $\lim_{i\to\infty} a_i (=: a)$.
(*c*) Show that

$$\lim_{n\to\infty} \frac{1}{n}\sum_{i=1}^{n} a_i = a.$$

Prob 3.9 (Source code design for a generalized Markov process) Let $\{X_i\}$ be a generalized Markov process with two memory states:

$$p(s_n|s_{n-1}, s_{n-2}, \ldots, s_1) = p(s_n|s_{n-1})$$

where $s_n := (x_{n-1}, x_n)$. Suppose that

$$\mathbb{P}(S_n = 00|S_{n-1} = 00) = \mathbb{P}(S_n = 11|S_{n-1} = 11) = p,$$

$$\mathbb{P}(S_n = 10|S_{n-1} = 01) = \mathbb{P}(S_n = 01|S_{n-1} = 10) = 0.5.$$

(*a*) Show that the minimum number of bits that represent the above information source per symbol is

$$H(\mathcal{X}) = \lim_{n\to\infty} H(X_n|X^{n-1}).$$

In Section 1.8, we introduced a terminology for the quantity $H(\mathcal{X})$. What is it? Also explain the rationale behind the naming.
(*b*) Using part (*a*), show that $H(\mathcal{X}) = H(X_3|X_1, X_2)$. Also compute $H(X_3|X_1, X_2)$.
(*c*) Define a typical set:

$$A_\epsilon^{(n)} := \{x^n : 2^{-n(H(X_3|X_1,X_2)+\epsilon)} \le p(x^n) \le 2^{-n(H(X_3|X_1,X_2)-\epsilon)}\}.$$

It has been verified that for any $\epsilon > 0$,

$$\mathbb{P}(A_\epsilon^{(n)}) := \mathbb{P}\left(X^n \in A_\epsilon^{(n)}\right) \longrightarrow 1 \text{ as } n \to \infty.$$

Using this together with part (*b*), construct a prefix-free code (i.e., draw a binary code tree) in which the expected codeword length (per symbol) approaches $H(X_3|X_1, X_2)$ as $n \to \infty$. Also show that the expected codeword length of your code indeed achieves the limit $H(X_3|X_1, X_2)$.

Prob 3.10 (WLLN of a stationary process) Let $\{X_i\}$ be a stationary process. Suppose that $\mu := \mathbb{E}[X_i]$; $\sigma_k^2 := \mathbb{E}[(X_i - \mu)(X_{i+k} - \mu)]$, $\forall k \in \mathbb{N}$; and $\sum_{k=0}^{\infty} \sigma_k^2 < \infty$.

(*a*) Prove that

$$S_n := \frac{X_1 + X_2 + \cdots + X_n}{n} \longrightarrow \mu \quad \text{in prob.} \tag{1.58}$$

(*b*) As mentioned in Section 1.8, (1.58) is a generalized version of the Weak Law of Large Numbers (WLLN). Using this, show that

$$\frac{1}{n}\log\frac{1}{p(X^n)} \longrightarrow H(\mathcal{X}) \quad \text{in prob.}$$

Here $X^n := (X_1, X_2, \ldots, X_n)$.

Prob 3.11 (Statistics of optimally compressed output) Consider an information source $S_1, S_2, \ldots, S_n$. Suppose we use an *optimal* source code that maps $(S_1, S_2, \ldots, S_n)$ into a binary string $(b_1, b_2, \ldots, b_m)$. Compute $H(b_1, b_2, \ldots, b_m)$. *Hint:* You may want to assume that n is sufficiently large.

Prob 3.12 (Entropy rate) Let $\mathbf{X} = \{X_n, n \in \mathbb{Z}\}$ be a stationary sequence of random variables taking values in a finite alphabet $\mathcal{X}$. Define conditional entropy:

$$H_{L|L}(\mathbf{X}) := \frac{1}{L}H(X_{2L}, \ldots, X_{L+1}|X_L, \ldots, X_1).$$

The confused information theorist claims that

$$\lim_{L\to\infty} H_{L|L}(\mathbf{X}) = H(\mathbf{X})$$

where $H(\mathbf{X})$ denotes the entropy rate of the source. Prove or disprove this claim.

Prob 3.13 (Entropy rate and Markov chain) The standard nomenclature designates the four quadrants of the Euclidean plane as NE, NW, SE, and SW (i.e., northeast, northwest, southeast, and southwest). The particle travels between these quadrants, moving equiprobably either vertically or horizontally at each time, which means that it switches quadrants with equal probability. For example, if it currently occupies the NE quadrant, there is an equal probability that it will be in the SE quadrant or the NW quadrant at the next time instant. The particle's movement is denoted by the standard directional labels N, S, W, or E, with the move from NE to SE labeled as S and the move from SW to SE labeled as E, for instance. At time 0, the particle is equally likely to be in any of the quadrants. The $\{X_n, n \geq 0\}$ defines the quadrant in which the particle is located at time n, whereas $\{Y_n, n \geq 0\}$ provides the label for the move made by the particle from time n to time $n + 1$.

(*a*) Is $\{X_n, n \geq 0\}$ a Markov chain? Either prove or disprove it.

(*b*) Is $\{X_n, n \geq 0\}$ stationary? Either prove or disprove it.
(*c*) Find the mutual information rate between $\{X_n, n \geq 0\}$ and $\{Y_n, n \geq 0\}$:

$$\lim_{L \to \infty} \frac{1}{L} I(X^L; Y^L).$$

(*d*) Is $\{Y_n, n \geq 0\}$ a Markov chain? Either prove or disprove it.
(*e*) Is $\{Y_n, n \geq 0\}$ stationary? Either prove or disprove it.

Prob 3.14 (Huffman code construction) Consider a random variable with the probability distribution:

$$X = \begin{pmatrix} x_1 & x_2 & x_3 & x_4 & x_5 & x_6 & x_7 \\ 0.5 & 0.25 & 0.2 & 0.02 & 0.01 & 0.01 & 0.01 \end{pmatrix}.$$

(*a*) Construct a binary Huffman code for X.
(*b*) Construct a ternary Huffman code for X.

Prob 3.15 (Optimality of the Huffman code) Let $X \in \{a_1, a_2, \ldots, a_M\}$ be a discrete random variable where $M \geq 3$ is an integer. Let p_i be a pmf of X where $i \in \{1, 2, \ldots, M\}$. Without loss of generality, assume that $p_1 \geq p_2 \geq \cdots \geq p_M$. Consider a source code which takes X as an input. Let ℓ_i be the codeword length w.r.t. a_i. Describe the Huffman algorithm. Prove the optimality of the algorithm.

Prob 3.16 (Principles of the Huffman code) A student claims that if the most probable letter in an alphabet has probability less than $\frac{1}{3}$, then any Huffman code will assign a codeword length of at least 2 to that letter. Prove or disprove the claim.

Prob 3.17 (Principles of the Huffman code) Let X be a random variable taking values in the finite set $\{1, 2, 3\}$. Let $\ell(X)$ denote the expected length of an optimal Huffman code for X. A student claims that if $\mathbb{P}(X = i) > 0$ for all $i = 1, 2, 3$, then $\ell(X) - H(X) \geq \frac{5}{3} - \log 3$. Either prove or disprove this statement.

Prob 3.18 (True or False?)

(*a*) Let $\{X_i\}$ be a random process. We say that X_n converges to X in probability if for any $\epsilon > 0$,

$$\mathbb{P}(|X_n - X| \leq \epsilon) \to 1 \quad \text{as } n \to \infty.$$

(*b*) For a discrete random variable X and $\epsilon > 0$, consider a typical set:

$$A_\epsilon^{(n)} := \{x^n : 2^{-n(H(X)+\epsilon)} \leq p(x^n) \leq 2^{-n(H(X)-\epsilon)}\}.$$

Then,

$$|A_\epsilon^{(n)}| \geq (1-\epsilon)2^{n(H(X)-\epsilon)}.$$

(*c*) Let $\{X_n\}$ be an i.i.d. sequence of binary random variables, each equally likely to be 0 or 1. Define

$$Y_n := |\{j : 1 \leq j \leq n, X_j = 1\}|, \quad n = 1, 2, \ldots$$

Then, the entropy rate of the process $\{Y_n\}$ is 1 bit/symbol.

(*d*) Consider discrete random variables X and Y. Suppose that X and Y are independent. Then, the expected codeword length of a Huffman code for the pair (X, Y) is at least as large as the sum of the expected codeword lengths of individual Huffman codes for X and Y.

(*e*) Let $X \in \mathcal{X}$ be a discrete random variable with pmf $p(x)$. In Section 1.5, we consider the following optimization problem:

$$\min_{\ell(x)} \sum_{x \in \mathcal{X}} p(x)\ell(x) :$$

$$\sum_{x \in \mathcal{X}} 2^{-\ell(x)} = 1$$

$$\ell(x) \in \mathbb{N}.$$

The closed form solution to the problem has been open thus far. Hence, no one knows how to construct an optimal $\ell^*(x)$ that minimizes the objective function.

DOI: 10.1561/9781638281153.ch2

Chapter 2

Channel Coding

2.1 Statement of Channel Coding Theorem

Recap In Part I, we delved into Shannon's communication architecture consisting of two stages. The first stage, source coding, aims to convert an information source, which may be of a different type, into bits, a common information currency. The second stage, channel coding, converts these bits into a signal that can be transmitted reliably over a channel to the receiver. We explored the source coding theorem, which determines the maximum compression rate of an information source, and learned how to design optimal codes that achieve this limit. Specifically, we investigated the Huffman code and practiced implementing it in Python.

Moving onto Part II, we shift our focus to the second block in Shannon's architecture. Here, we will examine the channel coding theorem, which delineates the fundamental limit on the number of bits that can be transmitted reliably over a channel.

Outline Recall the statement we made in Section 1.1 regarding the channel coding theorem. Similar to the laws of physics, there is a fundamental law in communication systems. It states that the maximum amount of information that can be transmitted over a channel is fixed, regardless of any operations performed by the transmitter or receiver. To put it differently, communication systems have a law similar to that of physics. The maximum number of bits that can be transmitted over a channel is determined, regardless of any operations performed at the transmitter and receiver. This means that there is a fundamental limit on the number of bits that allows communication and beyond which communication becomes impossible, no matter what we do. The limit is known as the channel capacity. This section aims to provide a more precise understanding of this statement by examining (i) a specific problem scenario that Shannon investigated, (ii) a mathematical representation of channels, and (iii) the mathematical definition of communication that is possible or impossible. Once we have a clear understanding of the statement, we will proceed to prove the theorem.

Problem setup The objective of communication is to transmit binary string information (bits) to the receiver as much as possible. Our investigation begins with analyzing the statistics of the binary string. Unlike the information source in the context of source coding, it is possible to assume simple statistics for the binary string. In the source coding scenario, the information source has arbitrary statistics that depend on the application of interest, and source code design is customized accordingly. In contrast, in channel coding, there is some good news; the input binary string's statistics are not arbitrary. Under a reasonable assumption, it follows a particular distribution. This reasonable assumption is that we use an optimal source code.

To determine the specific statistics, we can use the source coding theorem. Suppose the binary string $(b_1, b_2, \ldots, b_m)$ is the output of an optimal source encoder. In that case, we can apply the source coding theorem to obtain:

$$\begin{aligned} m &= H(\text{information source}) \\ &= H(b_1, b_2, \ldots, b_m) \end{aligned} \tag{2.1}$$

where the second equality follows from the fact that a source encoder is one-to-one mapping and the fact $H(X) = H(f(X))$ for an one-to-one mapping function f. Why? Now observe that

$$\begin{aligned} H(b_1, b_2, \ldots, b_m) &= H(b_1) + H(b_2|b_1) + \cdots + H(b_m|b_1, \ldots, b_{m-1}) \\ &\le H(b_1) + H(b_2) + \cdots + H(b_m) \\ &\le m \end{aligned} \tag{2.2}$$

where the first inequality is due to the fact that conditioning reduces entropy and the last inequality comes from the fact that b_i's are binary random variables. This together with (2.1) suggests that the inequalities in (2.2) are tight. This implies that b_i's are independent (from the first inequality) and identically

Figure 2.1. Channel coding problem setup.

distributed $\sim$ Bern$(\frac{1}{2})$ (from the second inequality). Hence, we can make the following assumption: a binary string, an input to a channel encoder, is i.i.d., each being according to Bern$(\frac{1}{2})$.

For notational simplicity, information theorists introduced a simple notation which indicates the i.i.d. random process $\{b_i\}$. They expressed it with a single random variable, say W, using the following mapping rule:

$$
\begin{aligned}
(b_1, \ldots, b_m) &= (0, 0, \ldots, 0) \longrightarrow W = 1; \\
(b_1, \ldots, b_m) &= (0, 0, \ldots, 1) \longrightarrow W = 2; \\
&\vdots \\
(b_1, \ldots, b_m) &= (1, 1, \ldots, 1) \longrightarrow W = 2^m.
\end{aligned}
$$

This allows us to express the input simply with W having the uniform distribution. Why uniform?

What to design The design of the digital communication system involves designing two components: (i) a channel encoder, say f; and (ii) a channel decoder, say g. See Fig. 2.1. One crucial consideration in the design process is the channel's behavior. As previously mentioned, the channel is an adversary that introduces errors, turning the system into a random function. To combat these errors, the encoder and decoder must be designed to provide protection. One way to accomplish this is by adding redundancy, which is akin to repeating the transmission when communication fails. Building on this idea, we can represent the channel encoder's output as a sequence of symbols $(X_1, X_2, \ldots, X_n)$, where n is the code length. This n is distinct from the super-symbol size utilized in the source coding context. We use the shorthand notation $X^n := (X_1, X_2, \ldots, X_n)$ to represent this sequence. We denote Y^n as the channel's output, and the decoder g's objective is to infer W from Y^n. The challenge here is that Y^n is not a deterministic function of X^n. If the channel were deterministic, g would be the inverse function of the concatenation of f and the channel. However, as the channel is not deterministic in reality, designing g is not straightforward. In fact, understanding the channel's behavior is critical in designing g. Therefore, before delving into details on f and g, we must discuss how the channel behaves and how to model it briefly.

Channel modeling Let X_i and Y_i be the input and the output of the channel at time i, respectively. One typical way to capture the randomness of the channel is via conditional distribution $p(y|x)$. To give you an idea, consider one concrete example: binary erasure channel (BEC). In the BEC, Y_i takes "erasure" (garbage) with probability, say p (called the erasure probability), while taking the input X_i cleanly otherwise. So the relation between X_i and Y_i is given by:

$$Y_i = \begin{cases} X_i, & \text{w.p. } 1-p; \\ \mathsf{e}, & \text{w.p. } p \end{cases}$$

where e stands for "erasure". Usually we consider a mememoryless channel in which this relationship is independent across different time instants.

Two performance metrics Let us investigate how to design f and g such that communication is possible. To this end, we first need to understand what it means by possible communication. To explain what it means, we will introduce two performance metrics.

The first performance metric captures the amount of bits that we wish to transmit. Suppose $W \in \{1, 2, \ldots, M\}$. Then, the total number of bits that we intend to send is $\log M$. Since code length is n, the number of channels (time instants) that we use is n and hence, the number of bits transmitted per channel use (called data rate) is

$$R = \frac{\log M}{n} \quad \text{bits/channel use.}$$

The second performance metric is w.r.t. the decoding quality. Since the channel is random, we cannot always guarantee the decoded message (say $\hat{W}$) to be the same as the original message W. Hence, the decoding quality can be captured by the probability of an error event:

$$P_e := \mathbb{P}(W \neq \hat{W}).$$

The smaller P_e, the better the decoding quality.

Optimization problem Shannon devised an optimization problem using the two performance metrics and introduced the concept of possible communication, which will be explained shortly. Let us first take a look at the optimization problem. One can expect that there should be tradeoff relationship between R and P_e. The larger R, the larger P_e and vice versa. So one natural optimization problem is the following. Given R and n:

$$P_e^* := \min_{f,g} P_e.$$

What Shannon realized is that unfortunately, this is again a *non-convex* optimization problem, being very difficult to solve. Shannon took a different approach, as he did for the source coding theorem. This is, to *approximate*. In other words, he attempted to develop upper and lower bounds on P_e^*.

In the process, he discovered something interesting, which then led to the concept of possible communication. What he found is that when R is below a certain threshold, an upper bound on P_e^* can be made arbitrarily close to 0 as n increases. This implies that in the case, the actual P_e^* can be made very small. He also found that when R is above the threshold, a lower bound on P_e^* cannot be made arbitrarily close to 0 no matter what we do. This implies that the actual P_e^* cannot be made arbitrarily small.

This observation led him to come up with the concept of possible communication. We say that communication is possible if one can make P_e^* arbitrarily close to 0 as $n \to \infty$; otherwise communication is said to be impossible. He also came up with the related concept of *achievable* rate. We say that data rate R is *achievable* if we can make $P_e^* \to 0$ as $n \to \infty$ given R; otherwise R is said to be not achievable.

Channel coding theorem Moreover, Shannon made an interesting observation. What he observed is that the threshold below which communication is possible and above which communication impossible is *sharp*. In other words, there is a sharp *phase transition* on the achievable rate. He then called the limit *channel capacity* and denoted it by C. This forms the channel coding theorem:

The maximum achievable rate is channel capacity C.

From the next section onwards, we will prove this theorem. Specifically we will prove the following two:

$$R \leq C \Longrightarrow P_e \to 0$$

$$R > C \Longrightarrow P_e \nrightarrow 0.$$

The first is called the *achievability* proof or the direct proof. The second is called the *converse* proof. Why converse? Notice that the contraposition of the second statement is:

$$P_e \to 0 \Longrightarrow R \leq C,$$

which is exactly the reverse of the first statement.

Look ahead In the upcoming section, we will prove the achievability for a simple example. By deriving insights from this example, we will subsequently prove the achievability for a broader class of channels, determined by any arbitrary conditional distribution $p(y|x)$.

2.2 Achievability Proof for the Binary Erasure Channel

Recap In the preceding section, we presented the framework for the channel coding problem. We formally stated the channel coding theorem, which was initially introduced in a vague manner at the start of this book. Two performance metrics were considered: (i) the data rate R, defined as $\frac{\log M}{n}$ (bits/channel use); and (ii) the probability of error P_e, defined as $\mathbb{P}(W \neq \hat{W})$. Here, W belongs to the set $\{1, 2, \ldots, M\}$, and n represents the code length (i.e., the number of channel uses). To investigate the tradeoff between these two metrics, we formulated the following optimization problem. Given R and n:

$$P_e^* = \min_{f,g} P_e$$

where f and g indicate channel encoder and decoder respectively. Unfortunately, Shannon could not solve this problem. Instead he looked into upper and lower bounds. In the process, he made an interesting observation. If R is less then a threshold, say C, then P_e can be made arbitrarily close to 0 as $n \to \infty$; otherwise, $P_e \nrightarrow 0$ no matter what we do. This leads to the natural concept of *achievable rate*. The data rate R is said to be *achievable* if we can make $P_e \to 0$ for a given R. This finally formed the channel coding theorem:

$$\text{Maximum achievable rate} = C.$$

The channel coding theorem requires the proof of two parts. The first is: $R \leq C \Longrightarrow P_e \to 0$. To this end, we need to come up with an *achievable* scheme which yields $P_e \to 0$ given $R \leq C$. This is called the *achievability* proof. The second part to prove is: $R > C \Longrightarrow P_e \nrightarrow 0$. The contraposition of this statement is: $P_e \to 0 \Longrightarrow R \leq C$, which is the opposite of the statement for the first part proof. Hence, it is called the *converse* proof.

Outline In this section, our goal is to prove the achievability of the channel coding theorem. To achieve this, we will examine a simple example of a channel known as the binary erasure channel (BEC). This example provides valuable insights into proving the achievability of more complex channels. Once we have acquired enough insights from the BEC, we will apply them to a fairly general channel model setting.

Binary erasure channel (BEC) Remember that the channel output reads:

$$Y_i = \begin{cases} X_i, & \text{w.p. } 1-p; \\ \mathsf{e}, & \text{w.p. } p \end{cases}$$

where e stands for "erasure" (garbage) and p indicates the erasure probability. As mentioned earlier, we consider a memoryless channel in which noises are independent across different time instances.

First of all, let us guess what the channel capacity C is. One can make a naive guess based on the following observation. The channel is perfect w.p. $1-p$; erased w.p. p; and the erasure (garbage) does not provide any information w.r.t. what we transmit. This naturally leads to: C is at most $1-p$. Then, a question arises: Is this achievable? Think about one extreme scenario in which the transmitter knows all erasure patterns across all time instances beforehand. Of course this is far from reality. But for simplicity, consider this case for the time being. In this case, one can readily achieve $1-p$. The achievable scheme is to transmit a bit whenever the channel is perfect. Since the perfect channel probability is $1-p$, by the WLLN, we can achieve $1-p$ as n tends to infinity.

Let us consider a realistic scenario in which we cannot predict the future events and thus the transmitter has no idea of erasure patterns. In this case, we cannot apply the above naive transmission scheme because each transmission of a bit is not guaranteed to be successful due to the lack of the knowledge of erasure patterns. One may imagine that it is impossible to achieve $1-p$. Interestingly, one can still achieve $1-p$ even in this realistic scenario.

How to encode? Here is a transmission scheme. Fix R arbitrarily close to $1-p$. Given R, what is M (the cardinality of the range of the message W)? Since $R := \frac{\log M}{n}$, one needs to set it as: $M = 2^{nR}$. The message W takes one of the values among $1, 2, \ldots, 2^{nR}$.

Next, how to encode the message W? In other words, what is a mapping rule between W and X^n? Here we call X^n codeword[1]. Shannon's encoding rule, to be explained in the sequel, enables us to achieve $R = 1-p$. Shannon's encoding is simple. The idea is to generate a random binary string given any $W = w$. In other words, for any $W = w$, every component of $X^n(w)$ follows a binary random variable with parameter $\frac{1}{2}$ and those are independent with each other, i.e., $X_i(w)$'s are i.i.d. $\sim \mathsf{Bern}(\frac{1}{2})$ across all i's. These are i.i.d. also across all w's. This looks like a dumb scheme. Surprisingly, Shannon showed that this dumb scheme can achieve the rate of $1-p$. We have a terminology indicating a collection (book) of $X_i(w)$'s. It is called a codebook.

How to decode? Let us move onto the decoder side. The decoder input is a received signal Y^n (channel output). One can make a reasonable assumption on

---

1. The terminology "codeword" was used to indicate the output of the source encoder in Part I. By convention, people employ the same terminology for the output of channel encoder.

the decoder. The codebook, a collection of $X_i(w)$'s, is known. This assumption is realistic because this information can be shared only at the beginning of communication. Once this information is shared, we can use this all the time until communication is terminated.

How to decode W from Y^n, assuming the knowledge of the codebook? What is an optimal way of decoding W? Remember the second performance metric: the probability of error $P_e := \mathbb{P}(W \neq \hat{W})$. Due to the stochastic nature of the channel, the best we can hope for is to minimize the probability of error. Thus, an optimal decoder is one that achieves the minimum possible probability of error. Alternatively, the optimal decoder can be defined as the one that maximizes the success probability, $\mathbb{P}(W = \hat{W})$. Since the received signal is known (as it is an input to the decoder), the success probability can be defined as the conditional probability of correctly decoding the message, given the received signal:

$$\mathbb{P}(W = \hat{W} | Y^n = y^n).$$

Then, a more formal definition of the optimal decoder is:

$$\hat{W} = \arg\max_w \mathbb{P}(W = w | Y^n = y^n).$$

We have a terminology for the optimal decoder. Notice that $\mathbb{P}(W = w|Y^n = y^n)$ is the probability *after* an observation y^n is made; and $\mathbb{P}(W = w)$ is called the *a priori* probability because it is the one known *beforehand*. Hence, $\mathbb{P}(W = w|Y^n = y^n)$ is called the *a posteriori* probability. Observe that the optimal decoder is the one that Maximizes A Posteriori (MAP) probability. So it is called the MAP decoder. The MAP decoder acts as an optimal decoder for many interesting problems not limited to this problem context. As long as a problem of interest is an *inference problem* (*infer* X from Y when X and Y are *probabilistically* related), the optimal decoder is always the MAP decoder.

In fact, this MAP decoder can be simplified further in many cases including ours as a special case. Here is how it is simplified. Using the definition of conditional probability, we get:

$$\begin{aligned}\mathbb{P}(W = w|Y^n = y^n) &= \frac{\mathbb{P}(W = w, Y^n = y^n)}{\mathbb{P}(Y^n = y^n)} \\ &= \frac{\mathbb{P}(W = w)}{\mathbb{P}(Y^n = y^n)} \cdot \mathbb{P}(Y^n = y^n|W = w).\end{aligned}$$

Notice that $\mathbb{P}(W = w) = \frac{1}{2^{nR}}$, as it is irrelevant to w. Also $\mathbb{P}(Y^n = y^n)$ is not a function of w. This implies that it suffices to consider only $\mathbb{P}(Y^n = y^n|W = w)$

in figuring out when the above probability is maximized. Hence, we obtain:

$$\begin{aligned}\hat{W} &= \arg\max_w \mathbb{P}(W = w | Y^n = y^n)\\ &= \arg\max_w \mathbb{P}(Y^n = y^n | W = w).\end{aligned}$$

Given $W = w$, X^n is known as $x^n(w)$. Hence,

$$\begin{aligned}\hat{W} &= \arg\max_w \mathbb{P}(Y^n = y^n | W = w)\\ &= \arg\max_w \mathbb{P}(Y^n = y^n | X^n = x^n(w), W = w).\end{aligned}$$

Also given $X^n = x^n(w)$, Y^n is a sole function of the channel, meaning the independence between Y^n and w. Hence,

$$\begin{aligned}\hat{W} &= \arg\max_w \mathbb{P}(Y^n = y^n | X^n = x^n(w), W = w)\\ &= \arg\max_w \mathbb{P}(Y^n = y^n | X^n = x^n(w)).\end{aligned}$$

Notice that $\mathbb{P}(Y^n = y^n | X^n = x^n(w))$ is nothing but conditional distribution, which is easy to compute. There is another name for the conditional distribution: *likelihood*. So the decoder is called the Maximum Likelihood (ML) decoder.

How to derive the ML decoder? The ML decoder is simple in our problem context. Let us see this through a simple example where $n = 4$ and $R = \frac{1}{2}$. In this example, $M = 2^{nR} = 4$. Consider a particular codebook:

$$\begin{aligned}X^n(1) &= 0000;\\ X^n(2) &= 0110;\\ X^n(3) &= 1010;\\ X^n(4) &= 1111.\end{aligned}$$

Suppose $y^n = 0e00$. Then, one can compute the likelihood $\mathbb{P}(y^n | x^n(w))$. For instance,

$$\mathbb{P}(y^n | x^n(1)) = (1-p)^3 p.$$

The channel is perfect three times (time 1, 3 and 4), while being erased at time 2. On the other hand,

$$\mathbb{P}(y^n | x^n(2)) = 0$$

because the third bit 1 does not match $y_3 = 0$, and this is the event that would never happen. In other words, the second message is *incompatible* with the received signal y^n. Then, what is the ML decoding rule? Here is what the rule says.

1. Eliminate all the messages which are incompatible with the received signal.
2. If there is only one message that survives, declare the survival as the correct message that the transmitter sent.

However, this procedure is not sufficient to describe the ML decoding rule. The reason is that we may have a different erasure pattern that confuses the rule. To see this clearly, consider the following example. Suppose the received signal is now $y^n = (0ee0)$. Then,

$$X^n(1) = (1-p)^2 p^2;$$
$$X^n(2) = (1-p)^2 p^2.$$

The two messages (1 and 2) are compatible and those likelihood functions are equal. In such cases, the best we can do is to randomly choose one out of the two options. This forms the ML decoding rule.

3. If there are multiple survivals, choose one randomly.

Look ahead In the next section, we will demonstrate that we can achieve the rate of $1-p$ under the optimal ML decoding rule.

2.3 Achievability Proof for the Binary Symmetric Channel

Recap In the prior section, we claimed that the capacity of the BEC with erasure probability p is

$$C_{\mathsf{BEC}} = 1 - p.$$

We then attempted to prove the achievability: if $R < 1-p$, we can make P_e arbitrarily close to 0 as $n \to \infty$. We employed a random codebook where each component $X_i(w)$ of the codebook follows $\mathsf{Bern}(\frac{1}{2})$ and is i.i.d. across all $i \in \{1, \ldots, n\}$ and $w \in \{1, \ldots, 2^{nR}\}$. We also employed the optimal decoder: the maximum likelihood (ML) decoder in our problem context where the message W is uniformly distributed. Next, we intended to complete the achievability proof, by showing $P_e \to 0$ under the problem setup.

Outline In this section, we will finish the proof and move on to another channel example that provides additional insights into the general channel setting. The channel we will focus on is called the Binary Symmetric Channel (BSC).

Probability of error The probability of error is a function of codebook $\mathcal{C}$. So we are going to investigate the average error probability $\mathbb{E}_{\mathcal{C}}[P_e(\mathcal{C})]$ taken over all possible random codebooks, and will demonstrate that $\mathbb{E}_{\mathcal{C}}[P_e(\mathcal{C})]$ approaches 0 as $n \to \infty$. This then implies the existence of an optimal deterministic codebook, say $\mathcal{C}^*$, such that $P_e(\mathcal{C}^*) \to 0$ as $n \to \infty$. Why? See Prob 4.3 for the proof of the existence. Consider:

$$\begin{aligned}
\mathbb{E}_{\mathcal{C}}[P_e(\mathcal{C})] &\stackrel{(a)}{=} \sum_{c} \mathbb{P}(\mathcal{C} = c)\mathbb{P}(\hat{W} \neq W | \mathcal{C} = c) \\
&\stackrel{(b)}{=} \mathbb{P}(\hat{W} \neq W) \qquad (2.3) \\
&\stackrel{(c)}{=} \sum_{w=1}^{2^{nR}} \mathbb{P}(W = w)\mathbb{P}\left(\hat{W} \neq w | W = w\right)
\end{aligned}$$

where (a) follows from the fact that $\mathbb{P}(\mathcal{C} = c)$ indicates the probability that codebook is chosen as a particular realization c; (b) and (c) are due to the total probability law.

Now consider $\mathbb{P}(\hat{W} \neq w | W = w)$. Notice that $[X_1(w), \ldots, X_n(w)]$'s are identically distributed over all w's. The way that the wth codeword is constructed is the same for all w's. This implies that $\mathbb{P}(\hat{W} \neq w | W = w)$ is irrelevant of what the particular value of w is, meaning that

$$\mathbb{P}(\hat{W} \neq w | W = w) \text{ is the same for all } w.$$

This together with (2.3) gives:

$$\begin{aligned} \mathbb{E}_{\mathcal{C}}[P_e(\mathcal{C})] &= \mathbb{P}(\hat{W} \neq 1 | W = 1) \\ &= \mathbb{P}\left(\bigcup_{w=2}^{2^{nR}} \{\hat{W} = w\} | W = 1 \right) \end{aligned} \tag{2.4}$$

where the second equality is due to the fact that $\{\hat{W} \neq 1\}$ means that $\hat{W} = w$ for some $w \neq 1$. In general, the probability of the union of multiple events is not that simple to compute. Rather it is quite complicated especially when it involves a large number of multiple events. Even for the three-event case (A, B, C), the probability formula is not simple:

$$\begin{aligned} \mathbb{P}(A \cup B \cup C) = \mathbb{P}(A) + \mathbb{P}(B) + \mathbb{P}(C) \\ - \mathbb{P}(A \cap B) - \mathbb{P}(A \cap C) - \mathbb{P}(B \cap C) + \mathbb{P}(A \cap B \cap C). \end{aligned}$$

Even worse, the number of associated multiple events in (2.4) is $2^{nR} - 1$ and this will make the probability formula very complicated. So the calculation of that probability is involved. To make some progress, Shannon took an indirect approach as he did in the proof of the source coding theorem. He did not attempt to compute the probability of error exactly. Instead, he intended to *approximate* it. He tried to derive an upper bound because what we want to show at the end of day is that the probability of error approaches 0. Note that if an upper bound goes to zero, the exact quantity also approaches 0. We have a very well-known upper bound w.r.t. the union of multiple events. That is, the *union bound*: for events A and B,

$$\mathbb{P}(A \cup B) \leq \mathbb{P}(A) + \mathbb{P}(B).$$

The proof is immediate. It is because $\mathbb{P}(A \cap B) \geq 0$.

Now applying the union bound to (2.4), we get:

$$\begin{aligned} \mathbb{E}_{\mathcal{C}}[P_e(\mathcal{C})] &\leq \sum_{w=2}^{2^{nR}} \mathbb{P}(\hat{W} = w | W = 1) \\ &= (2^{nR} - 1)\mathbb{P}(\hat{W} = 2 | W = 1) \end{aligned}$$

where the equality follows from the fact that the codebook employed is symmetric w.r.t message indices. The event of $\hat{W} = 2$ implies that message 2 is *compatible*;

otherwise, $\hat{W}$ cannot be chosen as 2. Using the fact that for two events A and B such that A implies B, $\mathbb{P}(A) \leq \mathbb{P}(B)$, we get:

$$\begin{aligned}\mathbb{E}_{\mathcal{C}}[P_e(\mathcal{C})] &\leq (2^{nR} - 1)\mathbb{P}(\text{message 2 is } \textit{compatible} | W = 1) \\ &\leq 2^{nR}\mathbb{P}(\text{message 2 is } \textit{compatible} | W = 1).\end{aligned}$$

Message 2 being compatible implies that for time i in which the transmitted signal is not erased, $X_i(2)$ must be the same as $X_i(1)$; otherwise, message 2 cannot be compatible. To see this clearly, let $\mathcal{B} = \{i : Y_i \neq \mathrm{e}\}$. Then, what the above means is that we get:

$$\begin{aligned}\mathbb{E}_{\mathcal{C}}[P_e(\mathcal{C})] &\leq 2^{nR}\mathbb{P}\left(\bigcap_{i\in\mathcal{B}}\{X_i(1) = X_i(2)\} | W = 1\right) \\ &= 2^{nR}\left(\frac{1}{2}\right)^{|\mathcal{B}|}\end{aligned} \tag{2.5}$$

where the equality is due to the fact that $\mathbb{P}(X_i(1) = X_i(2)) = \frac{1}{2}$ and the codebook is independent across all i's.

Observe that due to the WLLN, for sufficiently large n:

$$|\mathcal{B}| \approx n(1 - p).$$

This together with (2.5) yields:

$$\mathbb{E}_{\mathcal{C}}[P_e(\mathcal{C})] \lesssim 2^{n(R-(1-p))}.$$

Hence, $\mathbb{E}_{\mathcal{C}}[P_e(\mathcal{C})]$ can be made arbitrarily close to 0 as $n \to \infty$, as long as $R < 1 - p$. This completes the achievability proof.

Binary symmetric channel (BSC) Before moving on to arbitrary channels, let us delve deeper into the Binary Symmetric Channel (BSC) to gain more understanding. While the techniques used in proving achievability for the Binary Erasure Channel (BEC) are not sufficient for generalization, the techniques we will learn in the BSC achievability proof offer valuable insights that can be extended to other channels.

In the BSC, the channel output Y_i is a flipped version of X_i with probability, say p (called crossover probability): $p(y|x) = p$ when $y \neq x$. So the relation between X_i and Y_i is given by:

$$Y_i = \begin{cases} X_i, & \text{w.p. } 1 - p; \\ X_i \oplus 1, & \text{w.p. } p. \end{cases}$$

Another simpler way to represent this is:

$$Y_i = X_i \oplus Z_i$$

where Z_i's are i.i.d. $\sim$ Bern(p). Without loss of generality, assume $p \in (0, \frac{1}{2})$; otherwise, we can flip all 0's and 1's to 1's and 0's, respectively.

Let's start by claiming the channel capacity:

$$C_{\text{BSC}} = 1 - H(p)$$

where $H(p) := p \log \frac{1}{p} + (1-p) \log \frac{1}{1-p}$. For the rest of this section, we will prove the achievability: if $R < 1 - H(p)$, one can make the probability of error arbitrarily close to 0 as $n \to \infty$.

Encoder & decoder The encoder that we will employ is the same as before: the random code where $X_i(w)$'s are i.i.d. $\sim$ Bern($\frac{1}{2}$) across all i's and w's. We will also use the optimal decoder, which is the ML decoder:

$$\hat{W}_{\text{ML}} = \arg\max_{w} \mathbb{P}(y^n | x^n(w)).$$

The way that we compute the likelihood function is different from that in the BEC case. Let us see this difference through the following example. Suppose that $(n, R) = (4, \frac{1}{2})$ and the codebook is:

$$X^n(1) = 0000;$$
$$X^n(2) = 1110;$$
$$X^n(3) = 1010;$$
$$X^n(4) = 1111.$$

Suppose that the received signal y^n is (0100). Then, the likelihood functions are:

$$\mathbb{P}(y^n | x^n(1)) = (1-p)^3 p^1;$$
$$\mathbb{P}(y^n | x^n(2)) = (1-p)^2 p^2;$$
$$\mathbb{P}(y^n | x^n(3)) = (1-p) p^3;$$
$$\mathbb{P}(y^n | x^n(4)) = (1-p) p^3.$$

Unlike the BEC case, all the messages are *compatible* because all of the likelihood functions are strictly positive. So we need to compare all the functions to choose the message that maximizes the function. Since we assume $p \in (0, \frac{1}{2})$, in the above example, the first message is the maximizer. It has the minimum number of flips (marked in red) and the flipping probability is smaller than $\frac{1}{2}$, and hence, the corresponding likelihood function is maximized.

This reveals that the decision is heavily dependent on the number of non-matching bits (flips):

$$d(x^n, y^n) := |\{i : x_i \neq y_i\}|.$$

This is called the *Hamming* distance (Hamming, 1950). Using this, the ML decoder can be re-written as:

$$\hat{W}_{\mathsf{ML}} = \arg\min_{w} d(x^n(w), y^n).$$

Our remaining task is to demonstrate that the probability of error can be made arbitrarily close to zero as n approaches infinity when utilizing the ML decoder. However, we will not use the ML decoder for two reasons. Firstly, the analysis of error probability is somewhat intricate. Secondly, the proof method cannot be applied to arbitrary channels. How can we then prove the achievability without employing the ML decoder? Fortunately, there exists an alternative but suboptimal decoder that simplifies the proof of achievability significantly while still achieving $1 - H(p)$. Additionally, the suboptimal decoder is generalizable, which makes it suitable for proving the achievability. Therefore, we will utilize the suboptimal decoder to prove the achievability.

A suboptimal decoder The suboptimal decoder that we will employ is inspired by the following observation. Notice that the input to the decoder is Y^n and codeword is available at the decoder, i.e., $X^n(w)$ is known for all w's. Now observe that

$$Y^n \oplus X^n = Z^n$$

and we know the statistics of Z^n: i.i.d. $\sim \mathsf{Bern}(p)$.

Suppose that the actually transmitted signal is $X^n(1)$. Then,

$$Y^n \oplus X^n(1) = Z^n \sim \mathsf{Bern}(p).$$

On the other hand, for $w \neq 1$,

$$Y^n \oplus X^n(w) = X^n(1) \oplus X^n(w) \oplus Z^n.$$

The statistics of the resulting sequence is $\mathsf{Bern}(\frac{1}{2})$. Notice that the sum of any two independent Bernoulli random variables is $\mathsf{Bern}(\frac{1}{2})$, as long as at least one of them follows $\mathsf{Bern}(\frac{1}{2})$. Why? This motivates us to employ the following decoding rule.

1. Compute $Y^n \oplus X^n(w)$ for all w's.

2. Eliminate messages such that the resulting sequence is *not typical* w.r.t. Bern(p). More precisely, let

$$A_\epsilon^{(n)} := \{z^n : 2^{-n(H(p)+\epsilon)} \leq p(z^n) \leq 2^{-n(H(p)-\epsilon)}\}$$

be a typical set w.r.t. Bern(p). Eliminate all w's such that $Y^n \oplus X^n(w) \notin A_\epsilon^{(n)}$.

3. If there is only one message that survives, then declare the survival as the correct message. Otherwise, declare an error.

The error event is of two types. The first is the case where there are multiple survivals. The second refers to the scenario where there is no survival.

Probability of error We are ready to analyze the probability of error to complete the achievability proof. For notational simplicity, let $\bar{P}_e := \mathbb{E}_{\mathcal{C}}[P_e(\mathcal{C})]$. Using the same argument that we made in the BEC case, we get:

$$\bar{P}_e := \mathbb{E}_{\mathcal{C}}[P_e(\mathcal{C})] = \mathbb{P}(\hat{W} \neq 1 | W = 1).$$

As mentioned earlier, the error event is of the two types: (i) multiple survivals; and (ii) no survival. The multiple-survival event implies that there exists $w \neq 1$ such that $Y^n \oplus X^n(w) \in A_\epsilon^{(n)}$. The no-survival event implies that even $Y^n \oplus X^n(1)$ w.r.t. the correct message "1" is not a typical sequence, meaning that $Z^n \notin A_\epsilon^{(n)}$. Hence, we get:

$$\begin{aligned}
\bar{P}_e &= \mathbb{P}(\hat{W} \neq 1 | W = 1) \\
&\leq \mathbb{P}\left(\bigcup_{w \neq 1} \{Y^n \oplus X^n(w) \in A_\epsilon^{(n)}\} \cup \{Z^n \notin A_\epsilon^{(n)}\} | W = 1 \right) \\
&\overset{(a)}{\leq} \sum_{w=2}^{2^{nR}} \mathbb{P}(\{Y^n \oplus X^n(w) \in A_\epsilon^{(n)}\} | W = 1) + \mathbb{P}(Z^n \notin A_\epsilon^{(n)} | W = 1) \\
&\overset{(b)}{=} (2^{nR} - 1)\mathbb{P}(Y^n \oplus X^n(2) \in A_\epsilon^{(n)} | W = 1) + \mathbb{P}(Z^n \notin A_\epsilon^{(n)} | W = 1) \\
&\overset{(c)}{\approx} (2^{nR} - 1)\mathbb{P}(Y^n \oplus X^n(2) \in A_\epsilon^{(n)} | W = 1)
\end{aligned} \tag{2.6}$$

where (a) follows from the union bound; (b) is by symmetry of the codebook; and (c) follows from the fact that Z^n is a typical sequence w.h.p. due to the WLLN. Observe that

$$Y^n \oplus X^n(2) = X^n(1) \oplus X^n(2) \oplus Z^n \sim \text{Bern}\left(\frac{1}{2}\right).$$

How to compute $\mathbb{P}(Y^n \oplus X^n(2) \in A_\epsilon^{(n)} | W = 1)$? To this end, we need to consider two quantities: (i) the total number of possible sequences that $Y^n \oplus X^n(2) \sim \text{Bern}(\frac{1}{2})$ can take on; and (ii) the size of the typical set $|A_\epsilon^{(n)}|$. Specifically, we have:

$$\mathbb{P}(Y^n \oplus X^n(2) \in A_\epsilon^{(n)} | W = 1) = \frac{|A_\epsilon^{(n)}|}{\text{total number of Bern}(\frac{1}{2}) \text{ sequences}}$$

Note that the total number of $\text{Bern}(\frac{1}{2})$ sequences is 2^n and $|A_\epsilon^{(n)}| \leq 2^{n(H(p)+\epsilon)}$ (Why?). Hence,

$$\mathbb{P}(Y^n \oplus X^n(2) \in A_\epsilon^{(n)} | W = 1) \leq \frac{2^{n(H(p)+\epsilon)}}{2^n}.$$

This together with (2.6), we get:

$$\bar{P}_e \lesssim 2^{n(R-(1-H(p)-\epsilon))}.$$

Here ϵ can be made arbitrarily close to 0. Hence, if $R < 1 - H(p)$, $\bar{P}_e \to 0$ as $n \to \infty$. This completes the proof.

Look ahead Having established the achievability proof for the BSC, our next section will focus on expanding the proof of achievability to cover the general scenario in which the channel is characterized by any arbitrary conditional distribution $p(y|x)$.

Problem Set 4

Prob 4.1 (Concept of channel capacity) We wish to send a uniformly distributed message $W \in \{1, 2, \ldots, M\}$ from a transmitter to a receiver. Let f and g be channel encoder and decoder respectively. Let $R = \frac{\log M}{n}$ be data rate and $P_e = \mathbb{P}(\hat{W} \neq W)$ be the probability of error. Here n denotes code length and $\hat{W}$ indicates a decoded message. In an attempt to understand the fundamental tradeoff between the data rate R and the error probability P_e, Shannon considered the following optimization problem. Given R and n,

$$P_e^*(R, n) := \min_{f,g} P_e. \tag{2.7}$$

(*a*) Could Shannon solve the optimization problem (2.7)?
(*b*) State the definition of *possible (reliable) communication*.
(*c*) State the definition of an *achievable rate*.
(*d*) State the definition of *channel capacity* using the concept of an achievable rate.
(*e*) Consider a slightly different optimization problem: Given R,

$$P_e^*(R) := \min_{f,g,n} P_e.$$

Now the design variables that we optimize over include code length n. Let C be the channel capacity. For any $\epsilon > 0$, what are $P_e^*(C-\epsilon)$ and $P_e^*(C+\epsilon)$?

Prob 4.2 (Optimal decoding principle) Consider a binary random variable $X \sim \mathsf{Bern}(m)$. The signal X is transmitted over a binary symmetric channel with crossover probability p, yielding a channel output Y. Given $Y = y$, the receiver wishes to decode X so as to minimize the probability of error $P_e := \mathbb{P}(X \neq \hat{X})$. Here $\hat{X}$ denotes a decoder output. This problem explores the optimal way of decoding X.

(*a*) Compute the *a posteriori probability*: $\mathbb{P}(X = 1|Y = y)$.
(*b*) Derive the optimal decoder, i.e., derive $\hat{X}$ that yields the minimum P_e.
(*c*) The optimal decoder derived in part (*b*) is called the *Maximum A Posteriori probability (MAP)* decoder. Explain the rationale behind the naming.

Prob 4.3 (Existence of an optimal deterministic code) Consider a memoryless binary erasure channel with erasure probability p. In Section 2.3, we claimed the achievability of the data rate $1 - p$ when employing a random code. Given $R < 1 - p$, $\mathbb{E}_{\mathcal{C}}[P_e(\mathcal{C})]$ can be made arbitrarily close to 0 as $n \to \infty$. Here $P_e(\mathcal{C})$ indicates the probability of error when using a codebook $\mathcal{C}$ and the ML decoding, and the expression is over a random choice of the codebook. In this problem, you are asked to show that the above statement implies the existence of an optimal deterministic code, say $\mathcal{C}^*$: $\mathbb{P}(\mathcal{C}^*) \to 0$ as $n \to \infty$.

(*a*) Suppose X is a real-valued discrete random variable. Argue that there exists a value, say a such that $\mathbb{P}(X = a) \neq 0$ and $a \leq \mathbb{E}[X]$.
(*b*) Using part (*a*) or otherwise, show that the above statement for the achievability implies the existence of a deterministic code $\mathcal{C}^*$ such that given $R < 1 - p$, $P_e(\mathcal{C}^*) \to 0$ as $n \to \infty$.

Prob 4.4 (Chernoff bound)

(*a*) Show that for a random variable X and some constant a,

$$\mathbb{P}(X > a) \leq \min_{\lambda > 0} \frac{\mathbb{E}[e^{\lambda X}]}{e^{\lambda a}}.$$

(*b*) Suppose $X_1, \ldots, X_n$ are i.i.d., each being distributed according to $\mathsf{Bern}(m)$. Fix $\delta > 0$. Show that

$$\mathbb{P}(X_1 + \cdots + X_n \geq n(m + \delta)) \leq 2^{-n\mathsf{KL}(m+\delta \| m)}$$

where $\mathsf{KL}(m + \delta \| m)$ denotes the KL divergence between $\mathsf{Bern}(m + \delta)$ and $\mathsf{Bern}(m)$.
(*c*) Consider the same setup as that in part (*b*). Show that

$$\mathbb{P}(X_1 + \cdots + X_n \leq n(m - \delta)) \leq 2^{-n\mathsf{KL}(m-\delta \| m)}.$$

Prob 4.5 (Channel modeling & channel capacity) Consider a memoryless channel which concatenates the following two channels serially: (i) a binary symmetric channel with crossover probability p; and (ii) a binary erasure channel with erasure probability ϵ. Let X and Y be the input and the output of the channel.

(*a*) Derive the conditional distribution $p(y|x)$.
(*b*) Compute $\max_{p(x)} I(X; Y)$.

Prob 4.6 (Maximum Likelihood decoding) The achievability proof for BSC that we learned in Section 2.3 uses *joint typicality decoding*. As mentioned in the section, the joint typicality decoding is not optimal in terms of minimizing the probability of error. In this problem, you are asked to prove the achievability when using the optimal decoder: maximum likelihood decoder (MLD).

We consider a BSC with crossover probability $p < \frac{1}{2}$. Define the Hamming distance $d(x^n, y^n)$ between two binary sequences x^n and y^n as the number of positions where they differ, i.e., $d(x^n, y^n) = |\{i : x_i \neq y_i\}|$.

(*a*) Show that the MLD rule reduces to the minimum Hamming distance decoding rule – declare $\hat{w}$ is sent if $d(x^n(\hat{w}), y^n) < d(x^n(w), y^n)$ for all $w \neq \hat{w}$.

(*b*) Using the random code (that we learned in Section 2.2) and the minimum distance decoder, show that for every $\epsilon > 0$, the probability of error is upper bounded as

$$\begin{aligned} P_e &= \mathbb{P}\{\hat{W} \neq 1 | W = 1\} \\ &\leq \mathbb{P}\{d(X^n(1), Y^n) > n(p+\epsilon) | W = 1\} \\ &+ (2^{nR} - 1)\mathbb{P}\{d(X^n(2), Y^n) \leq n(p+\epsilon) | W = 1\}. \end{aligned}$$

(*c*) Show that the first term in the RHS in the above inequality tends to zero as $n \to \infty$. Using the Chernoff bound, show that

$$\mathbb{P}\{d(X^n(2), Y^n) \leq n(p+\epsilon) | W = 1\} \leq 2^{-n(1-H(p+\epsilon))}.$$

Using these results, show that any $R < C = 1 - H(p)$ is achievable.

Prob 4.7 (Maximum Likelihood decoding) We wish to transmit a message $W \in \{1, \ldots, M\}$ over a binary erasure channel with erasure probability p. Suppose we employ random encoding that we learned in Section 2.2.

(*a*) State the MAP decoding rule.

(*b*) Show that the MAP rule is optimal in a sense of minimizing the probability of error.

(*c*) Show that the MAP rule reduces the ML rule when the message W is uniformly distributed.

(*d*) A student claims that given $W = 1$, the following events are disjoint: $\{\hat{W} = 2\}, \ldots, \{\hat{W} = M\}$. Prove or disprove it. Here $\hat{W}$ indicates the output of the ML decoder.

Prob 4.8 (Random vs deterministic codes) Consider a channel coding problem setup in which code length $n = 4$ and data rate $R = \frac{1}{2}$. Let $W \in \{1, \ldots, 2^{nR}\}$ and X^n be the message and codeword respectively. Suppose that codeword X^n is transmitted over a BEC with erasure probability $p \in [0, 1]$, thus yielding a received signal Y^n. Assume we use an optimal decoder (i.e., the ML decoder in this problem setup). Let $\hat{W}$ be the decoded message.

(*a*) Suppose we employ the random code in which components of codewords $X_i(w)$'s are i.i.d. $\sim$ Bern($\frac{1}{2}$) across both $w \in \{1, \ldots, 2^{nR}\}$ and $i \in \{1, \ldots, n\}$. Show that $\mathbb{P}(\hat{W} = i | W = 1)$ is the same for all $i \neq 1$.

(*b*) Now we use a deterministic code, say $\mathcal{C}$, instead:

$$X^n(1) = 0000;$$
$$X^n(2) = 1100;$$
$$X^n(3) = 1110;$$
$$X^n(4) = 1111.$$

A student claims that $\mathbb{P}(\hat{W} = i|W = 1, \mathcal{C})$ is still the same for all $i \neq 1$. Prove or disprove this claim.

Prob 4.9 (Basic bounds) Let A and B be events.

(*a*) Show that $\mathbb{P}(A \cup B) \leq \mathbb{P}(A) + \mathbb{P}(B)$.
(*b*) Suppose that A implies B, i.e., whenever A occurs, B also occurs. Show that $\mathbb{P}(A) \leq \mathbb{P}(B)$.

Prob 4.10 (A chance of a random sequence being typical) Let $\{X_i\}$ be an i.i.d. binary random process, each being according to $\mathsf{Bern}(\frac{1}{2})$. Define a set

$$A_\epsilon^{(n)} := \{z^n : 2^{-n(H(p)+\epsilon)} \leq p(z^n) \leq 2^{-n(H(p)-\epsilon)}\}.$$

where $H(p) := p \log \frac{1}{p} + (1-p) \log \frac{1}{1-p}$ and $p \in (0, 0.5)$.

(*a*) Show that

$$\mathbb{P}(X^n \in A_\epsilon^{(n)}) = \frac{|A_\epsilon^{(n)}|}{2^n}.$$

(*b*) Show that

$$\mathbb{P}(X^n \in A_\epsilon^{(n)}) \leq 2^{-n(1-H(p)-\epsilon)}.$$

Prob 4.11 (Concept of jointly typical sequences) Consider an i.i.d. sequence pair (X^n, Y^n) with $p(x, y)$ where $x \in \mathcal{X}$ and $y \in \mathcal{Y}$. Here the i.i.d sequence pair means that (X_i, Y_i)'s are i.i.d. over i and each follows the identical distribution $p(x, y)$. Fix $\epsilon > 0$. Let

$$A_\epsilon^{(n)}(X) := \{x^n : 2^{-n(H(X)+\epsilon)} \leq p(x^n) \leq 2^{-n(H(X)-\epsilon)}\}$$
$$A_\epsilon^{(n)}(Y) := \{y^n : 2^{-n(H(Y)+\epsilon)} \leq p(y^n) \leq 2^{-n(H(Y)-\epsilon)}\}$$
$$A_\epsilon^{(n)}(X, Y) := \{(x^n, y^n) : 2^{-n(H(X,Y)+\epsilon)} \leq p(x^n, y^n) \leq 2^{-n(H(X,Y)-\epsilon)}\}.$$

Show that for any $\epsilon > 0$,

$$\mathbb{P}(X^n \in A_\epsilon^{(n)}(X)) \longrightarrow 1,$$
$$\mathbb{P}(Y^n \in A_\epsilon^{(n)}(Y)) \longrightarrow 1,$$
$$\mathbb{P}((X^n, Y^n) \in A_\epsilon^{(n)}(X, Y)) \longrightarrow 1,$$

as $n \to \infty$.

Prob 4.12 (Concept of jointly typical sequences) Consider an i.i.d. sequence pair (X^n, Y^n) with $p(x, y)$ where $x \in \mathcal{X}$ and $y \in \mathcal{Y}$. Fix $\epsilon > 0$. Let

$$A_\epsilon^{(n)}(X) := \{x^n : 2^{-n(H(X)+\epsilon)} \leq p(x^n) \leq 2^{-n(H(X)-\epsilon)}\}$$
$$A_\epsilon^{(n)}(Y) := \{y^n : 2^{-n(H(Y)+\epsilon)} \leq p(y^n) \leq 2^{-n(H(Y)-\epsilon)}\}$$
$$A_\epsilon^{(n)}(X, Y) := \{(x^n, y^n) : 2^{-n(H(X,Y)+\epsilon)} \leq p(x^n, y^n) \leq 2^{-n(H(X,Y)-\epsilon)}\}.$$

A student claims that there exists a sequence pair of $(\tilde{X}^n, \tilde{Y}^n)$ such that as $n \to \infty$,

$$\mathbb{P}(\tilde{X}^n \in A_\epsilon^{(n)}(X)) \to 1;$$
$$\mathbb{P}(\tilde{Y}^n \in A_\epsilon^{(n)}(Y)) \to 1;$$
$$\mathbb{P}((\tilde{X}^n, \tilde{Y}^n) \in A_\epsilon^{(n)}(X, Y)) \nrightarrow 1.$$

Prove or disprove this statement.

Prob 4.13 (Sum of Bernoulli random variables) Suppose that $X_1 \sim \text{Bern}(p)$ is independent of $X_2 \sim \text{Bern}(\frac{1}{2})$. What is the statistics of $X_1 \oplus X_2$?

Prob 4.14 (True or False?)

(*a*) Consider a memoryless binary erasure channel. Let X_i and Y_i be the input and the output of the channel at time i, respectively. Then,

$$H(Y_1, Y_2|X_1, X_2) = H(Y_1|X_1) + H(Y_2|X_2).$$

(*b*) Let $\{X_i\}$ be a binary random process such that $\mathbb{P}(X_1 = i_1, X_2 = i_2, \ldots, X_n = i_n) = \frac{1}{2^n}$ for all possible sequence patterns $(i_1, i_2, \ldots, i_n)$. Then, $\{X_i\}$'s are identically distributed, but not necessarily independent.

(*c*) The Shannon's landmark paper published in 1948 provides an explicit guideline as to how to design an optimal communication system.

(*d*) Consider a binary asymmetric channel in which the output Y is a flipped version of X with probability $p_1 \in [0, 1]$ when $X = 0$ (and with probability $p_2 \in [0, 1]$ when $X = 1$). The capacity is achieved when $X \sim \text{Bern}(1/2)$.

(*e*) For an inference problem, the optimal decoder is always the MAP decoder.

(*f*) Consider an inference problem in which we wish to decode $X \in \mathcal{X}$ from $Y \in \mathcal{Y}$. Given $Y = y$, the optimal decoder can be:

$$\hat{X} = \arg\max_{x \in \mathcal{X}} \mathbb{P}(Y = y | X = x).$$

(*g*) Suppose that $X_1 \sim \mathsf{Bern}(p), X_2 \sim \mathsf{Bern}(\frac{1}{2})$ and $S = X_1 \oplus X_2$. Then, S follows $\mathsf{Bern}(\frac{1}{2})$ no matter what p is.

(*h*) In the channel coding setup, we assumed that a message W is uniformly distributed. The rationale behind this assumption is that we use an optimal source code.

(*i*) In Section 2.1, we considered an optimization problem which aims to minimize the probability of error given data rate R and code length n. Denote by $P^*(R, n)$ the minimum probability of error. Instead of deriving the exact $P^*(R, n)$, Shannon developed a lower bound of $P^*(R, n)$ to show that for any $R \le C$, the probability of error can be made arbitrarily close to 0.

2.4 Achievability Proof for Discrete Memoryless Channels

Recap In the previous section, we demonstrated the achievability proof for BSC. However, we will postpone the converse proof because the technique for the converse proof, which we will discuss later, can be directly applied to a broad range of problem settings.

Outline This section will expand the proof of achievability to encompass the general scenario in which the channel is described by an arbitrary conditional distribution $p(y|x)$. Subsequently, in the following section, we will establish the converse proof for the general channel case.

Discrete memoryless channel (DMC) The general channel that we will consider is called the *discrete memoryless channel*, DMC for short. Let us start with the definition of the channel. We say that a channel is an DMC if input and output are on discrete alphabet sets and the following condition is satisfied:

$$p(y_i|x_i, x^{i-1}, y^{i-1}, W) = p(y_i|x_i).$$

This is called the memoryless property. Notice that given the current channel input x_i, the current output y_i is independent of the past input/output (x^{i-1}, y^{i-1}) and any other things including the message W. Here one key property that we need to keep in our mind is:

$$p(y^n|x^n) = \prod_{i=1}^{n} p(y_i|x_i). \tag{2.8}$$

This can be proved by using the memoryless property. Check in Prob 5.4. This property plays a crucial role in proving the achievability as well as the converse. This will be clearer later.

Guess on the capacity formula Let us guess the capacity formula for the DMC. Remember the capacity formulas of the BEC and BSC: $C_{\text{BEC}} = 1 - p$; $C_{\text{BSC}} = 1 - H(p)$. These capacities are closely related to the key notion that we introduced earlier: mutual information. Specifically what we can easily show is: when $X \sim \text{Bern}(\frac{1}{2})$,

$$C_{\text{BEC}} = 1 - p = I(X; Y);$$
$$C_{\text{BSC}} = 1 - H(p) = I(X; Y).$$

Also one can verify that for an arbitrary distribution of X,

$$C_{\text{BEC}} = 1 - p \geq I(X;Y);$$

$$C_{\text{BSC}} = 1 - H(p) \geq I(X;Y).$$

Check this in Prob 5.1. Hence, what one can guess on the capacity formula is:

$$C_{\text{DMC}} = \max_{p(x)} I(X;Y). \tag{2.9}$$

It turns out it is indeed the case. For the rest of this section, we will prove the achievability: if $R < \max_{p(x)} I(X;Y)$, the probability of error can be made arbitrarily close to 0 as $n \to \infty$.

Encoder The encoder that we will employ is a random code, meaning that $X_i(w)$'s are generated in an i.i.d. fashion according to some distribution $p(x)$. We know in the BEC and BSC that the input distribution is fixed as $\text{Bern}(\frac{1}{2})$. One can verify that $\text{Bern}(\frac{1}{2})$ is the maximizer of the above optimization problem (2.9). This motivates us to choose $p(x)$ as:

$$p^*(x) = \arg\max_{p(x)} I(X;Y).$$

Indeed, this choice enables us to achieve the capacity (2.9).

Decoder Assume that the codebook is known at the decoder. We use the suboptimal decoder employed in the BSC case. Remember that the suboptimal decoder is based on a typical sequence and the fact that $Y^n \oplus X^n = Z^n$. One significant distinction in the general DMC case is that Y^n is an *arbitrary random function* of X^n. So what we can do is to take a look at pairs of $(Y^n, X^n(w))$ and to check if we can see a particular behavior of the pair associated with the true codeword, compared to the other pairs w.r.t. the wrong codewords.

To illustrate this, consider the following situation. Suppose that $X^n(1)$ is transmitted, i.e., message 1 is the correct one. Then, the joint distribution of the correct pair $(x^n(1), y^n)$ would be:

$$\begin{aligned}
p(x^n(1), y^n) &= p(x^n(1))p(y^n|x^n(1)) \\
&\stackrel{(a)}{=} \prod_{i=1}^{n} p(x_i(1))p(y_i|x_i(1)) \\
&= \prod_{i=1}^{n} p(x_i(1), y_i)
\end{aligned}$$

where (a) follows from the key property (2.8). This implies that the pairs $(x_i(1), y_i)$'s are i.i.d over i's. Then, using the WLLN on the i.i.d. sequence of pairs, one can show that

$$\frac{1}{n}\log\frac{1}{p(X^n(1), Y^n)} \longrightarrow H(X,Y) \quad \text{in prob.}$$

Check in Prob 5.4. From this, one can say that

$$p(x^n(1), y^n) \approx 2^{-nH(X,Y)}$$

for sufficiently large n. Why?

On the other hand, the joint distribution of the wrong pair $(x^n(w), y^n)$ for $w \neq 1$ would be:

$$p(x^n(w), y^n) = p(x^n(w))p(y^n).$$

This is because y^n is associated only with $(x^n(1), \text{channel noise})$, which is independent of $x^n(w)$. Remember that we use a random code in which codewords are independent with each other. Also one can verify that y^n is i.i.d. Check in Prob 5.4. Again using the WLLN on $x^n(w)$ and y^n, one can get:

$$p(x^n(w), y^n) \approx 2^{-n(H(X)+H(Y))}$$

for sufficiently large n. This motivates the following decoder. Let

$$\begin{aligned} A_\epsilon^{(n)} = \Big\{(x^n, y^n) : 2^{-n(H(X)+\epsilon)} &\leq p(x^n) \leq 2^{-n(H(X)-\epsilon)} \\ 2^{-n(H(Y)+\epsilon)} &\leq p(y^n) \leq 2^{-n(H(Y)-\epsilon)} \\ 2^{-n(H(X,Y)+\epsilon)} &\leq p(x^n, y^n) \leq 2^{-n(H(X)+H(Y)-\epsilon)}\Big\}. \end{aligned}$$

The decoding rule is as follows:

1. Eliminate all the messages w's such that $(x^n(w), y^n) \notin A_\epsilon^{(n)}$.
2. If there is only one message that survives, then declare the survival as the correct message.
3. Otherwise, declare an error.

Similar to the previous case, the error event is of two types: (i) multiple survivals (or one wrong survival); and (ii) no survival.

Probability of error We analyze the probability of error to complete the achievability proof. Using the same argument that we made in the BEC case, we get:

$$\bar{P}_e := \mathbb{E}_{\mathcal{C}}[P_e(\mathcal{C})] = \mathbb{P}(\hat{W} \neq 1 | W = 1).$$

As mentioned earlier, the error event is of the two types: (i) multiple survivals; and (ii) no survival. The multiple-survival event implies the existence of the wrong pair being a jointly typical pair, meaning that there exists $w \neq 1$ such that $(X^n(w), Y^n) \in A_\epsilon^{(n)}$. The no-survival event implies that even the correct pair is not jointly typical, meaning that $(X^n(1), Y^n) \notin A_\epsilon^{(n)}$. Hence, we get:

$$\begin{aligned}
\bar{P}_e &= \mathbb{P}(\hat{W} \neq 1 | W = 1) \\
&\leq \mathbb{P}\left(\bigcup_{w \neq 1} \{(X^n(w), Y^n) \in A_\epsilon^{(n)}\} \cup \{(X^n(1), Y^n) \notin A_\epsilon^{(n)}\} | W = 1 \right) \\
&\overset{(a)}{\leq} \sum_{w=2}^{2^{nR}} \mathbb{P}(\{(X^n(w), Y^n) \in A_\epsilon^{(n)}\} | W = 1) + \mathbb{P}((X^n(1), Y^n) \notin A_\epsilon^{(n)} | W = 1) \\
&\overset{(b)}{\leq} 2^{nR} \mathbb{P}((X^n(2), Y^n) \in A_\epsilon^{(n)} | W = 1) + \mathbb{P}((X^n(1), Y^n) \notin A_\epsilon^{(n)} | W = 1) \\
&\overset{(c)}{\approx} 2^{nR} \mathbb{P}((X^n(2), Y^n) \in A_\epsilon^{(n)} | W = 1)
\end{aligned}$$

where (*a*) follows from the union bound; (*b*) is by symmetry of the codewords w.r.t. message indices; and (*c*) follows from the fact that $(X^n(1), Y^n))$ is jointly typical for sufficiently large n w.h.p due to the WLLN. Observe that $X^n(2)$ and Y^n are independent. So the total number of pair patterns w.r.t. $(X^n(2), Y^n)$ would be:

$$\begin{aligned}
\text{total number of } (X^n(2), Y^n) \text{ pairs} &\approx 2^{nH(X)} \cdot 2^{nH(Y)} \\
&= 2^{n(H(X)+H(Y))}.
\end{aligned}$$

On the other hand, the cardinality of the jointly typical pair set $A_\epsilon^{(n)}$ is:

$$|A_\epsilon^{(n)}| \approx 2^{nH(X,Y)}$$

Why? Hence, we get:

$$\begin{aligned}
\mathbb{P}(Y^n \oplus X^n(2) \in A_\epsilon^{(n)} | W = 1) &= \frac{|A_\epsilon^{(n)}|}{\text{total number of } (X^n(2), Y^n) \text{ pairs}} \\
&\approx \frac{2^{n(H(X,Y)-H(X)-H(Y))}}{2^n} \\
&= \frac{2^{-nI(X;Y)}}{2^n}.
\end{aligned}$$

Using this, we get:

$$\bar{P}_e \lesssim 2^{n(R-I(X;Y))}.$$

Hence, if $R < I(X;Y)$, then $\bar{P}_e \to 0$ as $n \to \infty$. Since we choose $p(x)$ such that $I(X;Y)$ is maximized, $\max_{p(x)} I(X;Y)$ is achievable. This completes the proof.

Look ahead So far we have proved the achievability for discrete memoryless channels. In the next section, we will prove the converse to complete the proof of channel coding theorem.

2.5 Converse Proof for Discrete Memoryless Channels

Recap In the previous section, we have proven the achievability for discrete memoryless channels which are described by conditional distribution $p(y|x)$:

$$R < \max_{p(x)} I(X;Y) \Longrightarrow P_e \to 0.$$

Outline In this section, we will prove the converse to complete the channel coding theorem:

$$P_e \to 0 \Longrightarrow R < \max_{p(x)} I(X;Y).$$

The proof consists of three parts. Firstly, we will examine Fano's inequality, a fundamental inequality that plays a crucial role in the converse proof. Secondly, we will explore another critical inequality known as the data processing inequality (DPI). Lastly, utilizing both these inequalities, we will present the final proof of the converse.

Fano's inequality Fano's inequality is a significant inequality in the context of inference problems, which involve inferring an input from an output that is probabilistically related to the input. In such problems, the only possible action with respect to the input is to make an inference or guess. The communication problem we have been studying can also be viewed as an inference problem, where the goal is to infer the message W from the received signal Y^n, which is stochastically related to W. In this context, Fano's inequality relates the following two quantities:

$$P_e := \mathbb{P}(\hat{W} \neq W) \;\&\; H(W|\hat{W}).$$

One can expect that the smaller P_e, the smaller $H(W|\hat{W})$. Fano's inequality presents how one affects the other in a precise manner. In the converse proof, we need to establish a lower bound on P_e to show that P_e does not converge to zero when $R > C$. Fano's inequality plays a crucial role in providing this lower bound by giving an upper bound on $H(W|\hat{W})$ expressed in terms of P_e. The formula for this upper bound is well-known, and we will focus on its expression:

$$\textit{Fano's inequality: } H(W|\hat{W}) \leq 1 + P_e \cdot nR. \tag{2.10}$$

This indeed captures the intimate relationship between P_e and $H(W|\hat{W})$. The smaller P_e, the more contracted $H(W|\hat{W})$. The proof of this is simple. Let

$$E = \mathbf{1}\{\hat{W} \neq W\}.$$

By the definition of the error probability and the fact that $\mathbb{E}[\mathbf{1}\{\hat{W} \neq W\}] = \mathbb{P}(\hat{W} \neq W)$, we see that $E \sim \text{Bern}(P_e)$. Starting with the fact that E is a function of $(W, \hat{W})$, we have:

$$
\begin{aligned}
H(W|\hat{W}) &= H(W, E|\hat{W}) \\
&\overset{(a)}{=} H(E|\hat{W}) + H(W|\hat{W}, E) \\
&\overset{(b)}{\leq} 1 + H(W|\hat{W}, E) \\
&\overset{(c)}{=} 1 + \mathbb{P}(E = 1) \cdot H(W|\hat{W}, E = 1) \\
&= 1 + P_e \cdot H(W|\hat{W}, E = 1) \\
&\overset{(d)}{\leq} 1 + P_e \cdot nR
\end{aligned}
$$

where (a) is due to a chain rule; (b) follows from the cardinality bound on $H(E|\hat{W})$; (c) comes from the definition of conditional entropy; and (d) follows from the cardinality bound on $H(W|\hat{W}, E = 1)$.

Data processing inequality (DPI) The next inequality we will examine is called the data processing inequality (DPI). Essentially, DPI states that any processing of data cannot enhance the quality of inference beyond what was originally available. In the case of our problem, this means that the quality of inference of W based on X^n (which we can view as the original data) cannot be inferior to that based on processed data, such as Y^n. DPI is a mathematical statement formulated in the context of a Markov process. Therefore, let us first study what a Markov process is. We say that a random process, say $\{X_i\}$, is a Markov process if

$$p(x_{i+1}|x_i, x_{i-1}, \ldots, x_1) = p(x_{i+1}|x_i).$$

The meaning of this condition is that, given the current state x_i, the future state x_{i+1} and the past states x^{i-1}'s are independent of each other. A Markov process is often represented graphically using a well-known diagram. For example, if (X_1, X_2, X_3) form a Markov process, it can be represented as:

$$X_1 - X_2 - X_3.$$

The reason for representing a Markov process in this way is as follows. If we are given X_2, we can remove it from the graph since it is already known. This removal results in X_3 and X_1 being disconnected, indicating that they are statistically independent. Since the resulting graph resembles a chain, it is referred to as a Markov chain.

Our problem context exhibits a Markov chain. One can show that (W, X^n, Y^n) forms a Markov chain:

$$W - X^n - Y^n.$$

The proof is straightforward. Given X^n, Y^n is a sole function of the noise induced by the channel. Since the noise has nothing to do with the message W, (Y^n, W) are independent of each other.

DPI is defined w.r.t. the Markov chain. It captures the relationship between the following quantities: $I(W;X^n), I(W;Y^n), I(X^n;Y^n)$. In fact, $I(W;X^n)$ represents the common information shared between W and X^n, which can be seen as the quality of inference on W based on X^n. Similarly, $I(W;Y^n)$ represents the quality of inference on W based on Y^n. The verbal statement of DPI states that the quality of inference on W cannot be improved by processing the original data X^n to obtain Y^n. In terms of mutual information, this can be expressed as follows:

$$I(W;Y^n) \leq I(W;X^n).$$

The proof of this is simple. Starting with a chain rule and applying non-negativity of mutual information, we have:

$$\begin{aligned} I(W;Y^n) &\leq I(W;Y^n,X^n) \\ &= I(W;X^n) + I(W;Y^n|X^n) \\ &\overset{(a)}{=} I(W;X^n) \end{aligned} \tag{2.11}$$

where (a) follows from the fact that $W - X^n - Y^n$. There is another DPI w.r.t. $I(W;Y^n)$ and another mutual information $I(X^n;Y^n)$. Note that X^n is closer to Y^n relative to the distance between W and Y^n. Hence, one can guess:

$$I(W;Y^n) \leq I(X^n;Y^n).$$

It turns out this is indeed the case. The proof is also simple.

$$\begin{aligned} I(W;Y^n) &\leq I(W,X^n;Y^n) \\ &= I(X^n;Y^n) + I(W;Y^n|X^n) \\ &= I(X^n;Y^n). \end{aligned} \tag{2.12}$$

From (2.11) and (2.12), one can summarize that mutual information between two ending terms in the Markov chain does not exceed mutual information between any two terms that lie in-between the two ends.

Looking at our problem setting, there is another term: $\hat{W}$. This together with the above Markov chain ($W - X^n - Y^n$) forms a longer Markov chain:

$$W - X^n - Y^n - \hat{W}.$$

Given Y^n, $\hat{W}$ is completely determined regardless of (W, X^n) because it is a function of Y^n. Now applying DPI, one can verify that

$$I(W;\hat{W}) \leq I(W;Y^n); \tag{2.13}$$

$$I(W;\hat{W}) \leq I(W;X^n); \tag{2.14}$$

$$I(W;\hat{W}) \leq I(X^n;\hat{W}); \tag{2.15}$$

$$I(W;\hat{W}) \leq I(X^n;Y^n); \tag{2.16}$$

$$I(W;\hat{W}) \leq I(Y^n;\hat{W}). \tag{2.17}$$

Converse proof We are ready to prove the converse with the two inequalities. Starting with the fact that $nR = H(W)$, we have:

$$\begin{aligned} nR &= H(W) \\ &= I(W;\hat{W}) + H(W|\hat{W}) \\ &\overset{(a)}{\leq} I(W;\hat{W}) + 1 + P_e \cdot nR \\ &\overset{(b)}{=} I(W;\hat{W}) + n\epsilon_n \\ &\overset{(c)}{\leq} I(X^n;Y^n) + n\epsilon_n \\ &= H(Y^n) - H(Y^n|X^n) + n\epsilon_n \\ &\overset{(d)}{=} H(Y^n) - \sum_{i=1}^{n} H(Y_i|X_i) + n\epsilon_n \end{aligned}$$

$$\overset{(e)}{\leq} \sum_{i=1}^{n} [H(Y_i) - H(Y_i|X_i)] + n\epsilon_n$$

$$= \sum_{i=1}^{n} I(X_i; Y_i) + n\epsilon_n$$

$$\overset{(f)}{\leq} nC + n\epsilon_n$$

where (a) follows from Fano's inequality (2.10); (b) comes from the definition of ϵ_n that we set as:

$$\epsilon_n := \frac{1}{n}(1 + nP_e R);$$

(c) is due to DPI (2.16); (d) follows from the memoryless channel property ($p(Y^n|X^n) = \prod_{i=1}^{n} p(Y_i|X_i)$ and hence $H(Y^n|X^n) = \sum_{i=1}^{n} H(Y_i|X_i)$); (e) conditioning reduces entropy; and (f) is due to the definition $C := \max_{p(x)} I(X; Y)$. Dividing by n on both sides, we get:

$$R \leq C + \epsilon_n.$$

If $P_e \to 0$, $\epsilon_n := \frac{1}{n}(1 + P_e nR)$ tends to 0 as $n \to \infty$. Hence, we get:

$$R \leq C.$$

This completes the proof.

Look ahead The source and channel coding theorems have been proven within Shannon's two-stage communication architecture, which does not permit interaction between the source and channel codes. However, a question arises as to whether this separation approach is optimal and can achieve the same performance as the general architecture that allows for cooperation between the two codes. Surprisingly, the answer is yes, and we will prove this in the next section.

2.6 Source-Channel Separation Theorem and Feedback

Recap So far, we have studied Shannon's two fundamental theorems: the source and channel coding theorems. These theorems are established under a specific two-stage architecture, as shown in Fig. 2.2. Some readers may be curious about what would happen if the architecture was arbitrary and allowed for any interaction between the source and channel codes. Can we achieve better performance with potential cooperation between the two codes? This was a question that Shannon himself raised in his landmark paper of 1948 (Shannon, 2001). Surprisingly, he answered this question negatively by establishing the source-channel separation theorem. This result is both surprising and important. It is surprising because the simple separation approach is optimal, and it is important because it forms the foundation of the digital communication architecture, where the digital interface operates independently with the source code block.

Outline In this section, we will present the proof for the source-channel separation theorem. The proof consists of three parts. Firstly, we will identify a condition that the separation approach relies upon for reliable transmission of an information source, which will serve as a sufficient condition for reliable transmission. Secondly, using the two critical inequalities, Fano's inequality and data processing inequality, introduced in the previous section, we will demonstrate that the condition is also necessary, thus establishing the optimality of the separation approach. Finally, we will explore a distinct topic that has a close technical connection with the two inequalities, namely the role of channel output feedback, and provide a detailed analysis of this topic.

What to prove for the optimality of the separation approach? As the separation approach is a specific communication scheme, a condition that guarantees reliable transmission of an information source using this approach can serve as a sufficient condition. Therefore, proving that this condition is also necessary would establish the optimality of the separation approach. In the sequel, we will

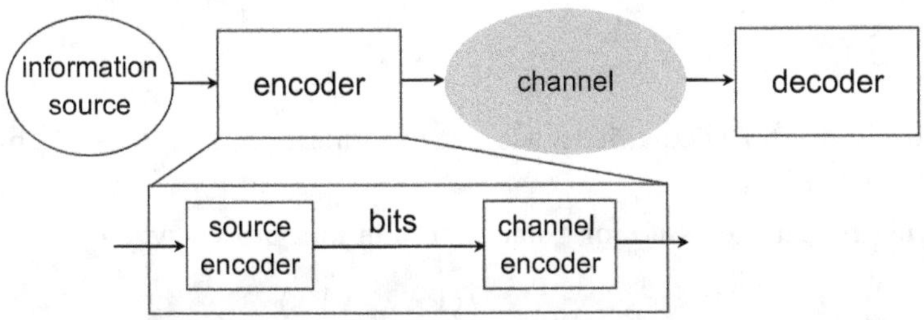

Figure 2.2. Shannon's two-staged architecture for communication systems.

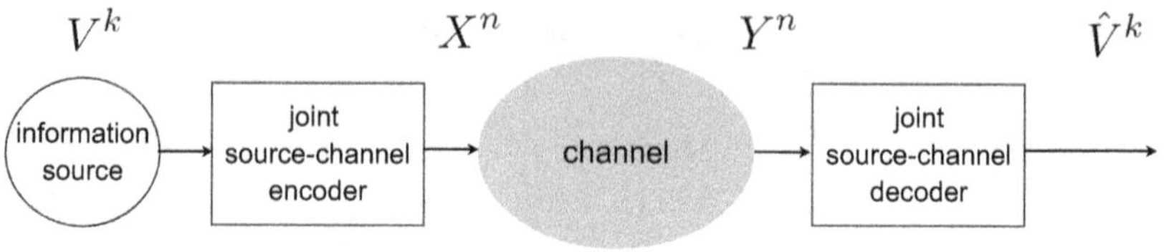

Figure 2.3. Joint source-channel encoder and decoder.

first come up with a sufficient condition due to the separation approach, and then prove its necessity accordingly.

A sufficient condition due to the separation approach Suppose we want to transmit k source symbols using n channels, as shown in Fig. 2.3. An important question is: under what conditions can the separation approach reliably transmit k symbols? This condition can be identified using the source and channel coding theorems. The source coding theorem states that the entropy rate H (in bits per symbol) is the minimum number of bits required to represent the information source per symbol. The channel coding theorem states that the capacity C (in bits per channel use) is the maximum number of bits that can be reliably transmitted over a channel. We can represent the entropy of the k source symbols as kH, and the total number of bits that can be transmitted using n channels as nC. If we apply the source and channel coders separately, a sufficient condition would be:

$$kH < nC. \tag{2.18}$$

Necessity of the sufficient condition (2.18) We will prove that the condition (2.18) is also necessary by using Fano's inequality and DPI. To determine what needs to be proven precisely, let us first express several quantities that arise in the problem setup. We denote by V^k the k source symbols. X^n is the sequence fed into the channel, and Y^n is the channel output. $\hat{V}^k$ is the decoded source. In contrast to the separation approach setup, $\hat{V}^k$ is the output of a joint source-channel decoder that allows for any interaction across source and channel codes. This is because we intend to prove the necessity of the condition (2.18) under any arbitrary scheme. The probability of error is defined as:

$$P_e = \mathbb{P}(V^k \neq \hat{V}^k).$$

What we wish to show is that for reliable communication, i.e., $P_e \to 0$, the condition (2.18) must hold.

We will prove its necessity for a stationary random process where

$$H = \lim_{m \to \infty} \frac{H(V_1, \ldots, V_m)}{m}.$$

A stationary process has an interesting property on $\frac{H(V_1,\ldots,V_m)}{m}$. To see this, let us first massage this term as:

$$\frac{H(V_1,\ldots,V_m)}{m}=\frac{1}{m}\sum_{i=1}^{m}H(V_i|V^{i-1}).$$

This is because of a chain rule. Let $a_i := H(V_i|V^{i-1})$. Using the stationarity and the fact that conditioning reduces entropy,

$$a_i \geq a_{i-1}.$$

Keeping this in our mind, compare the following two terms:

$$\frac{1}{m}\sum_{i=1}^{m}a_i \qquad \text{vs} \qquad \frac{1}{m+1}\sum_{i=1}^{m+1}a_i.$$

Since a_i is non-increasing in i, $\frac{1}{m}\sum_{i=1}^{m}a_i$ is also non-increasing in m:

$$\frac{1}{m}\sum_{i=1}^{m}a_i \geq \frac{1}{m+1}\sum_{i=1}^{m+1}a_i.$$

Hence, for any positive integer k,

$$\lim_{m\to\infty}\frac{H(V_1,\ldots,V_m)}{m}\leq\frac{H(V_1,\ldots,V_k)}{k}.$$

Using this property, we have:

$$\begin{aligned}kH &= k\lim_{m\to\infty}\frac{H(V_1,\ldots,V_m)}{m}\\ &\leq k\cdot\frac{H(V_1,\ldots,V_k)}{k}\\ &= H(V^k)\end{aligned}$$

Starting with this and applying the definition of $I(V^k;\hat{V}^k) := H(V^k) - H(V^k|\hat{V}^k)$, we then get:

$$\begin{aligned}kH &\leq H(V^k)\\ &\leq I(V^k;\hat{V}^k)+H(V^k|\hat{V}^k).\end{aligned}$$

We are now ready to employ Fano's inequality and data processing inequality. To be specific, Fano's inequality reads:

$$H(V^k|\hat{V}^k) \leq 1 + kP_e \log|\mathcal{V}| =: k\epsilon_k.$$

DPI that we want to use in this context is:

$$I(V^k; \hat{V}^k) \leq I(X^n; Y^n).$$

Applying these into the above, we obtain:

$$\begin{aligned} kH &\leq H(V^k) \\ &\leq I(V^k; \hat{V}^k) + H(V^k|\hat{V}^k) \\ &\leq I(X^n; Y^n) + k\epsilon_k \\ &\leq H(Y^n) - H(Y^n|X^n) + k\epsilon_k. \end{aligned}$$

Manipulating further, we get:

$$\begin{aligned} kH &\leq H(Y^n) - H(Y^n|X^n) + k\epsilon_k \\ &\overset{(a)}{=} \sum H(Y_i|Y^{i-1}) - H(Y^n|X^n) + k\epsilon_k \\ &\overset{(b)}{\leq} \sum [H(Y_i) - H(Y_i|X_i)] + k\epsilon_k \\ &= \sum I(X_i; Y_i) + k\epsilon_k \\ &\overset{(c)}{\leq} nC + k\epsilon_k \end{aligned}$$

where (a) comes from a chain rule; (b) follows from the memoryless property of DMC and the fact that conditioning reduces entropy; and (c) is due to the definition of $C := \max_{p(x)} I(X; Y)$. As $k \to \infty$, $\epsilon_k \to 0$. Hence, we get:

$$H \leq \frac{n}{k} C.$$

This completes the proof.

Discrete memoryless channel with feedback Next, we will discuss another topic that is related to Fano's inequality and DPI in a technical sense. The topic we will cover is the role of channel output feedback, which was studied by Shannon previously. Feedback has proven to be valuable in numerous areas, particularly in control. Feedback is known to have a significant role in stabilizing systems. In communication, however, feedback has not been very useful. This is mainly due

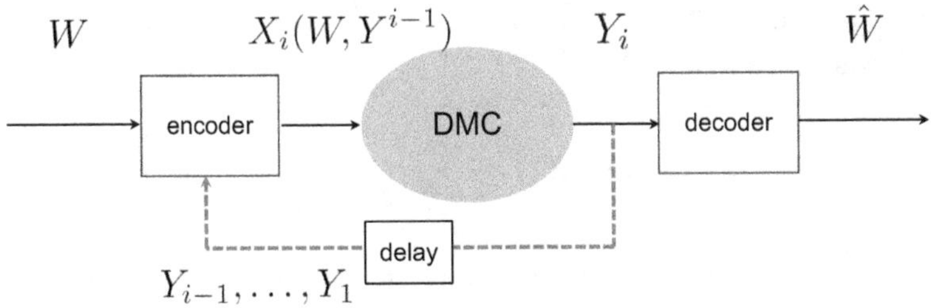

Figure 2.4. A discrete memoryless channel with feedback.

to Shannon's original result in the 1950s (Shannon, 1956). What Shannon showed is that feedback cannot increase the capacity. Hence, feedback has been used to verify the success of transmission. For the rest of this section, we will explore this counter-intuitive result in depth.

We first introduce a channel model that Shannon considered. It is based on the memoryless channel which respects:

$$p(y_i|x_i, x^{i-1}, y^{i-1}, W) = p(y_i|x_i).$$

We consider channel output feedback where the past channel output is fed back to the encoder. See Fig. 2.4. Hence, a transmitted signal X_i at time i is a function of the message W and the past channel output $Y^{i-1} := (Y_1, \ldots, Y_{i-1})$. Under this model, what Shannon showed is that the feedback capacity C_{FB} is the same as the non-feedback capacity:

$$C_{\mathsf{FB}} = C_{\mathsf{NO}}, \tag{2.19}$$

meaning that feedback cannot increase the capacity.

Proof of (2.19) The initial procedure for the converse proof is the same as that of the non-feedback case. We start with:

$$\begin{aligned} nR &= H(W) \\ &= I(W; \hat{W}) + H(W|\hat{W}) \\ &\overset{(a)}{\leq} I(W; \hat{W}) + n\epsilon_n \end{aligned}$$

where (a) is due to Fano's inequality and $\epsilon_n := \frac{1}{n}(1 + nRP_e)$. The next step is to employ DPI to make a progress w.r.t. $I(W; \hat{W})$. Remember in the non-feedback case that we used the following DPI:

$$I(W; \hat{W}) \leq I(X^n; Y^n).$$

We then expressed $I(X^n; Y^n)$ as $H(Y^n) - H(Y^n|X^n)$. The conditional entropy $H(Y^n|X^n)$ was expressed as:

$$H(Y^n|X^n) = \sum_{i=1}^{n} H(Y_i|X_i), \tag{2.20}$$

and it played a crucial role to prove the converse. The above (2.20) was because of the key memoryless property:

$$p(y^n|x^n) = \prod_{i=1}^{n} p(y_i|x_i).$$

It holds in the non-feedback case. See Prob 5.4 for the proof. In the feedback case, however, it does not hold any more. Hence, we should take a different approach to prove the converse.

Even in the feedback case, what we know for sure is about a Markov chain relationship, which reads:

$$(W, X^n) - Y^n - \hat{W}.$$

This then yields the following DPI:

$$I(W; \hat{W}) \leq I(W; Y^n).$$

Applying this different DPI to $I(W; \hat{W})$ in the above, we get:

$$\begin{aligned}
nR &\leq I(W; Y^n) + n\epsilon_n \\
&\overset{(a)}{=} \sum [H(Y_i|Y^{i-1}) - H(Y_i|Y^{i-1}, W)] + n\epsilon_n \\
&\overset{(b)}{=} \sum [H(Y_i|Y^{i-1}) - H(Y_i|Y^{i-1}, W, X_i)] + n\epsilon_n \\
&\overset{(c)}{=} \sum [H(Y_i|Y^{i-1}) - H(Y_i|X_i)] + n\epsilon_n \\
&\overset{(d)}{\leq} \sum [H(Y_i) - H(Y_i|X_i)] + n\epsilon_n \\
&= \sum I(X_i; Y_i) + n\epsilon_n \\
&\overset{(e)}{\leq} nC_{\mathsf{NO}} + n\epsilon_n
\end{aligned}$$

where (a) is due to the definition of mutual information and a chain rule; (b) follows from the fact that X_i is a function of (W, Y^{i-1}) and adding a function in conditioning does not alter entropy; (c) follows from the Markov property

$(W, Y^{i-1}) - X_i - Y_i$; (d) comes from the fact that conditioning reduces entropy; and (e) is because of the definition $C_{\mathsf{NO}} := \max_{p(x)} I(X;Y)$.

The above gives

$$R \leq C_{\mathsf{NO}} + \epsilon_n.$$

Under reliable communication, $\epsilon_n \to 0$. Hence, we prove:

$$R \leq C_{\mathsf{NO}}.$$

Look ahead Up to this point, we have proven the source coding theorem, the channel coding theorem, and the source-channel separation theorem for a broad range of channels, specifically the discrete memoryless channel. We also explored the technical connection between the role of feedback and the converse proof. However, there is a topic that has not been thoroughly addressed in the channel coding theorem, which pertains to the achievable scheme. During the achievability proof, we established the existence of an optimal code that can achieve P_e close to 0 as n approaches ∞, but we did not discuss how to construct the optimal code. Moving forward, we will delve into deterministic codes that are explicit and approach or even reach the capacity.

Problem Set 5

Prob 5.1 (Channel capacity) Let X and Y be the input and the output of a discrete memoryless channel, respectively.

(*a*) Suppose the channel is a BEC with erasure probability p. Show that

$$I(X;Y) \le 1 - p.$$

Also derive the condition under which the equality in the above holds.

(*b*) Suppose the channel is now a BSC with crossover probability p. Show that

$$I(X;Y) \le 1 - H(p).$$

Also derive the condition under which the equality in the above holds.

(*c*) In Section 2.4, we learned that the capacity of a discrete memoryless channel is

$$C = \max_{p(x)} I(X;Y).$$

Show that $I(X;Y)$ is a concave function in $p(x)$.

Prob 5.2 (Computation of channel capacity) Let $S_1 \sim \mathsf{Bern}(q)$ and $S_2 \sim \mathsf{Bern}(\frac{1}{2})$. Let $X = S_1 \oplus S_2$ be an input to a BSC with crossover probability p and Y denote an output of the channel. Suppose S_1 and S_2 are independent of each other. Compute $I(X;Y)$.

Prob 5.3 (Typical versus non-typical sequence pairs) Consider a discrete memoryless channel. Suppose X^n are i.i.d., each being generated as per $p_X(x)$. Let T^n be another i.i.d. random process, being independent of X^n yet each being generated according to the same $p_X(\cdot)$. Let Y^n be the output of the channel when X^n is fed into. Explain the limiting behavior of $\lim_{n\to\infty} \frac{1}{n} \log \frac{1}{p_{X^n,Y^n}(T^n,Y^n)}$.

Prob 5.4 (Jointly typical sequence) Consider a discrete memoryless channel. Suppose that the encoder uses a random code in which $X_i(w)$'s are i.i.d. $\sim p(x)$ across $i \in \{1,\ldots,n\}$ and $w \in \{1,\ldots,2^{nR}\}$.

(*a*) Show that for $w \in \{1,\ldots,2^{nR}\}$,

$$p(y^n|x^n(w)) = \prod_{i=1}^{n} p(y_i|x_i(w)).$$

(*b*) Suppose that $X^n(1)$ is transmitted. Show that

$$\frac{1}{n}\log\frac{1}{p(X^n(1),Y^n)} \xrightarrow{\text{in prob.}} H(X,Y) \quad \text{as } n\to\infty.$$

Also show that for $\epsilon > 0$, as $n\to\infty$,

$$\mathbb{P}(2^{-n(H(X,Y)+\epsilon)} \le p(x^n(1),y^n) \le 2^{-n(H(X,Y)-\epsilon)}) \longrightarrow 1.$$

(*c*) Suppose that $X^n(1)$ is transmitted. A student claims that as in part (*b*),

$$\frac{1}{n}\log\frac{1}{p(X^n(2),Y^n)} \xrightarrow{\text{in prob.}} H(X,Y) \quad \text{as } n\to\infty.$$

Prove or disprove this statement.

(*d*) Show that Y^n is i.i.d.

Prob 5.5 (Fano's inequality) We wish to transmit a message $W \in \{1,\ldots,2^{nR}\}$ over a discrete memoryless channel. Let $\hat{W}$ be a decoded message at a receiver. Let $E = \mathbf{1}\{\hat{W} \neq W\}$ where $\mathbf{1}\{\cdot\}$ denotes an indicator function which returns 1 if the argument is true; 0 otherwise. Let $P_e := \mathbb{P}(\hat{W} \neq W)$. Show that

$$H(W|\hat{W}) = H(E|\hat{W}) + P_e \cdot H(W|\hat{W}, E=1).$$

Prob 5.6 (Data processing inequality) Suppose a random process $\{X_i\}$ satisfies: for all $i \ge 1$,

$$p(x_{i+2}|x_{i+1},x_i,\ldots,x_1) = p(x_{i+2}|x_{i+1},x_i).$$

Let $S_i := (X_{i+1}, X_i)$. A student claims that $I(S_1;S_2) \ge I(S_1;S_3)$. Prove or disprove the claim.

Prob 5.7 (Source-channel separation theorem) We wish to transmit an information source of a stationary process V^k over a DMC. Let X^n be the output of an encoder fed by V^k, and Y^n be the output of the DMC when X^n is fed into. Let $\hat{V}^k$ be the decoded information source at a receiver. Let $C := \max_{p(x)} I(X;Y)$ and $P_e := \mathbb{P}(\hat{V}^k \neq V^k)$.

(*a*) Using the source coding and channel coding theorems, show that if

$$kH(\mathcal{V}) < nC, \tag{2.21}$$

then P_e can be made arbitrarily close to 0 as $n\to\infty$. Here $H(\mathcal{V})$ denotes the entropy rate:

$$H(\mathcal{V}) := \lim_{k\to\infty}\frac{H(V^k)}{k}.$$

(*b*) Show that for an integer $m \geq 0$,

$$\frac{H(V^m)}{m} \geq \frac{H(V^{m+1})}{m+1}.$$

Hint: $H(V^m) = \sum_{i=1}^{m} H(V_i|V^{i-1})$ and $H(V_i|V^{i-1})$ is a non-increasing sequence in i.

(*c*) Prove: (i) Fano's inequality $H(V^k|\hat{V}^k) \leq 1 + P_e \cdot k \log |\mathcal{V}|$; and (ii) data processing inequality $I(V^k; \hat{V}^k) \leq I(X^n; Y^n)$. Here $\mathcal{V}$ denotes the range of V.

(*d*) Using parts (*b*) and (*c*), show that if we want to make $P_e \to 0$, then $kH(\mathcal{V}) < nC$ must hold.

(*e*) Does the condition (2.21) serve as the sufficient and necessary condition for reliable communication, even when using an encoder/decoder that doesn't follow Shannon's two-stage architecture? Based on the answer, can we determine the optimality of the two-stage architecture? Is the two-stage architecture lossless in terms of optimality, meaning that any data rate achieved with an arbitrary encoder/decoder can also be achieved using the two-stage architecture?

Prob 5.8 (Source-channel separation theorem) We wish to encode i.i.d. $V^n \sim \text{Bern}(\frac{1}{4})$ for transmission over a binary erasure channel with erasure probability ϵ. Find the necessary and sufficient condition under which the probability of error $\mathbb{P}(\hat{V}^n \neq V^n)$ can be made arbitrarily close to 0 as $n \to \infty$.

Prob 5.9 (Capacity of a composite channel) Suppose that two binary symmetric channels (BSCs) with crossover probabilities p_1 and p_2 respectively are connected end-to-end to form a composite channel. Let X_1^n and Y_1^n (or X_2^n and Y_2^n) indicate the input and output of the first (or second) BSC. Here n denotes the code length.

(*a*) Suppose no operation is allowed between the two BSCs, i.e., X_2^n is simply set as Y_1^n. Compute the capacity of this composite channel.

(*b*) Suppose any operation is allowed between the two BSCs, i.e., X_2^n can now be an arbitrary function of Y_1^n. Compute the capacity of this composite channel.

Prob 5.10 (Capacity of the union of two channels) Consider a discrete memoryless channel which is the union of the following two channels: (i) a binary symmetric channel with crossover probability p; and (ii) an erasure channel with erasure probability ϵ. At each time, one can send a symbol channel 1 or channel 2 but not both. Find the capacity of this channel.

Prob 5.11 (Capacity of a composite channel) Let X, Y, Z be three discrete random variables defined on $\mathcal{X}, \mathcal{Y}$ and $\mathcal{Z}$ respectively. Suppose that $p(y|x)$ is a DMC with input in $\mathcal{X}$, output in $\mathcal{Y}$ and capacity C_1; and $q(z|y)$ is another DMC with input in $\mathcal{Y}$, output in $\mathcal{Z}$, and capacity C_2. Consider the following DMC

$$r(z|x) := \sum_{y \in \mathcal{Y}} q(z|y)p(y|x)$$

with input in $\mathcal{X}$ and output in $\mathcal{Z}$. A confused information theorist claims that the capacity of this DMC is $\min(C_1, C_2)$. Either prove or disprove this statement.

Prob 5.12 (Channel capacity and achievability) We wish to transmit a uniformly distributed message $W \in \{1, \ldots, 2^{nR}\}$ over a memoryless binary erasure channel with erasure probability p. Here n denotes the code length and $R := \log \frac{M}{n}$ where M indicates the cardinality of the range of W. Let f and g be channel encoder and decoder respectively. Let $P_e = \mathbb{P}(\hat{W} \neq W)$ be the probability of error where $\hat{W}$ indicates a decoded message. Let C be the capacity of this channel.

(*a*) Given R, define:

$$P_e^*(R) := \min_{f,g,n} P_e.$$

For $\epsilon > 0$, compute $P_e^*(C - \epsilon)$ and $P_e^*(C + \epsilon)$.

(*b*) Show that $C \leq 1 - p$.

(*c*) In Section 2.2, we intended to prove the achievability of $1 - p$. To this end, we employed a random encoder in which $X_i(w)$'s are generated i.i.d. according to $p(x)$ for $i \in \{1, 2, \ldots, n\}$ and $w \in \{1, 2, \ldots, 2^{nR}\}$. What was the choice of $p(x)$?

(*d*) Given $Y^n = y^n$, derive an optimal decoder. What is the name of the optimal decoder? Also explain the rationale behind the naming. We say that a decoder is optimal if it minimizes the probability of error.

(*e*) It turns out that under the random encoding in part (*c*) and the optimal decoder in part (*d*), P_e can be made arbitrarily close to 0 as $n \to 0$. Show that this implies the existence of an optimal deterministic code, say C^* such that given $R < 1 - p$, $P_e(C^*) \to 0$ as $n \to \infty$.

Prob 5.13 (Role of feedback) This problem explores the role of the channel output feedback in a DMC. We wish to transmit a message $W \in \{1, 2, \ldots, 2^{nR}\}$ over the DMC. Let X^n be a transmitted signal and Y^n be the channel output. Unlike the conventional non-feedback setting, we assume that the past channel output Y^{i-1} is available at the encoder at time i; hence, X_i is a function of (W, Y^{i-1}). Let $\hat{W}$ be the decoded message.

(*a*) Show that $(W, Y^n, \hat{W})$ form a Markov chain, i.e., $W - Y^n - \hat{W}$.
(*b*) Using part (*a*), show that $I(W; \hat{W}) \leq I(W; Y^n)$ (data processing inequality).
(*c*) Prove that $H(W|\hat{W}) \leq 1 + P_e \cdot nR$ (Fano's inequality).
(*d*) Using the definition of memoryless channels and the fact that X_i is a function of (W, Y^{i-1}), show that

$$H(Y^n|W) = \sum_{i=1}^{n} H(Y_i|X_i).$$

(*e*) Using parts (*b*), (*c*), (*d*), prove that the capacity of this feedback channel is still $C := \max_{p(x)} I(X; Y)$, meaning that feedback cannot increase capacity. Since the achievability readily comes from the nonfeedback scheme, you only need to prove the converse: if $P_e \to 0$, then $R \leq C$.

Prob 5.14 (Capacity of a cascaded channel) Consider a cascade of n identical binary symmetric channels (BSCs), each with crossover probability $p \in (0, 1)$. Assume that no operation is allowed in between any two BSCs. Compute the capacity of this cascaded channel.

Prob 5.15 (Capacity of a cascaded channel) Consider a discrete memoryless channel which concatenates the following two channels serially: (i) a binary symmetric channel with crossover probability p; and (ii) an erasure channel with erasure probability ϵ. Let X and Y be the input and output of the channel.

(*a*) Derive the conditional distribution $p(y|x)$.
(*b*) Find the capacity of this channel.

Prob 5.16 (Markov chain) Suppose

$$X - Y - (Z, W).$$

(*a*) Show that

$$X - (Y, Z) - W.$$

(*b*) Find $I(X; W|Y)$.

Prob 5.17 (Capacity of the union of two channels) Define the probability transition matrix P_i for a discrete memoryless channel (DMC) as a matrix whose (x, y) entry is the probability that the output of the channel is y given the input x. Consider two DMCs, called DMC_1 and DMC_2, with transition matrices P_1 and P_2 respectively. Consider a third DMC, say DMC_3, whose transition matrix is given by

$$\begin{pmatrix} P_1 & 0 \\ 0 & P_2 \end{pmatrix}.$$

Define a selector S that selects the channel to which a symbol is transmitted.

(*a*) Show that

$$I(X;Y) = I(X,S;Y).$$

(*b*) Show that the capacity of DMC$_3$ is given by

$$C_3 = \log(2^{C_1} + 2^{C_2}),$$

where C_i is the capacity of DMC$_i$.
Hint: Use part (*a*).

Prob 5.18 (Capacity of an erasure channel with two erasures) Consider a DMC with binary input $X \in \{0,1\}$, output $Y \in \{0, e_0, e_1, 1\}$, and channel probabilities:

$$p(0|0) = p(1|1) = 1 - p - q,$$

$$p(e_0|0) = p(e_1|1) = p,$$

$$p(e_1|0) = p(e_0|1) = q$$

where $p > q > 0$. Find the capacity of this DMC.

Prob 5.19 (Quantum channel) Alice wishes to transmit a single bit $X \sim \text{Bern}(\frac{1}{2})$ to Bob. There is an intruder Eve who intends to interfere with the communication. With probability $1-p$, Eve does nothing, so Bob receives X. With probability p, however, Eve intervenes in the communication between Alice and Bob. With probability p, Eve intercepts the transmitted bit in a possibly noisy manner. The intercepted bit Z is:

$$Z = \begin{cases} X, & \text{w.p. } \frac{1}{2}; \\ X+N, & \text{w.p. } \frac{1}{2} \end{cases}$$

where $N \sim \text{Bern}(\frac{1}{2})$, independent of X. Eve then resends Z to Bob. Hence, a received signal Y at Bob reads:

$$Y = \begin{cases} X, & \text{w.p. } 1-p; \\ Z, & \text{w.p. } p. \end{cases}$$

(*a*) Eve wishes to decode X given its received signal. Remember that Eve gets nothing w.p. $1-p$ and gets Z w.p. p. Let $\hat{X}_E$ be the optimal detector output w.r.t. X. Compute $\mathbb{P}(\hat{X}_E \neq X)$. Also compute $I(X;\hat{X}_E)$.

(*b*) Similarly Bob wants to decode X given Y. Let $\hat{X}_B$ be the corresponding optimal detector output. Compute $\mathbb{P}(\hat{X}_B \neq X)$. Compute $I(X;\hat{X}_B)$.

(*c*) Find a condition on $\mathbb{P}(\hat{X}_B \neq X)$ such that $I(X;\hat{X}_B) \geq I(X;\hat{X}_E)$.

Prob 5.20 (True or False?)

(*a*) Consider a discrete memoryless channel with input X_i and output Y_i. Then,

$$H(Y_1, Y_2|X_1, X_2) = H(Y_1|X_1) + H(Y_2|X_2).$$

(*b*) Consider a channel coding problem setup where the channel is a BEC with erasure probability 0.11. A hard-working student claims that she can come up with a transmission scheme that achieves 100 bits of reliable transmission with code length 1000. Does the claim make sense?

(*c*) Let $V_1, V_2, \ldots, V_k$ be a finite alphabet i.i.d source. The information source is encoded as a sequence of n input symbols X^n of a memoryless binary erasure channel. Let Y^n be the output of the channel. Given $Y^n = y^n$, we wish to decode X^n. The optimal decoder is:

$$\hat{X}^n = \arg\max_{x^n} \mathbb{P}(Y^n = y^n | X^n = x^n).$$

(*d*) Consider a discrete memoryless channel (DMC) with probability transition probability:

$$p(y|x) = \begin{bmatrix} p & 1-p & 0 & 0 \\ 1-p & p & 0 & 0 \\ 0 & 0 & q & 1-q \\ 0 & 0 & 1-q & q \end{bmatrix}$$

where rows and columns correspond to values of x and y, respectively. The capacity of this channel is $C = 2 - H(p) - H(q)$ where $H(p) := p \log \frac{1}{p} + (1-p) \log \frac{1}{1-p}$.

(*e*) Consider an additive channel whose input alphabet $\mathcal{X} = \{0, \pm 2, \pm 4\}$ and whose output $Y = X + Z$. Here Z is distributed uniformly over the interval $[-1, 1]$. The capacity of this channel is $\log 3$.

(*f*) Consider a DMC with feedback in which the past channel output is available at encoder: an encoded signal X_i at time i is a function of (W, Y^{i-1}). Here W indicates a message $W \in \{1, 2, \ldots, 2^{nR}\}$, Y_i denotes channel output at time i, and $Y^{i-1} := (Y_1, \ldots, Y_{i-1})$. Then, $W - X^n - Y^n$.

(*g*) Suppose a random process $\{X_i\}$ satisfies: for all $i \geq 1$,

$$p(x_{i+2}|x_{i+1}, x_i, x_{i-1}, \ldots, x_1) = p(x_{i+2}|x_{i+1}, x_i).$$

Let $S_i := (X_{i+1}, X_i)$. Then, $I(S_1; S_3) \geq I(S_2; S_4)$.

(*h*) Suppose that two binary symmetric channels with crossover probabilities p_1 and p_2 respectively are connected end-to-end to form a composite channel. Suppose that any operation is allowed between the two channels. Then, the capacity of the composite channel is $1 - \max\{H(p_1), H(p_2)\}$.

(*i*) Suppose random variables $(X_1, X_2, X_3, X_4, X_5)$ form a Markov chain. Then, $I(X_1; X_2) \geq I(X_2; X_5)$.

(*j*) Shannon created a communication architecture that separates source coding and channel coding, allowing them to be performed independently of each other. This independent design facilitates the standardization of the digital interface and simplifies implementation. However, this comes at the expense of performance degradation resulting from the separation of the architecture.

(*k*) Let X, Y, Z be three discrete random variables defined on $\mathcal{X}, \mathcal{Y}, \mathcal{Z}$ respectively. Suppose that $p(y|x)$ is a discrete memoryless channel (DMC) with input in $\mathcal{X}$, output in $\mathcal{Y}$ and capacity C_1; and $q(z|y)$ is another DMC with input in $\mathcal{Y}$, output in $\mathcal{Z}$, and capacity C_2. Consider the following DMC

$$r(z|x) := \sum_{y \in \mathcal{Y}} q(z|y)p(y|x)$$

with input in $\mathcal{X}$ and output in $\mathcal{Z}$. This DMC can have capacity greater than $\max(C_1, C_2)$.

(*ℓ*) Let Y_1 and Y_2 be conditionally independent and conditionally identically distributed given X. The capacity of a channel with input X and output (Y_1, Y_2) is twice the capacity of a channel with input X and output Y_1.

2.7 Polar Code: Polarization

Recap In Part II, we proved the channel coding theorem and source-channel separation theorem for discrete memoryless channels. To achieve this, we used a random coding argument in the proof of the channel coding theorem to establish the existence of an optimal code. However, we did not discuss how to construct an explicit and practical code that guarantees optimality.

Outline In the remainder of Part II, we will delve into a specific deterministic code called the *polar code*, which achieves channel capacity for a class of channels. This section is divided into four parts. Firstly, we will share an interesting backstory about the random code utilized in the achievability proof of the channel coding theorem. Following that, we will discuss two major endeavors focused on explicit code constructions, with emphasis on the polar code. Then, we will highlight a crucial feature of the polar code known as polarization, which brings about a fascinating phenomenon. Finally, we will examine the polarization-inspired encoding and decoding methods to describe how the polar code operates.

Initial reactions on Shannon's channel coding theorem When Shannon first introduced his channel coding theorem, it was met with mixed reactions, particularly from communication systems engineers. There were three main reasons for this. Firstly, many engineers did not comprehend the concepts of reliable communication, achievable data rate, and capacity, which made it difficult for them to understand the theorem. Secondly, even those who understood the theorem had a negative outlook on the development of an optimal code. The achievability proof of the theorem suggested that the optimal code required a lengthy code-length to attain capacity, which seemed complex and unfeasible given the technology of the time. Lastly, even the optimistic engineers who saw the potential for implementation with future technology were not confident because Shannon did not provide a concrete method for designing an optimal code. The proof only established the existence of optimal codes without specifying how to construct them.

Two major efforts Due to these reasons, Shannon and his supporters, including some intelligent MIT folks, made significant efforts to develop explicit and deterministic optimal codes with potentially low implementation complexity. Unfortunately, Shannon himself failed to develop a good deterministic code. Instead, his MIT supporters came up with some successful codes, and in this section, we will discuss two of their major efforts.

One of the major efforts was made by Robert G. Gallager, who developed the "Low Density Parity Check code" (LDPC code for short) in 1960 (Gallager, 1962). It is an explicit and deterministic code, which provides a detailed guideline on how to design such a code. The code's performance is remarkable, as it approaches capacity as the code length tends to infinity, although it does not match the capacity precisely. However, the LDPC code was not initially given enough credit since it was still of high implementation complexity given the digital signal processing (DSP) technology of the day.[2] However, the code was later revived 30 years later, as it became an efficient code when the DSP technology evolved, finally enabling the code to be implemented. Currently, it is widely being employed in a variety of systems, such as LTE, WiFi, DMB,[3] and storage systems.

Gallager was not entirely satisfied with his result, as his code was not guaranteed to achieve capacity precisely, even in the limit of code length. This motivated one of his PhD students, Erdal Arikan, to develop a capacity-achieving deterministic code. Arikan developed the first capacity-achieving deterministic code, called polar code (Arikan, 2009). Interestingly, he could develop the code in 2007, 30+ years later than the motivation for Gallager's work. Due to its excellent performance and low-complexity nature, the polar code is being seriously considered for implementation in a variety of systems. We will dedicate the next three sections to study the polar code.

The encoder structure that Arikan imagined Arikan focused on a simple channel in which an input to the channel is binary-valued. Such examples are BEC and BSC that we learned earlier. He proceeded to examine the statistical characteristic that the channel input X^n must possess to achieve the capacity. This brought to mind the random code utilized by Shannon, where X^n is independent and identically distributed (i.i.d.). This served as a source of inspiration for Arikan, leading to the development of the encoding structure shown in Fig. 2.5.

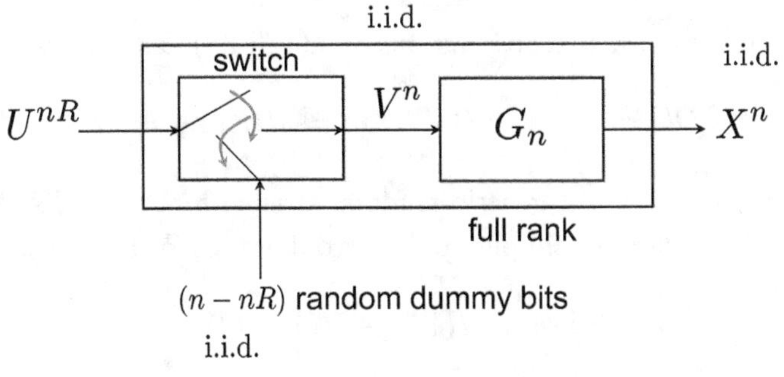

Figure 2.5. The encoder structure of the polar code.

2. Despite the lack of immediate impact on the world, his remarkable work earned him a position as a faculty member at MIT right after graduation. It is fortunate that there are scholars who are patient enough to recognize the potential of such groundbreaking work.

3. It is the name of the technology for a digital broadcast system, standing for Digital Multimedia Broadcasting

To explain the structure in detail, we first introduce some notations. We denote the message W by binary string $U^{nR} := (U_1, U_2, \ldots, U_{nR})^T$ where T denotes a transpose. Note that the message can be represented as nR i.i.d. bits. Why? We intend to generate another sequence X^n from U^{nR}. Remember that we consider a binary-input channel where the capacity cannot exceed 1 (why?). Hence, X^n is of a longer length relative to U^{nR}. We introduce additional number $(n - nR)$ bits that we call dummy bits. Since we want to have i.i.d. sequence X^n, we induce the random dummy bits which are independent of U^{nR} and also i.i.d.

The way of constructing such X^n takes two steps. First we combine U^{nR} and the dummy bits to generate a length-n sequence, say V^n. Here V_i takes either a component of U^{nR} or that of the dummy bits. For instance, when $n = 6$ and $R = \frac{1}{2}$, we may have $V^n = (U_1, U_2, \text{dummy}_1, \text{dummy}_2, U_3, \text{dummy}_3)^T$. Whether we choose the one from U^{nR} or from the dummy bits for the value of V_i depends on a particular rule that will be specified later on. We then pass V^n through a full-rank matrix of size n-by-n, say G_n, yielding:

$$X^n = G_n V^n.$$

One can verify that X^n is also i.i.d. (the property that we wished to obtain) as long as G_n has full rank. Check in Prob 6.3. It will be clearer soon as to why the full-rank matrix is employed here.

This is the encoder structure that Arikan imagined. Under this structure, what he observed is that for some particular G_n, an interesting phenomenon occurs. He called that phenomenon "polarization" – the rationale behind the naming will be clearer later. In fact, he discovered the phenomenon in the process of manipulating the following quantity: $I(V^n; Y^n)$.

Polarization He made two observations on $I(V^n; Y^n)$. The first is:

$$\textit{Observation } \#1 : I(V^n; Y^n) = nI(X; Y) =: nI \tag{2.22}$$

where X (or Y) indicates a generic random variable for X_i (or Y_i). We denote $I(X; Y)$ by I for notational simplicity. The proof of Observation #1 is as follows:

$$\begin{aligned} I(V^n; Y^n) &= H(Y^n) - H(Y^n|V^n) \\ &\overset{(a)}{=} H(Y^n) - H(Y^n|X^n, V^n) \\ &\overset{(b)}{=} H(Y^n) - H(Y^n|X^n) \\ &\overset{(c)}{=} H(Y^n) - \sum_{i=1}^{n} H(Y_i|X_i) \end{aligned}$$

$$\overset{(d)}{=} \sum_{i=1}^{n} H(Y_i) - \sum_{i=1}^{n} H(Y_i|X_i)$$
$$= nI(X; Y) = nI$$

where (a) follows from the fact that X^n is a function of V^n; (b) is due to the Markov chain of $V^n - X^n - Y^n$ (why?); (c) comes from the memoryless property of the channel; and (d) follows from the fact that Y^n is i.i.d. (why?).

The second observation is:

$$\textit{Observation \#2} : I(V^n; Y^n) = \sum_{i=1}^{n} I(V_i; Y^n, V^{i-1}). \tag{2.23}$$

This is due to the chain rule w.r.t. mutual information and the fact that V_i is independent of V^{i-1}: $I(V_i; Y^n|V^{i-1}) = I(V_i; Y^n, V^{i-1})$. If you are not convinced, please check in Prob 6.2. Arikan viewed $I(V_i; Y^n, V^{i-1})$ as a quantity that indicates the data rate w.r.t. the ith *virtual* subchannel (say p_i) with input V_i and output (Y^n, V^{i-1}):

$$V_i \rightarrow \text{(virtual subchannel } i) \rightarrow (Y^n, V^{i-1}). \tag{2.24}$$

He then made an interesting phenomenon for the quantity $I(V_i; Y^n, V^{i-1})$ under some particular choice of G_n. To illustrate this, let us plot an empirical CDF (cumulative density function) of the quantity:

$$\frac{|\{i : I(V_i; Y^n, V^{i-1}) \leq x|}{n}$$

as a function of a dummy variable x.

You will soon understand why we plot the empirical CDF to see the interesting phenomenon. To understand this, let us first consider the case of $n = 1$. In this case, we have only one virtual subchannel where the data rate is $I(V_1; Y_1)$, which is I due to (2.22). So the empirical CDF is a dirac-delta function jumping at I. See the first subfigure in Fig. 2.6.

When $n = 2$, the summation includes:

$$\begin{aligned} I_1 &:= I(V_1; Y_1, Y_2); \\ I_2 &:= I(V_2; Y_1, Y_2, V_1). \end{aligned} \tag{2.25}$$

Due to Observations #1 and #2 in (2.22) and (2.23):

$$I_1 + I_2 = 2I. \tag{2.26}$$

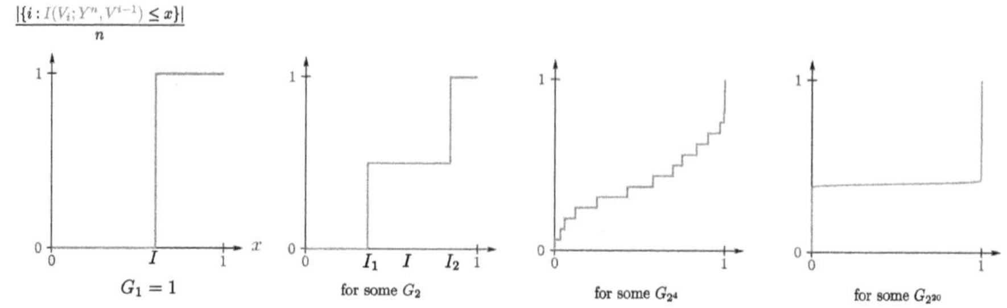

Figure 2.6. Polarization: $I(V_i; Y^n, V^{i-1})$ takes either 1 or 0 in the limit of n.

Suppose that $I_1 \leq I \leq I_2$ under some G_2. Then, we would have two jumps at I_1 and I_2 as plotted in the second subfigure in Fig. 2.6. What Arikan found is that under some G_n the empirical CDF is of a very interesting shape in the limit of n: the jumps occur only at two extreme points, which are 0 and 1. This implies:

$$I(V_i; Y^n, V^{i-1}) = 1 \text{ or } 0 \text{ as } n \to \infty, \tag{2.27}$$

meaning that the data rate of the ith subchannel is *polarized* (perfect or completely noisy). Also (2.22) together with (2.23) suggests that the fraction of $I(V_i; Y^n, V^{i-1})$ being 1 (or 0) approaches I (or $1 - I$):

$$\begin{aligned} \frac{|\{i : I(V_i; Y^n, V^{i-1}) \approx 1|}{n} &\to I; \\ \frac{|\{i : I(V_i; Y^n, V^{i-1}) \approx 0|}{n} &\to 1 - I. \end{aligned} \tag{2.28}$$

Encoding & decoding The polarization reflected in (2.27) and (2.28) immediately suggests the following encoding rule:

$$\text{Set } V_i = \begin{cases} \text{information bit (from } U^{nR}), & \text{if } I(V_i; Y^n, V^{i-1}) \approx 1; \\ \text{dummy bit,} & \text{otherwise,} \end{cases};$$

$$\text{Set } R \approx I.$$

You may wonder when $I(V_i; Y^n, V^{i-1})$ is close to 1 or 0. It depends on the structure of G_n that we will investigate in detail later on.

Now what about decoding? As we learned earlier, the optimal decoding rule is ML: Choosing v^n such that the corresponding likelihood $p(y^n|V^n = v^n)$ is maximized. But we are not going to employ the ML rule since the complexity of the rule is prohibitive. It requires an exhaustive search over all possible choices of U^{nR} (why?), i.e., the complexity scales like 2^{nR}. Instead we will use a suboptimal yet intuitive and low-complexity rule, so called *successive cancellation decoding*. This is inspired by the subchannel representation as illustrated in Fig. 2.7.

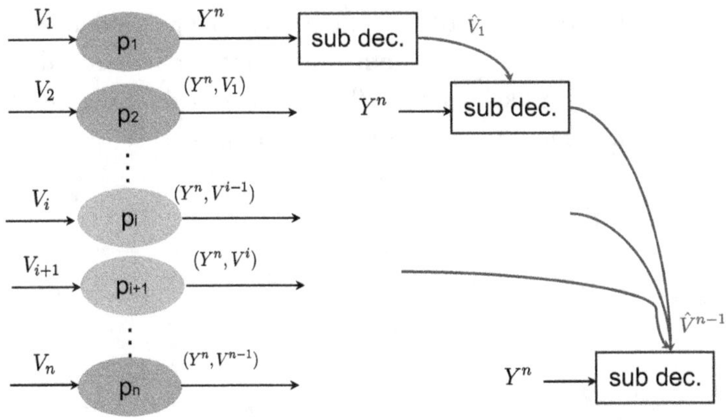

Figure 2.7. Successive cancellation decoding.

The blue-colored (or red-colored) virtual subchannels indicate the perfect (or completely noisy) channels. Note in the first subchannel, say p_1, that the output contains only the received signal Y^n. So one can decode V_1 by employing the following ML rule associated with that particular subchannel:

$$p(y^n|V_1 = 1) \geq p(y^n|V_1 = 0) \longrightarrow \hat{V}_1 = 1;$$
$$p(y^n|V_1 = 1) < p(y^n|V_1 = 0) \longrightarrow \hat{V}_1 = 0.$$

On the other hand, in the second subchannel p_2, the output contains V_1 (in addition to Y^n) which is not available at the decoder. But the good news is that the *estimate* of V_1 is available instead once we perform the above operation regarding p_1. This suggests a successive way of decoding. We first decode V_1; we then use this to decode V_2; and all the way up to decode V_n with $\hat{V}^{n-1}$. To be specific, for the ith subchannel, the decoding rule is:

$$p(y^n, \hat{v}^{i-1}|V_i = 1) \geq p(y^n, \hat{v}^{i-1}|V_i = 0) \longrightarrow \hat{V}_i = 1;$$
$$p(y^n, \hat{v}^{i-1}|V_i = 1) < p(y^n, \hat{v}^{i-1}|V_i = 0) \longrightarrow \hat{V}_i = 0$$

where the estimates $\hat{v}^{i-1}$ are available from the earlier steps. For time indices i in which dummy bits are transmitted, we do not need to decode such bits. Hence we simply ignore them while using them instead to decode other information bits. It turns out that this suboptimal (yet intuitive) decoding rule enables the error probability to be arbitrarily close to 0 as long as $R < I$. The proof of this is not that simple. So we omit the detailed proof in this book.

Look ahead What follows are two questions that need to be addressed. The first question is: How can we create G_n to induce polarization? The second question is: How can we calculate the likelihood $p(y^n|\hat{v}^{i-1}|V_i)$ needed for successive cancellation decoding? These questions will be answered in the next section.

2.8 Polar Code: Implementation of Polarization

Recap In the previous section, we began exploring the polar code, which is the first code to achieve capacity with low complexity and explicit construction. The encoder structure of the code consists of two stages. In the first stage, nR information bits (U^{nR}) are combined with $n - nR$ dummy bits according to a certain rule to construct a longer i.i.d. sequence V^n. In the second stage, V^n is converted into X^n by multiplying it with a full-rank matrix G_n to achieve polarization.

Arikan observed that when converting n independent copies of a channel $p(y|x)$ into n virtual subchannels (each having input V_i and output (Y^n, V^{i-1})) under a particular choice of G_n, the data rate of each virtual subchannel $I(V_i; Y^n, V^{i-1})$ takes either 1 or 0 in the limit of n, indicating complete polarization of the subchannels. The polarization phenomenon led naturally to the encoding and decoding strategies. The encoding rule assigns information bits to good channels (with a data rate of 1) and dummy bits to bad channels (with a data rate of 0). The decoding rule follows successive cancellation decoding, where V_i is decoded in a step-by-step manner from $i = 1$ to $i = n$ with the aid of previously decoded bits $\hat{V}^{i-1}$.

Outline This section will cover two topics that were not previously discussed. Firstly, we will investigate the construction of G_n, which is necessary to achieve polarization. Secondly, we will delve into the computation of the likelihood functions $p(y^n, v^{i-1}|v_i)$, which are required for successive cancellation decoding of the virtual subchannels.

Case of $n = 2$: Choice of G_2 & likelihood computation Let us start with the simplest case of $n = 2$. In this case, one obvious yet non-trivial choice for G_2 is:

$$G_2 = \begin{bmatrix} 1 & 1 \\ 0 & 1 \end{bmatrix}. \tag{2.29}$$

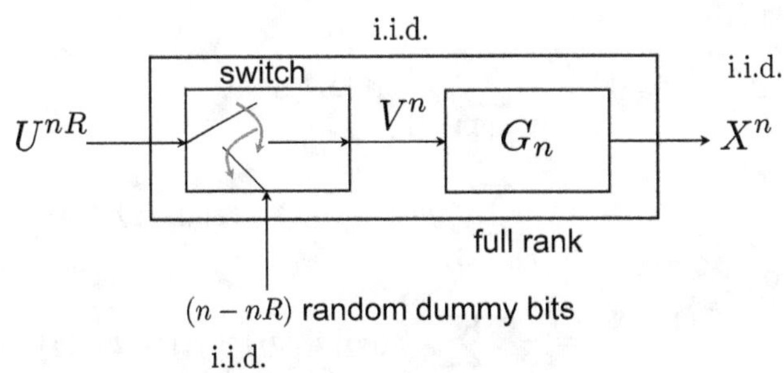

Figure 2.8. The encoder structure of the polar code.

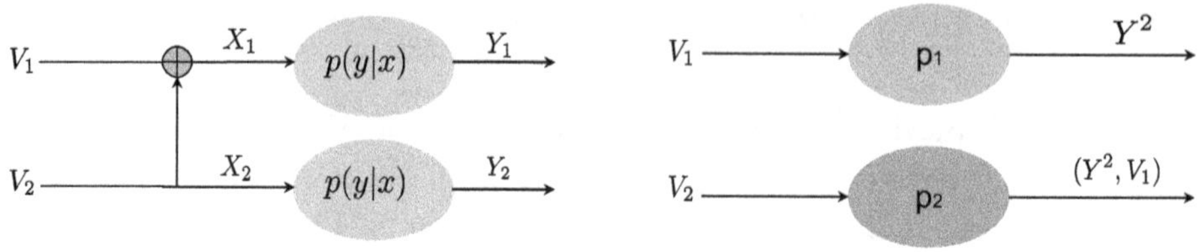

Figure 2.9. (Left): Two independent channels and the mapping between (V_1, V_2) and (X_1, X_2); (Right): Two converted virtual subchannels (denoted by p_1 and p_2 respectively).

Clearly, no polarization occurs when G_2 is either

$$\begin{bmatrix} 1 & 0 \\ 0 & 1 \end{bmatrix} \text{ or } \begin{bmatrix} 0 & 1 \\ 1 & 0 \end{bmatrix}.$$

One can readily verify that in the case, the data rate of the first virtual subchannel $I_1 := I(V_1; Y^2)$ coincides with that of the second subchannel $I_2 := I(V_2; Y^2, V_1)$, meaning no polarization between I_1 and I_2 (please check). The choice of (2.29) is one of the two remaining non-trivial candidates. This choice yields: $X_1 = V_1 \oplus V_2$ and $X_2 = V_2$; see the left in Fig. 2.9. In the previous section, we converted these two independent copies of $p(y|x)$ into two virtual subchannels, inspired by the following formula:

$$I(V^2; Y^2) = I(V_1; Y^2) + I(V_2; Y^2, V_1).$$

See the right in Fig. 2.9 for illustration of each virtual subchannel, say p_i where $i \in \{1, 2\}$.

Now recall one of the two questions raised earlier: how to compute (p_1, p_2) which are required to perform successive cancellation decoding? Here the key is to represent p_i in terms of the one that is known: the conditional distribution of the DMC $p(y|x)$. First we get:

$$\begin{aligned}
p_1(y_1, y_2|v_1) &\overset{(a)}{=} \sum_{v_2 \in \{0,1\}} p(y_1, y_2, v_2|v_1) \\
&\overset{(b)}{=} \sum_{v_2 \in \{0,1\}} p(v_2|v_1) p(y_1, y_2|v_1, v_2) \\
&\overset{(c)}{=} \frac{1}{2} \sum_{v_2 \in \{0,1\}} p(y_2|v_1, v_2) p(y_1|v_1, v_2, y_2)
\end{aligned}$$

$$\stackrel{(d)}{=} \frac{1}{2} \sum_{v_2 \in \{0,1\}} p(y_2|v_2)p(y_1|v_1, v_2, y_2)$$

$$\stackrel{(e)}{=} \frac{1}{2} \sum_{v_2 \in \{0,1\}} p(y_2|v_2)p(y_1|v_1 \oplus v_2)$$

where (*a*) follows from the total probability law; (*b*) is due to the definition of conditional distribution; (*c*) comes from the definition of conditional distribution and the fact that $p(v_2|v_1) = \frac{1}{2}$; (*d*) follows from the Markov chain: $V_1 - X_2(= V_2) - Y_2$ (why?); and (*e*) is due to the Markov chain: $(V_1, V_2, Y_2) - V_1 \oplus V_2 - Y_1$ (why?). Notice that $p_1(y_1, y_2|v_1)$ is represented in terms of $p(y_2|v_2)$ and $p(y_1|v_1 \oplus v_2)$ which are known and therefore can be computed. Similarly we get:

$$\begin{aligned} p_1(y_1, y_2, v_1|v_2) &= p(v_2|v_1)p(y_1, y_2|v_1, v_2) \\ &= \frac{1}{2}p(y_2|v_2)p(y_1|v_1 \oplus v_2). \end{aligned}$$

Example: Binary Erasure Channel To give you a concrete feel as to how to compute p_1 and p_2 derived as above, let us give you an example of a binary erasure channel. Let α denote erasure probability of the BEC. Note in the BEC that

$$(Y_1, Y_2) = \begin{cases} (V_1 \oplus V_2, V_2), & \text{w.p. } (1-\alpha)^2; \\ (e, V_2) & \text{w.p. } \alpha(1-\alpha); \\ (V_1 \oplus V_2, e) & \text{w.p. } (1-\alpha)\alpha; \\ (e, e) & \text{w.p. } \alpha^2. \end{cases}$$

One can decode V_1 only when there is no erasure (the first case). Hence, one can view p_1 as another BEC yet having different erasure probability: $1-(1-\alpha)^2$. Since $I_1 = (1-\alpha)^2$ is smaller than the capacity of the original BEC ($C_{\mathsf{BEC}} = 1-\alpha$), p_1 is sort of a bad channel.

On the other hand,

$$(Y_1, Y_2, V_1) = \begin{cases} (V_1 \oplus V_2, V_2, V_1), & \text{w.p. } (1-\alpha)^2; \\ (e, V_2, V_1) & \text{w.p. } \alpha(1-\alpha); \\ (V_1 \oplus V_2, e, V_1) & \text{w.p. } (1-\alpha)\alpha; \\ (e, e, V_1) & \text{w.p. } \alpha^2. \end{cases}$$

One can decode V_2 except for one case in which both channels are erased (the last case); hence, p_2 can be viewed as another BEC with erasure probability α^2. Since $I_2 = 1-\alpha^2$ is larger than $C_{\mathsf{BEC}} = 1-\alpha$, p_2 is sort of a good channel.

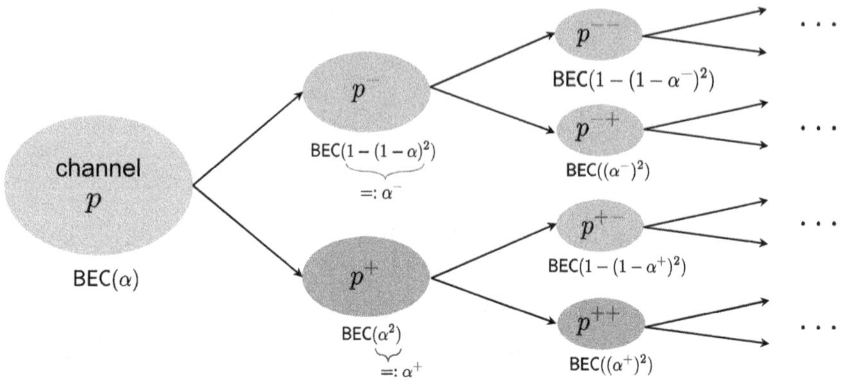

Figure 2.10. Channel splitting for arbitrary n.

The above two interpretations suggest that the two independent copies of $p(y|x)$ can be split into two virtual subchannels: one with a smaller capacity (the bad channel); the other with a larger capacity (the good channel). So we can view this split as partial polarization.

Idea: Channel splitting for an arbitrary n This observation leads to a natural way of polarizing n independent copies of $p(y|x)$. The idea is to repeat the same for the two split subchannels. See Fig. 2.10.

We first split the original BEC (with erasure probability α) into two subchannels, say p^- and p^+ to indicate the bad and good subchannels respectively. Then, the erasure probability of the bad (or good) channel would be $\alpha^- := 1-(1-\alpha)^2$ (or $\alpha^+ := \alpha^2$). Similarly we split p^- into another set of two split subchannels, say p^{--} and p^{-+} where $\alpha^{--} := 1-(1-\alpha^-)^2$ and $\alpha^{-+} := (\alpha^-)^2$. To this end, we should first construct two p^-'s from four p's. Similarly p^+ is split into (p^{+-}, p^{++}). We repeat this until we get n split subchannels. This way, one may imagine that the subchannels are completely polarized: the data rate of a split subchannel in the final stage is either 1 or 0 in the limit of n. It turns out this is indeed the case.

How to implement the idea? Before proving the complete polarization of the subchannels, let us explore how to implement the channel-splitting idea, i.e., how to construct G_n that yields the splitting? Consider a case in which n is of the form 2^k. Let us start with $n = 2^2$. First of all, we merge the first and second copies of $p(y|x)$ to construct (p^-, p^+), also merging the third and fourth copies to construct another (p^-, p^+). We then combine the two p^-'s to construct (p^{--}, p^{-+}). The way to construct is the same as before: adding two inputs of (p^-, p^-) to yield the first output while simply passing the second input to yield the second output. See the modulo addition on top in the left of Fig. 2.11. We add V_1 and V_3 (inputs

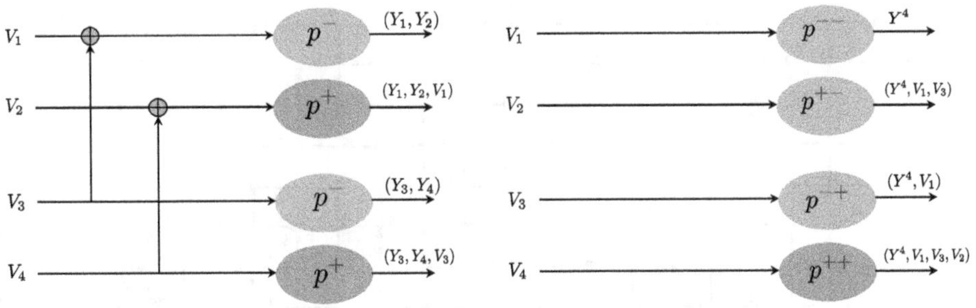

Figure 2.11. Channel splitting for $n = 4$.

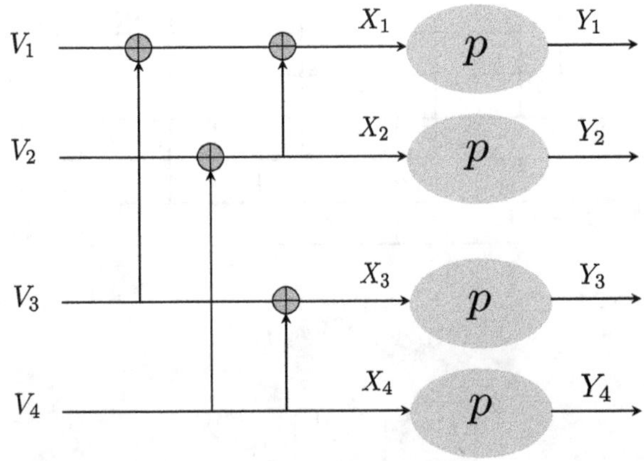

Figure 2.12. Relationship between V^4 and X^4, reflected in G_4.

of p^- subchannels) while simply passing V_3. We do the same for the remaining p^+'s, thus yielding (p^{+-}, p^{++}). Now remember how we constructed outputs of polarized virtual subchannels. The output of p^{--} should read a collection of the outputs of the p^-'s: $Y^4 = (Y_1, Y_2, Y_3, Y_4)$; and the output of p^{-+} should read a collection of Y^4 and the input V_1 of the first p^-. Similarly the outputs of p^{+-} and p^{++} should be: (Y^4, V_1, V_3) and (Y^4, V_1, V_3, V_2) respectively.

Then, what is G_4 that implements the mapping? To see this, let us represent the four virtual subchannels in terms of the four independent copies of $p(y|x)$. See Fig. 2.12.

From this, we see that

$$X_1 = V_1 + V_2 + V_3 + V_4;$$

$$X_2 = V_2 + V_4;$$

$$X_3 = V_3 + V_4;$$

$$X_4 = V_4.$$

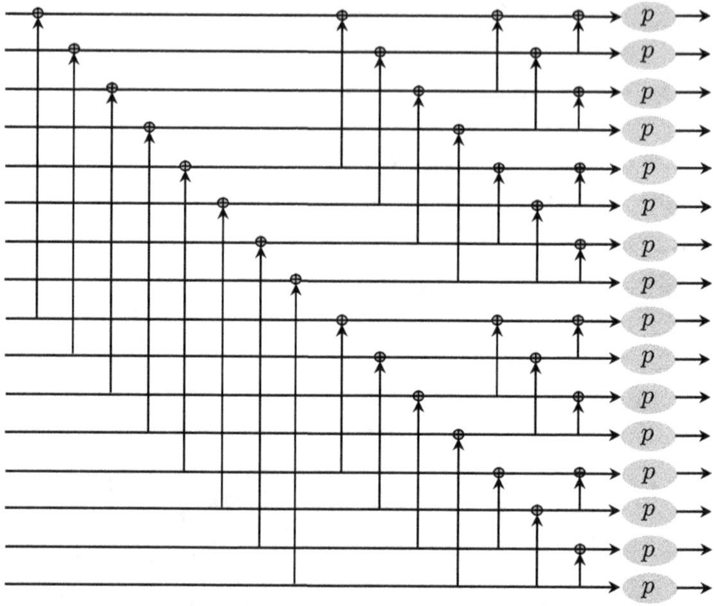

Figure 2.13. Relationship between V^{16} and X^{16}, reflected in G_{16}.

Hence, we get:

$$G_4 = \left[\begin{array}{cc|cc} 1 & 1 & 1 & 1 \\ 0 & 1 & 0 & 1 \\ \hline 0 & 0 & 1 & 1 \\ 0 & 0 & 0 & 1 \end{array}\right] = \left[\begin{array}{c|c} G_2 & G_2 \\ \hline 0 & G_2 \end{array}\right].$$

This observation leads to the following for the case of $n = 2^k$:

$$G_{2^k} = \left[\begin{array}{c|c} G_{2^{k-1}} & G_{2^{k-1}} \\ \hline 0 & G_{2^{k-1}} \end{array}\right]. \tag{2.30}$$

See Fig. 2.13 for $n = 2^4$.

Mathematical statement of polarization As claimed earlier, the construction (2.30) enables the complete polarization as n tends to infinity, meaning that $I(V_i; Y^n, V^{i-1})$ is either 0 or 1 in the limit of n. To prove this, let us first introduce some notations. Let I, I^-, I^+ be the data rate associated with p, p^-, p^+ respectively:

$$I = I(X; Y);$$

$$I^- = I(V_1; Y_1, Y_2);$$

$$I^+ = I(V_2; Y_1, Y_2, V_1).$$

As we verified earlier, $2I = I^- + I^+$; see (2.23). Similarly we define $I^{--}, I^{-+}, I^{+-}, I^{++}$:

$$\begin{aligned}
I^{--} &= I(V_1; Y_1, Y_2, Y_3, Y_4);\\
I^{-+} &= I(V_3; Y_1, Y_2, Y_3, Y_4, V_1);\\
I^{+-} &= I(V_2; Y_1, Y_2, Y_3, Y_4, V_1, V_3);\\
I^{++} &= I(V_4; Y_1, Y_2, Y_3, Y_4, V_1, V_3, V_2).
\end{aligned}$$

You may wonder why I^{-+} (or I^{+-}) is not of the form $I_2 = I(V_2; Y^4, V_1)$ (or $I_3 = I(V_3; Y^4, V^2)$). However, swapping the role of V_2 and V_3, we see that the above is equivalent to the form of (I_1, I_2, I_3, I_4). This swapping corresponds to changing the positions of V_2 and V_3 in Fig. 2.12. As before, one can show that $2I^- = I^{--} + I^{-+}$ and $2I^+ = I^{+-} + I^{++}$. Let B_i be a polarization sign in the ith layer where $i \in \{1, 2, \ldots, k\}$. Then, for general $n = 2^k$, we have $I_k := I^{B_1 B_2 \cdots B_k}$. Similarly we can show that

$$2I_k = I_k^- + I_k^+, \ \forall k \tag{2.31}$$

where I_k^- (or I_k^+) indicates $I^{B_1 B_2 \cdots B_k -}$ (or $I^{B_1 B_2 \cdots B_k +}$).

What the complete polarization means is: in the limit of k,

$$I_k \longrightarrow 1 \text{ or } 0. \tag{2.32}$$

This is the one that we intend to prove. In addition to this, we need to show that the fraction of the good subchannels is I. The second is much easier to prove if we make some assumption on B_i's. Suppose that B_i's are i.i.d., each being according to

$$B_i = \begin{cases} -, & \text{w.p. } \frac{1}{2};\\ +, & \text{w.p. } \frac{1}{2}. \end{cases}$$

Then, we get:

$$I_k = \begin{cases} I^{--\cdots--}, & \text{w.p. } \frac{1}{2^k};\\ I^{--\cdots-+}, & \text{w.p. } \frac{1}{2^k};\\ I^{--\cdots+-}, & \text{w.p. } \frac{1}{2^k};\\ \vdots & \\ I^{++\cdots+-}, & \text{w.p. } \frac{1}{2^k};\\ I^{++\cdots++}, & \text{w.p. } \frac{1}{2^k}. \end{cases}$$

This yields: $\forall k$,

$$\begin{aligned}\mathbb{E}[I_k] &= \frac{1}{2^k}\left(I^{--\cdots--} + I^{--\cdots-+} + \cdots + I^{++\cdots++}\right)\\ &\overset{(a)}{=} \frac{1}{2^k}\cdot(2^k I)\\ &= I\end{aligned}$$

where (a) follows from $\sum_{i=1}^{n} I(V_i; Y^n, V^{i-1}) = nI$. This is what we already proved in the previous section; see (2.22) and (2.23). This then immediately implies that the fraction of the good subchannels is I:

$$\mathbb{P}(I_\infty = 1) = \mathbb{E}[I_\infty] = I.$$

Look ahead Inspired by a simple observation made in $n = 2$, we could come up with G_n as in (2.30). We then claimed that under the G_n, the polarization formally stated in (2.32) occurs. In the next section, we will prove this claim.

2.9 Polar Code: Proof of Polarization and Python Simulation

Recap In the last section, we came up with an encoding strategy that enables the perfect polarization. We designed the full-rank matrix G_n, which relates V^n (consisting of nR information bits and $n - nR$ dummy bits) to X^n (channel input) as follows. For $n = 2^k$,

$$G_{2^k} = \left[\begin{array}{c|c} G_{2^{k-1}} & G_{2^{k-1}} \\ \hline 0 & G_{2^{k-1}} \end{array}\right] \tag{2.33}$$

where $G_2 = [1 \ 1; 0 \ 1]$. The key idea behind this matrix structure is to split the original channel into bad and good subchannels infinitely many. We first split the original channel p (with data rate $I := I(X;Y)$) into two subchannels: (1) p^- channel with a smaller data rate, say I^-; and (2) p^+ channel with a larger data rate, say I^+. From $I^- \le I \le I^+$, we see some partial polarization, i.e., data rates of the two split channels are being apart, being polarized to some extent. Repeating this, we obtain four subchannels $(p^{--}, p^{-+}, p^{+-}, p^{++})$. From $I^{--} \le I^- \le I \le I^+ \le I^{++}$, we see a larger difference between I^{--} and I^{++}, meaning further polarization. Repeating this k times, we get 2^k subchannels $(p^{-\cdots-}, \ldots, p^{+\cdots+})$ in the end. We claimed that the complete polarization on those subchannels occurs in the limit of k:

$$I^{+-\cdots-+} \text{ converges either to 1 or to 0.} \tag{2.34}$$

Outline In this section, we will bring the polar code story to a conclusion by proving the claim (2.34) mentioned earlier. To do so, we will take the following steps. First, we will introduce some mathematical notations to clearly state what needs to be proved. We can actually prove the polarization by relying on a prominent theorem in the random process literature. In the second part, we will delve into this theorem, which is known as the bounded martingale theorem. Then, we will use the theorem to prove the polarization. Lastly, we will provide numerical evidence of polarization through a Python simulation.

A simpler notation for $I^{+-\cdots-+}$ Recall the simpler notation for $I^{+-\cdots-+}$. Let B_i indicate the polarization sign in the ith layer where $i \in \{1, 2, \ldots, k\}$ and $n = 2^k$. Then, such data rate can be represented as $I^{B_1\cdots B_k}$. For notational simplicity, let $I_k := I^{B_1\cdots B_k}$. In terms of this notation, what we intend to prove is then:

$$I_k \longrightarrow 1 \text{ or } 0. \tag{2.35}$$

Observation The converging value in the above is either 1 or 0, meaning there is uncertainty in the quantity. So we can view it as a *random variable*. This implies that I_k is a *random process*, so is B_i.

Then, what is the statistics of the random process $\{B_i\}$? There is one key constraint that the statistics of $\{B_i\}$ needs to satisfy. Since I_k indicates one of the data rates of the n split subchannels, the aggregation of all the possible values should be the same as nI (remember Observation#1 that we made in Section 2.7):

$$I^{-\cdots-} + I^{-\cdots+} + \cdots + I^{+\cdots-} + I^{+\cdots+} = nI,$$

which is equivalent to:

$$\frac{1}{n}\left(I^{-\cdots-} + I^{-\cdots+} + \cdots + I^{+\cdots-} + I^{+\cdots+}\right) = I. \tag{2.36}$$

Suppose $\{B_i\}$'s are i.i.d. $\sim$ Bern$(\frac{1}{2})$. Then, the above (2.36) implies that

$$\mathbb{E}\left[I_k\right] = I. \tag{2.37}$$

If we can prove the perfect polarization (2.35) in the end, then this together with (2.37) yields:

$$\mathbb{P}(I_k = 1) = I. \tag{2.38}$$

Here (2.38) means the fraction of the perfect channels is I, which is what we need to satisfy. Hence this leads us to assume that B_i's are i.i.d., each being according to:

$$B_i = \begin{cases} -, & \text{w.p. } \frac{1}{2}; \\ +, & \text{w.p. } \frac{1}{2}. \end{cases}$$

Convergence of the random process I_k The proof of (2.35) requires the proof of the convergence of I_k. It turns out that I_k belongs to a special class of random processes and there is a key theorem for the special class that serves to prove the convergence.

To figure out what the special class is, let us make some observations. One observation is:

$$0 \leq I_k \leq 1.$$

This is obvious as we consider a binary-input channel where the capacity cannot exceed 1. This means that the random process I_k is *bounded*. Another observation can be made on the following quantity: $\mathbb{E}[I_{k+1}|B_1, B_2, \ldots, B_k]$. Viewing k as the current time index, one can interpret this quantity as the expected future outcome given the current & past knowledge. Observe that given $(B_1, \ldots, B_k)$, I_{k+1} is a sole function of B_{k+1}:

$$I_{k+1} = \begin{cases} I_k^-, & \text{w.p. } \mathbb{P}(B_{k+1} = -) = \frac{1}{2}; \\ I_k^+, & \text{w.p. } \mathbb{P}(B_{k+1} = +) = \frac{1}{2} \end{cases} \tag{2.39}$$

where I_k^- (or I_k^+) indicates $I^{B_1B_2\cdots B_k-}$ (or $I^{B_1B_2\cdots B_k+}$).

This then yields:

$$\begin{aligned} \mathbb{E}[I_{k+1}|B_1, B_2, \ldots, B_k] &= \mathbb{P}(B_{k+1} = -)I_k^- + \mathbb{P}(B_{k+1} = +)I_k^+ \\ &= \frac{1}{2}(I_k^- + I_k^+) \\ &= I_k \end{aligned} \tag{2.40}$$

where the last equality follows from the fact that $2I_k = I_k^- + I_k^+$ (why?). Notice that the expected future outcome, reflected in $\mathbb{E}[I_{k+1}|B_1, \ldots, B_k]$, is the same as the *current* outcome I_k. This is the key property that characterizes one of the well-known random processes, called the *martingale*. We say that a random process is a martingale if (2.40) holds.[4]

Bounded martingale theorem There is a well known result as to the *convergence* of such a *bounded martingale*: for a bounded martingale I_k,

$$I_k \longrightarrow I_\infty \qquad \textit{almost surely} \tag{2.41}$$

where I_∞ indicates a random variable that represents the limit of I_k. You may wonder what the "almost surely" means. Remember in Part I that we learned about one type of convergence w.r.t. a random process. That is, the convergence *in probability*. There are a couple of more types of convergence regarding a random process. One such type is the convergence *almost surely*. Mathematically, it means:

$$\mathbb{P}\left(\lim_{k \to \infty} I_k = I_\infty\right) = 1. \tag{2.42}$$

What it means is that the limit of I_k is almost surely I_∞ as the name of the convergence suggests. As the expression of (2.42) indicates, another name of it is the

4. The term "martingale" has an interesting historical origin in the gambling realm where it was used to describe a particular betting tactic. In the context of probability theory, however, it refers to a stochastic process that models a fair game of chance. For example, suppose we have a game where B_i represents the amount of money gained or lost in the ith round of play. In this scenario, the values of B_i are independent and identically distributed, taking either -1 or 1 with equal probability, ensuring that the game is unbiased. Consider $I_k = \sum_{i=1}^{k} B_i$ which denotes the capital after k games. Note that $\{I_k\}$ is a martingale as it satisfies the property (2.40): $\mathbb{E}[I_{k+1}|B_1, \ldots, B_k] = I_k$. We see that this property is a consequence of the fair game.

convergence *w.p. 1*. Actually this is a *stronger* type of convergence relative to the one *in probability* because:

$$(2.42) \Longrightarrow \lim_{k\to\infty} \mathbb{P}\left(|I_k - I_\infty| \le \epsilon\right) = 1 \quad \text{for any } \epsilon > 0.$$

It turns out that the other way around does not necessarily hold, i.e., there are examples in which only the convergence in probability holds (think about such examples). Hence, the convergence almost surely is of a stronger type.

In fact, the proof of the bounded martingale theorem (2.41) is not that simple. This book does not cover the proof of the theorem as it requires a background in probability theory, which is not covered here. If you are interested in the proof, you may refer to a graduate-level textbook on probability theory, such as (Grimmett and Stirzaker, 2020).

Proof of polarization: $I_\infty = 1$ or 0 The key property on $\{I_k\}$ reflected in (2.41) means that I_k actually converges to I_∞. For the proof of polarization, it suffices to show that

$$I_\infty = 1 \text{ or } 0.$$

The proof of this requires the following two: (i) the bounded martingale theorem; and (ii) the recursive relationship between I_k and I_{k+1}, reflected in (2.39). Using (2.39), we get:

$$\begin{aligned}
&\mathbb{E}[|I_{k+1} - I_k||B_1, \ldots, B_k] \\
&\overset{(a)}{=} \frac{1}{2}(I_k - I_k^-) + \frac{1}{2}(I_k^+ - I_k) \\
&= \frac{1}{2}(I_k^+ - I_k^-)
\end{aligned} \tag{2.43}$$

where (a) is due to (2.39). On the other hand, the bounded martingale theorem (2.41) yields:

$$\begin{aligned}
&I_k \longrightarrow I_\infty \text{ almost surely \&} \\
&I_{k+1} \longrightarrow I_\infty \text{ almost surely.}
\end{aligned}$$

This implies that:

$$|I_{k+1} - I_k| \longrightarrow 0 \quad \text{almost surely.}$$

This then yields:

$$\mathbb{E}[|I_{k+1} - I_k|] \longrightarrow 0 \quad \text{almost surely.} \tag{2.44}$$

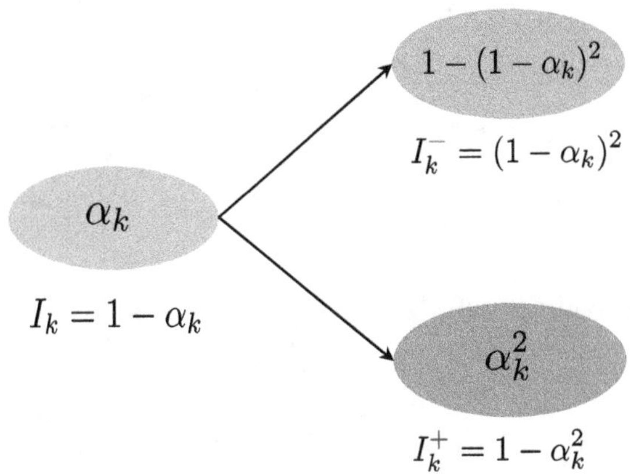

Figure 2.14. Channel-splitting from I_k to (I_k^-, I_k^+).

Now using the *tower property*,[5] we get:

$$\mathbb{E}[|I_{k+1} - I_k|] = \mathbb{E}_{B_1,\ldots,B_k}[\mathbb{E}_{B_{k+1}}[|I_{k+1} - I_k||B_1, \ldots, B_k]].$$

For a *non-negative* random process $|I_{k+1} - I_k|$, if its mean converges to 0, such random process also converges to 0. Hence, the above tower property together with (2.44) yields:

$$\mathbb{E}_{B_{k+1}}[|I_{k+1} - I_k||B_1, \ldots, B_k] \longrightarrow 0 \quad \text{almostly surely.} \tag{2.45}$$

Applying (2.43) to the above (2.45), we get:

$$I_k^+ - I_k^- \longrightarrow 0 \quad \text{almostly surely.} \tag{2.46}$$

Note that I_k is a random process which takes one of the virtual subchannels. Such virtual subchannels are of the same type as the original channel: BEC (remember what we learned in Section 2.8). So one can think of erasure probability for I_k, say α_k. Then, $I_k = 1 - \alpha_k$. Applying the channel-splitting idea that we came up with earlier, the bad channel split from the α_k-channel would have $\alpha_k^- = 1-(1-\alpha_k)^2$; on the other hand, the good channel would have $\alpha_k^+ = \alpha_k^2$. See Fig. 2.14.

This then gives:

$$\begin{aligned} I_k^- &= (1-\alpha_k)^2; \\ I_k^+ &= 1-\alpha_k^2. \end{aligned} \tag{2.47}$$

---

5. The tower property is a useful and well-known property that arises in the random process context. It says: for random variables (or vectors), say X and Y, $\mathbb{E}[X] = \mathbb{E}_Y[\mathbb{E}_X[X|Y]]$. One can readily prove this using the definition of conditional probability. Try Prob 6.1.

This together with (2.46) yields:

$$1 - \alpha_k^2 - (1 - \alpha_k)^2 \longrightarrow 0, \tag{2.48}$$

which is equivalent to:

$$\alpha_k \longrightarrow 0 \text{ or } 1. \tag{2.49}$$

Hence, we complete the proof:

$$I_k \longrightarrow 1 \text{ or } 0. \tag{2.50}$$

Extension Thus far, we have outlined the narrative of the polar code, including the encoder structure that facilitates polarization and the proof of polarization using the bounded martingale theorem. We have primarily focused on the binary erasure channel (BEC) for the sake of simplicity, but it should be noted that the polar code has been shown to achieve capacity for all binary-input channels. The expansion to the general case is explored in Prob 6.5 and Prob 6.7. If you wish to demonstrate polarization for binary symmetric channels, follow the guidelines in Prob 6.5. If you prefer to show polarization for binary-input memoryless channels, refer to the instructions in Prob 6.7.

Python simulation of polarization We will use a Python simulation to confirm the polarization of I_k. We will simulate the same scenario we have been focusing on: $n = 2^k, I_k := I^{B_1 B_2 \cdots B_k}$, and the channel is the BEC with an erasure probability α. Since I_{k+1} takes either I_k^{-1} or I_k^{+1}, which are generated recursively from I_k, we will use a binary tree class with top and bottom children.

```
class TreeNode:
    def __init__(self, val, top=None, bottom=None):
        self.val = val
        self.top = top
        self.bottom = bottom
```

We wish to construct a binary code tree as illustrated in Fig. 2.14. We assign I_k to node.val and associate the top and bottom trees with I_k^- and I_k^+, respectively. The relationship between I_k and I_k^- (or I_k^+) reads:

$$\begin{aligned} I_k^- &= I_k^2; \\ I_k^+ &= 1 - (1 - I_k)^2. \end{aligned}$$

We iterate this procedure until we reach all the leaves. See below for code implementation.

```
k=20
n=2**k
```

```
res=[] # an array for containing all the I_k's
       # k=2 --> "res" is of the following structure
       #       res=[[X],[X,X],[X,X,X,X]]
alpha = 0.4 # erasure probablity of BEC
root=TreeNode(1-alpha) # I=1-alpha (root node)

# Recursive function for tree generation
def rec_treeGen(node,depth):
    # initialization of the resulting array
    if len(res)<=depth: res.append([])
    # Append node.val in the depth level
    res[depth].append(node.val)
    # If reaching a leaf node, terminate the process
    if depth==k: return
    else: # for an internal node, iterate further
        # Construct the top node and go deeper
        node.top=TreeNode(node.val**2)
        rec_treeGen(node.top,depth+1)
        # Construct the bottom node and go deeper
        node.bottom=TreeNode(1-(1-node.val)**2)
        rec_treeGen(node.bottom,depth+1)
rec_treeGen(root,0)
```

Using the resulting array res, we can then plot an empirical CDF of I_k as in Fig. 2.6:

$$\frac{|\{i : \text{res[k][i]} \leq x|}{n}.$$

```
import numpy as np

# range of x
x_grid=np.arange(0,1,0.0001)
# initialization of cdf
cdf_1=[0]*len(x_grid)
cdf_2=[0]*len(x_grid)
cdf_2_4=[0]*len(x_grid)
cdf_2_20=[0]*len(x_grid)
# Sorting I_k
# Why? To ease the computation of empirical cdf
sres_1 = sorted(res[0])
sres_2 = sorted(res[1])
sres_2_4 = sorted(res[4])
sres_2_20 = sorted(res[20])

for i,x in enumerate(x_grid):
```

```
        # Case n=1
        for j in range(len(sres_1)):
            if sres_1[j]>x:
                cdf_1[i]=j # because sres_1 is sorted
                break
            # if sres_1[j]<=x for all j, set cdf(x)=1
            if j==len(sres_1)-1: cdf_1[i]=1
        # Case n=2
        for j in range(len(sres_2)):
            if sres_2[j]>x:
                cdf_2[i]=j/2 # divided by n=2
                break
            if j==len(sres_2)-1: cdf_2[i]=1
        # Case n=2^4
        for j in range(len(sres_2_4)):
            if sres_2_4[j]>x:
                cdf_2_4[i]=j/(2**4) # divided by n=2^4
                break
            if j==len(sres_2_4)-1: cdf_2_4[i]=1
        # Case n=2^20
        for j in range(len(sres_2_20)):
            if sres_2_20[j]>x:
                cdf_2_20[i]=j/(2**20)
                break
            if j==len(sres_2_20)-1: cdf_2_20[i]=1
    # Set cdf_2_20[-1]=1 since x_grid is not dense enough
    cdf_2_20[-1]=1
```

Here is a code for plotting the empirical CDFs for four cases: $n = 1, 2, 2^4, 2^{20}$.

```
import matplotlib.pyplot as plt
import matplotlib

plt.figure(figsize=(10,2),dpi=500)
# Adjust the font size of axis tick values
matplotlib.rc('xtick', labelsize=7)
matplotlib.rc('ytick', labelsize=7)

plt.subplot(1,4,1)
plt.plot(x_grid,cdf_1,color='red')
plt.xlabel('x', fontsize=10)
plt.ylabel('empirical CDF', fontsize=8)
plt.title('$n=1$',fontsize=10)
plt.subplot(1,4,2)
plt.plot(x_grid,cdf_2,color='red')
plt.xlabel('x', fontsize=10)
plt.title('$n=2$',fontsize=10)
```

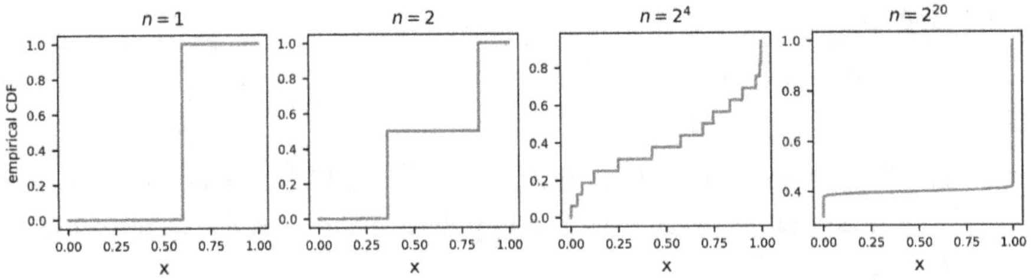

Figure 2.15. Python simulation for verifying polarization. As n increases, the number of perfect subchannels converges to $I = 1 - \alpha$ (=0.6 in this simulation).

```
plt.subplot(1,4,3)
plt.plot(x_grid,cdf_2_4,color='red')
plt.xlabel('x', fontsize=10)
plt.title('$n=2^4$',fontsize=10)
plt.subplot(1,4,4)
plt.plot(x_grid,cdf_2_20,color='red')
plt.xlabel('x', fontsize=10)
plt.title('$n=2^{20}$',fontsize=10)
plt.show()
```

Note that the obtained curves follow the same trend as those in Fig. 2.6:

Look ahead In Parts I and II, we delved into the source and channel coding theorems, respectively, and also explored important information-theoretic measures like entropy, mutual information, and KL divergence. Additionally, we placed significant emphasis on the concept of phase transition, which resembles a fundamental law in physics. These information-theoretic tools and the phase transition concept are instrumental in addressing critical issues that emerge in various fields beyond communication. Part III aims to showcase their role in data science applications, ranging from social networks, ranking, computational biology, machine learning, and deep learning. The following section will commence our investigation into their role within the context of social networks.

Problem Set 6

Prob 6.1 (Basics)

(*a*) Let X_1 and X_2 be Bern($\frac{1}{2}$) random variables which are independent with each other. Let $Z = X_1 \oplus X_2$. Show that X_1 and Z are independent, i.e., $I(X_1; Z) = 0$.

(*b*) A curious student claims that X_1 and Z (in part (*a*)) are independent even if X_2 is given, i.e., $I(X_1; Z|X_2) = 0$. Prove or disprove it.

(*c*) Let X and Y be random variables (or random vectors). Prove the tower property:

$$\mathbb{E}[X] = \mathbb{E}_Y[\mathbb{E}_X[X|Y]].$$

Prob 6.2 (Chain rule for mutual information) Consider two random processes: $\{X_i\}$ and $\{Y_i\}$.

(*a*) Show that

$$I(X^i; Y_i|Y^{i-1}) = \sum_{j=1}^{i} I(X_j; Y_i|X^{j-1}, Y^{i-1})$$

where $X^i := (X_1, \ldots, X_i)$ and $Y^i := (Y_1, \ldots, Y_i)$.

(*b*) Let $f(i,j)$ be a function of i and j where $i, j \in \mathbb{N}$. Show that

$$\sum_{i=1}^{n} \sum_{j=1}^{i} f(i,j) = \sum_{j=1}^{n} \sum_{i=j}^{n} f(i,j).$$

(*c*) Using parts (*a*) and (*b*), show that

$$\sum_{i=1}^{n} I(X^i; Y_i|Y^{i-1}) = \sum_{j=1}^{n} I(X_j; Y_j^n|X^{j-1}, Y^{j-1})$$

where $Y_j^n := (Y_j, Y_{j+1}, \ldots, Y_n)$.

Prob 6.3 (Useful facts in the polar code) Let $\{V_i\}$ be an i.i.d. random process $\sim$ Bern($\frac{1}{2}$). Let $G_n = [G_{ij}]$ be a full rank matrix of size n-by-n where each entry G_{ij} is a binary value. Let $X^n = G_n V^n$ where $V^n := [V_1, V_2, \ldots, V_n]^T$ denotes an n-by-1 column vector.

(*a*) Consider a case in which $n = 2$ and

$$G_2 = \begin{pmatrix} 1 & 1 \\ 0 & 1 \end{pmatrix}.$$

Show that X_1 and X_2 are i.i.d.

(*b*) Consider $X_i = \sum_{k=1}^{n} G_{ik}V_k$ and $X_j = \sum_{k=1}^{n} G_{jk}V_k$ where $\sum$ indicates the modulo-2 addition and $i \neq j$. Argue that there exists some component V_k which appears in X_i but not in X_j (or vice versa). Show that $I(X_i; X_j) = 0$, i.e., X_i and X_j are independent. Also show that X_i's are i.i.d.
Hint: The full rank condition on G_n implies that its row vectors are linearly independent.

(*c*) Let Y^n be the output of a BEC when X^n is fed into. Show that

$$I(V^n; Y^n) = I(X^n; Y^n) \text{ and } I(X^n; Y^n) = nI$$

where $I := I(X; Y)$ and X (or Y) denotes a generic random variable for X_i (or Y_i).

(*d*) Show that

$$I(V^n; Y^n) = \sum_{i=1}^{n} I(V_i; Y^n, V^{i-1}).$$

Prob 6.4 (Polarization in BEC) Suppose that V_1 and V_2 are i.i.d. $\sim$ Bern($\frac{1}{2}$), and (X_1, X_2) are constructed through $G_2 = [1, 1; 0, 1]$:

$$\begin{bmatrix} X_1 \\ X_2 \end{bmatrix} = G_2 \begin{bmatrix} V_1 \\ V_2 \end{bmatrix} = \begin{bmatrix} 1 & 1 \\ 0 & 1 \end{bmatrix} \begin{bmatrix} V_1 \\ V_2 \end{bmatrix}.$$

We pass (X_1, X_2) through two independent copies of a BEC, say p, with erasure probability α, thus yielding the channel output (Y_1, Y_2). As we learned in Section 2.7, we convert the collection of the two independent copies into two virtual subchannels:

$$p^- : V_1 \to (Y_1, Y_2); \tag{2.51}$$

$$p^+ : V_2 \to (Y_1, Y_2, V_1). \tag{2.52}$$

Let (I, I^-, I^+) be the data rates associated with (p, p^-, p^+) respectively:

$$2I = I(X_1, X_2; Y_1, Y_2);$$

$$I^- = I(V_1; Y_1, Y_2);$$

$$I^+ = I(V_2; Y_1, Y_2, V_1).$$

(*a*) Compute (I, I^-, I^+). Express them in terms of α.

(*b*) Can p^- be interpreted as another BEC? If so, explain why and indicate the corresponding erasure probability.

(*c*) Repeat part (*b*) for p^+.
(*d*) Compare I^- and I^+. Which one is larger?

Prob 6.5 (Polarization in BSC) Consider the same problem setup as that in Prob 6.4. The only distinction is that the channel is a BSC with crossover probability α.

(*a*) Compute $H(V_1|Y_1, Y_2)$ and $I(V_1; Y_1, Y_2)$.
(*b*) Compute $H(V_1|Y_1 \oplus Y_2)$ and $I(V_1; Y_1 \oplus Y_2)$.
(*c*) Using the above, show that $I(V_1; Y_1, Y_2|Y_1 \oplus Y_2) = 0$, i.e., $V_1 - (Y_1 \oplus Y_2) - (Y_1, Y_2)$.
(*d*) Can p^- be interpreted as another BSC? If so, explain why and indicate the corresponding crossover probability.
(*e*) Show that $2I = I^- + I^+$. Using this and part (*d*), compute I^+.
(*f*) Compare I^- and I^+. Which one is larger or same?
(*g*) Show that when $I^+ - I^- = 0$, $H(\alpha)$ is either 0 or 1.
(*h*) The result of part (*g*) proves the perfect polarization for BSC. Explain why.

Prob 6.6 (Python simulation for BSC polarization) Consider a setting where $n = 2^k$, $I_k := I^{B_1 \cdots B_k}$, and the channel is a BSC with crossover probability α. Let I be the capacity of the BSC: $I = 1 - H(\alpha)$.

(*a*) Using the result in Prob 6.5, express I^- and I^+ in terms of α.
(*b*) Using part (*a*) and the skeleton Python code in Section 2.9, construct a code yielding all the possible values (say res[k][i]) that I_k can take on. Here res[k] is a 2^k-sized array that contains all the values for I_k.
(*c*) Set $\alpha = 0.3$. Using part (*b*) and the skeleton code in Section 2.9, plot empirical CDFs for four cases: $n = 1, 2, 2^4, 2^{20}$:

$$\frac{|\{i : \texttt{res[k][i]} \leq x|}{n}.$$

Prob 6.7 (Polarization in B-DMC) Consider the same problem setup as that in Prob 6.4 and Prob 6.5. The only distinction is that the channel is a B-DMC where the channel input is binary-valued. Let $\mathcal{Y}$ be the range of the channel output.

(*a*) Argue that given $Y_1 = y_1$, $X_1 \sim \text{Bern}(q_1)$ where q_1 is a function of y_1. Using this, show that

$$I := I(X_1; Y_1) = 1 - \sum_{y_1 \in \mathcal{Y}} p(y_1) H(q_1).$$

(*b*) Argue that given $(Y_1, Y_2) = (y_1, y_2)$, channel inputs (X_1, X_2) are two independent binary random variables, say $X_i \sim \text{Bern}(q_i)$ where q_i is a sole function of y_i, $i \in \{1, 2\}$. Using this, show that

$$I^- = 1 - \sum_{y_1 \in \mathcal{Y}} \sum_{y_2 \in \mathcal{Y}} p(y_1)p(y_2)H\left(q_1(1-q_2) + (1-q_1)q_2\right).$$

(*c*) Using $2I = I^- + I^+$, show that

$$I^+ = 1 + \sum_{y_1 \in \mathcal{Y}} \sum_{y_2 \in \mathcal{Y}} p(y_1)p(y_2)H\left(q_1(1-q_2) + (1-q_1)q_2\right) - 2\sum_{y_1 \in \mathcal{Y}} p(y_1)H(q_1).$$

(*d*) Show that when $I^+ - I^- = 0$, I is either 0 or 1.

(*e*) The result of part (*d*) proves the perfect polarization for B-DMC. Explain why.

Prob 6.8 (Python simulation for B-DMC polarization) Consider a setting where $n = 2^k$, $I_k := I^{B_1 \cdots B_k}$, and the channel is a B-DMC. Let I, I^-, I^+ be the quantities derived in Prob 6.7. Assume that $\mathcal{Y} = \{0, 1, 2\}$.

(*a*) Using the skeleton Python code in Section 2.9, construct a code yielding all the possible values (say `res[k][i]`) that I_k can take on. Here `res[k]` is a 2^k-sized array that contains all the values for I_k.

(*b*) Set $q_i = \frac{y_i}{4}$ for $i \in \{1, 2\}$. Using part (*a*) and the skeleton code in Section 2.9, plot empirical CDFs for four cases: $n = 1, 2, 2^4, 2^{20}$:

$$\frac{|\{i : \texttt{res[k][i]} \leq x|}{n}.$$

Prob 6.9 (True or False?)

(*a*) Let X, Y, Z be Bernoulli random variables. Suppose $I(X; Y|Z) = 0$. Then, $I(X; Y) = 0$.

(*b*) Let (X_1, X_2, Y_1, Y_2) be discrete random variables. Suppose

$$I(X_1; X_2|Y_1) = 0;$$
$$I(X_1; X_2|Y_2) = 0.$$

A curious student claims that

$$I(X_1; X_2|Y_1, Y_2) = 0.$$

Either prove or disprove the statement.

DOI: 10.1561/9781638281153.ch3

Chapter 3

Data Science Applications

3.1 Social Networks: Fundamental Limits

Concepts and tools learned from Parts I and II In Parts I and II, we investigated the well-known theorems in information theory: the source coding theorem, channel coding theorem, and source-channel separation theorem. In Part I, we introduced important concepts in information theory such as entropy, mutual information, and KL divergence. We also examined the idea of prefix-free codes, typical sequences, and some lower and upper bounding techniques. With these tools, we proved the source coding theorem and analyzed a specific code, the Huffman code, along with its implementation in Python.

In Part II, we explored various concepts and techniques related to channel codes. One crucial idea we discussed was the phase transition, where there is a critical data rate below which we can make the error probability arbitrarily close to zero and above which no matter what we do, the error probability is not zero. We used techniques like random coding, maximum likelihood decoding, joint typicality decoding, and union bound for the achievability proof, and Fano's inequality and data processing inequality for the converse proof. We also proved the source-channel separation theorem and demonstrated that feedback cannot increase capacity in a DMC. Moreover, we examined the polar code, an explicit channel code that achieves the capacity of a specific type of memoryless channel, binary-input DMCs.

The concepts and information-theoretic notions we have explored are applicable in various domains beyond communication. In particular, these concepts are instrumental in data science. One example is the occurrence of phase transition in various inference problems within data science. Examples include: (i) community recovery of social networks (Girvan and Newman, 2002; Fortunato, 2010; Abbe, 2017; Chen *et al.*, 2016b); (ii) DNA sequencing in computational biology (Browning and Browning, 2011; Das and Vikalo, 2015; Chen *et al.*, 2016a; Si *et al.*, 2014); (iii) ranking in search engine (Negahban *et al.*, 2012; Chen and Suh, 2015); and (iv) matrix completion in recommender systems (Candès and Tao, 2010; Keshavan *et al.*, 2010; Candes and Recht, 2012; Ahn *et al.*, 2018; Elmahdy *et al.*, 2020; Zhang *et al.*, 2021). The KL divergence is utilized in the development of a loss function for optimizing supervised learning, which is one of the most significant frameworks in machine learning. Mutual information is used to enhance powerful unsupervised learning frameworks such as generative adversarial networks (GANs) (Goodfellow *et al.*, 2014). Recently, mutual information has also been applied in the design of new machine learning models known as fair prediction models. These models not only ensure accurate predictions but also guarantee fairness in prediction statistics for different groups and individuals (Larson *et al.*, 2016; Zafar *et al.*, 2017; Cho *et al.*, 2020; Roh *et al.*, 2020).

Goal of Part III In Part III, our aim is to illustrate the significance of the phase transition concept and fundamental notions through the exploration of three inference problems and three machine learning models:

(1) Inference problem #1: Community detection;
(2) Inference problem #2: DNA sequencing;
(3) Inference problem #3: Ranking;
(4) Supervised learning: Design of a loss function;
(5) Unsupervised learning: Generative Adversarial Networks (GANs);
(6) Fair machine learning: Design of fair classifiers.

Outline This section will examine the first inference problem: Community detection. It consists of four parts. Initially, we will study what community detection is and discuss why it is essential in data science. Next, we will establish a related mathematical problem and address a crucial issue regarding phase transition. We will then establish a connection to the communication problem we have previously studied. Finally, we will investigate how phase transition manifests in the problem context.

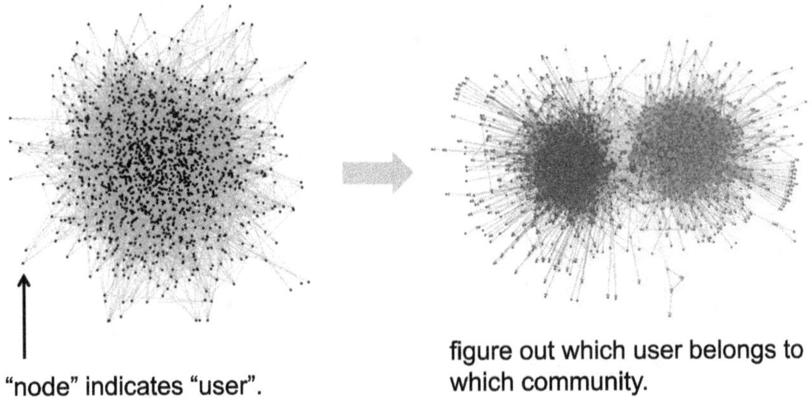

Figure 3.1. Community detection in picture.

Community detection (Girvan and Newman, 2002; Fortunato, 2010; Abbe, 2017) Community refers to a collection of individuals with shared interests or residing in the same locality. The objective of community detection is to identify similar groups, as illustrated in Fig. 3.1. Suppose users indicate nodes in a graph (shown on the left in Fig. 3.1), and there are two communities: blue and red communities. The aim of the problem is to determine which user (node) belongs to which community among the blue and red communities (shown on the right in Fig. 3.1). Clustering is another term for this problem (Bansal *et al.*, 2004; Jalali *et al.*, 2011). Note that the output nodes are clustered into either a blue or a red community.

One may wonder why we should be concerned with this problem. The problem arises in several critical domains, such as social networks (Facebook, LinkedIn, Twitter, etc.), and biological networks (Chen and Yuan, 2006). In social networks, identifying community memberships can assist in identifying target groups for product advertisements. In biological networks, the problem is relevant to DNA sequencing for cancer detection and personalized medicine. In Section 3.5, we will discuss how this problem is connected to DNA sequencing. The problem has applications beyond these examples.

Problem formulation First consider the type of information that can be accessed in many applications. One common type of information that is easily accessible is relationship information, such as friendship in Meta's social network, connections in LinkedIn, and followers in Twitter. However, community memberships are often not revealed in practice. For example, in Meta's social network, only the friendship information is available, and it is unknown whether a user belongs to a particular community. In privacy-concerned contexts, this information is prohibited from being made public by law, even if Meta has access to it.

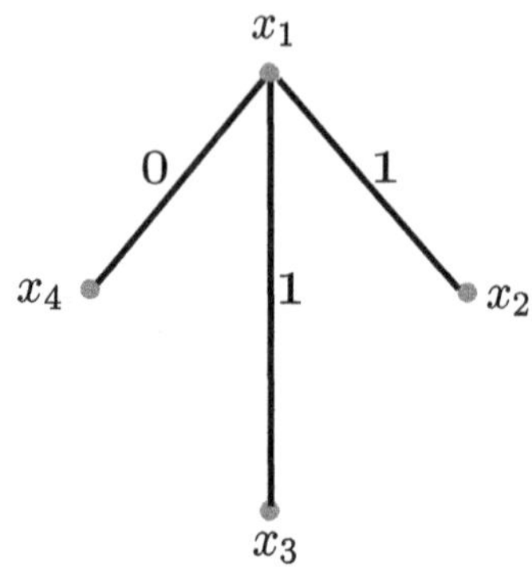

Figure 3.2. An example of community detection.

To give you a concrete feel as to what the relationship information looks like, let us give you an example. Suppose that x_i indicates community membership of user i, and we assign x_i to node i. Then, one natural function is a parity function between the values assigned to two nodes, e.g., x_i and x_j. For instance, when $x_1 = x_2$, we get, say $y_{12} = 0$; otherwise $y_{12} = 1$.

Given a collection of y_{ij}'s, the goal of the problem is then to decode $\boldsymbol{x} := [x_1, x_2, \ldots, x_n]$. Upon reflection, this objective is unattainable as decoding $\boldsymbol{x}$ with precision is not feasible based solely on the parities y_{ij}'s. To illustrate this point, consider the scenario depicted in Fig. 3.2. Suppose that $n = 4$ and (y_{12}, y_{13}, y_{14}) are given as $(0, 1, 1)$. If $x_1 = 0$, $\boldsymbol{x} = [0, 1, 1, 0]$. However, an ambiguity arises on the value of x_1 because there is no way to infer x_1 only from the parities. The other solution $\boldsymbol{x} = [1, 0, 0, 1]$ is also valid. We have always two valid solutions: (i) the correct one; and (ii) its flipped counterpart. To resolve this ambiguity, we should relax the goal as follows: decoding $\boldsymbol{x}$ or $\boldsymbol{x} \oplus \mathbf{1}$ from y_{ij}'s. Here $\oplus$ indicates the bit-wise modulo sum.

Two challenges Given the relaxed goal, solving the problem may appear to be not too difficult, as suggested by the example in Fig. 3.2. However, in the era of big data, two challenges arise. Firstly, in many applications, such as social networks, the number of nodes (i.e., users) can be very large. For example, as of December 2022, the number of Facebook users has reached 2.96 billion (Meta, 2022). Therefore, we may only have access to a portion of the parities. Note that the number of all possible pairs is huge: $\binom{n}{2} \approx 4.38 \times 10^{18}$ in the above example.

In situations where we are allowed to choose any pair of two nodes, community detection is straightforward. For instance, by selecting a set of consecutive pairs like $(y_{12}, y_{23}, y_{34}, \ldots, y_{(n-1)n})$, we can easily decode $\boldsymbol{x}$ up to a global shift. However, this approach can still be challenging due to the second challenge. In many applications, such as Meta's social network, similarity relationships are passively given according

Figure 3.4. Translation to a communication problem.

Let us first translate the problem as below. See Fig. 3.4. One can view $\boldsymbol{x}$ as a message that we wish to send and $\hat{\boldsymbol{x}}$ as a decoded message. What we are given are pairwise measurements. A block diagram at the transmitter converts $\boldsymbol{x}$ into pairwise measurements, say x_{ij}'s. Here $x_{ij} := x_i \oplus x_j$. One can view this as an encoder. Since we assume that only part of the pairwise measurements are observed at random, we have another block which implements the partial & random measurements to extract a subset of x_{ij}'s. One can view this processing as the one that behaves like a channel where the output y_{ij} admits:

$$y_{ij} = \begin{cases} x_{ij}, & \text{w.p. } p; \\ \mathsf{e}, & \text{w.p. } 1-p \end{cases}$$

where p denotes the observation probability and e indicates empty information (erasure). In other words, the measurement process can be modeled as an *erasure channel* with erasure probability $1 - p$. These y_{ij}'s are then fed into an algorithm block, thus yielding $\hat{\boldsymbol{x}}$. The algorithm block can be interpreted as a decoder.

Performance metrics & an optimization problem As in the communication setting, we can think of two performance metrics. The first refers to a quantity that we are interested in characterizing the limit on. That is, the number of pairwise measurements that are observed, namely *sample complexity*. In this problem context, the sample complexity would be concentrated around:

$$\text{sample complexity} \longrightarrow \binom{n}{2} p \quad \text{as } n \to \infty.$$

This is because of the WLLN. The second refers to a metric conventionally employed in the context of inference problems. That is, the probability of error defined as:

$$P_e := \mathbb{P}\left(\hat{\boldsymbol{x}} \notin \{\boldsymbol{x}, \boldsymbol{x} \oplus \mathbf{1}\}\right).$$

An error occurs when $\hat{\boldsymbol{x}} \neq \boldsymbol{x}$ and $\hat{\boldsymbol{x}} \neq \boldsymbol{x} \oplus \mathbf{1}$.

There must be a tradeoff relationship between the sample complexity and P_e. The larger the sample complexity, the smaller P_e and vice versa. Hence, as Shannon did in the communication problem, we can formulate the following optimization

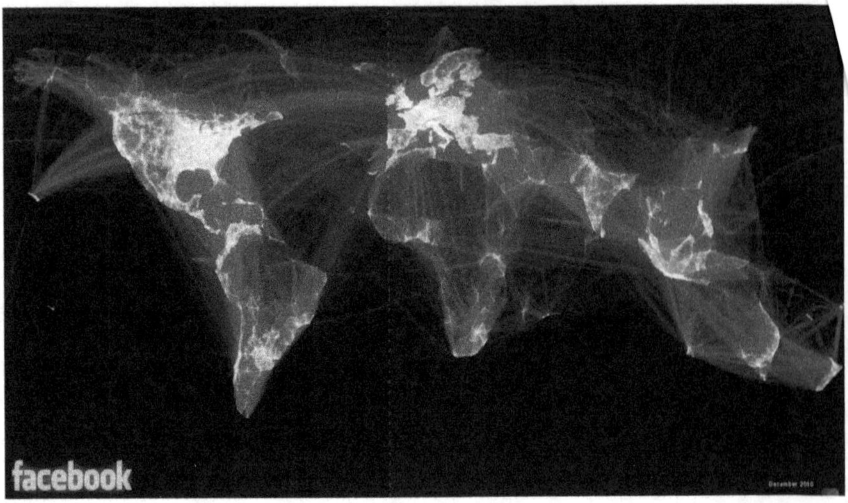

Figure 3.3. The number of facebook users is ≈2.96 billion as of December 2022.

to a context. This means that such information is not obtained by our own choice. For instance, the friendship information in Meta's social network is simply given by the context. It is not possible to ask an arbitrary pair of two users whether they are friends or not. Therefore, one natural assumption is that the similarity information is given in a probabilistic manner. For example, the parity of a pair of any two users is given with probability p, independently across all the other pairs.

An information-theoretic question Although community detection is a challenging task due to the limited and probabilistic nature of pairwise measurements, there is good news. The good news is that, as demonstrated in the example in Fig. 3.2, it may not be necessary to observe every measurement pair in order to decode $\boldsymbol{x}$. Since the pairs are highly dependent on each other, partial pairs might be enough to achieve this goal. An information-theoretic question arises: is there a fundamental limit on the number of measurement pairs required to enable detection? Interestingly, similar to the channel capacity in communication, a phase transition occurs. There exists a sharp threshold on the number of pairwise measurements above which reliable community detection is possible and below which it is impossible, regardless of any method used. For the rest of this section, we will investigate this threshold in detail.

Translation to a communication problem Under the partial and random observation setting, y_{ij} is statistically related to x_i and x_j. Hence, community detection is an inference problem, suggesting an intimate connection with a communication problem. Translating community detection into a communication problem, we can come up with a mathematical statement on the limit.

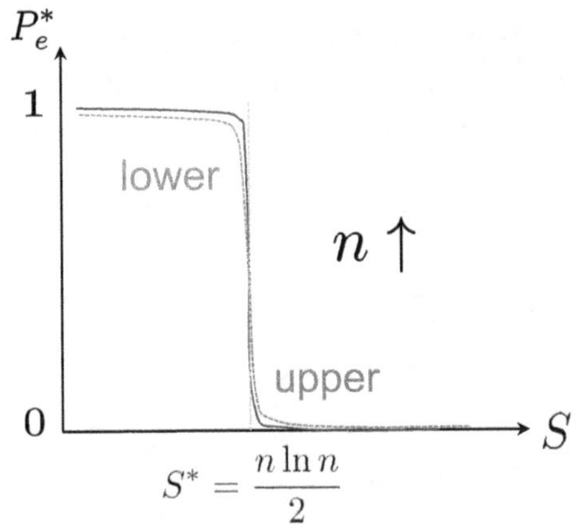

Figure 3.5. Phase transition in community detection. If S is above the minimal sample complexity $S^* = \frac{n \ln n}{2}$, we can make P_e arbitrarily close to 0 as n tends to infinity. If $S < S^*$, P_e cannot be made arbitrarily close to 0 no matter what we do and whatsoever.

from which the tradeoff relationship can be characterized. Given p and n,

$$P_e^*(p, n) := \min_{\text{algorithm}} P_e. \tag{3.1}$$

Remember the optimization problem that Shannon formulated earlier. It was a difficult non-convex problem which has been open even thus far. Here we see the same thing. The above problem (3.1) is very difficult; hence, the exact error probability $P_e^*(p, n)$ is still open.

Phase transition But we have a good news. The good news is that as in the communication setting, phase transition occurs w.r.t. the sample complexity in the limit of n. If sample complexity is above a threshold, one can make P_e arbitrarily close to 0 as $n \to \infty$; otherwise (i.e., if it is below the threshold), $P_e \nrightarrow 0$ no matter what we do and whatsoever. In other words, there exists a sharp threshold on the sample complexity which determines the boundary between possible vs. impossible detection. This sharp threshold is called the *minimum sample complexity* S^*. See Fig. 3.5.

It turns out the minimal sample complexity reads:

$$S^* = \frac{n \ln n}{2}.$$

Notice that S^* is much smaller than the total number of possible pairs: $\binom{n}{2} \approx \frac{n^2}{2}$. This result implies that the limit on the observation probability is

$$p^* = \frac{S^*}{\binom{n}{2}} = \frac{\ln n}{n}$$

where the second equality is because $\lim_{n\to\infty} \frac{n(n-1)}{n^2} = 1$. Notice that $p^* = \frac{\ln n}{n}$ vanishes as $n \to \infty$, meaning that community detection requires only a negligible fraction of pairwise measurements for successful community detection.

Look ahead In the next section, we will prove the achievability of the limit:

$$p > \frac{\ln n}{n} \Longrightarrow P_e \to 0 \text{ as } n \to \infty. \tag{3.2}$$

3.2 Social Networks: Achievability Proof

Recap In the previous section, we embarked on Part III which focuses on the application of information theory to data science. We discussed a specific application known as community detection, which involves identifying similar communities. Given two communities and a set of users with community memberships $x_i \in \{0, 1\}$, the goal is to decode the community membership vector $\boldsymbol{x} = [x_1, \ldots, x_n]$. We used parity information, i.e., whether two users belong to the same community, as the measurement information. However, since $\boldsymbol{x}$ and its flipped version $\boldsymbol{x} \oplus \mathbf{1}$ are indistinguishable from the parities y_{ij}'s alone, we included decoding both as a success event. We considered a setting where only a subset of pairwise measurements y_{ij}'s are accessible and the pairs are chosen randomly without our control, motivated by big data applications such as Meta's social networks. This led us to ask whether there exists a fundamental limit on the number of pairwise measurements needed to make community detection possible. We claimed that there is indeed a limit, and we explored what this means preciesely by translating it into a communication problem, as shown in Fig. 3.6.

We introduced two performance metrics: (i) sample complexity (concentrated around $\binom{n}{2}p$); and (ii) the probability of error $P_e := \mathbb{P}(\hat{\boldsymbol{x}} \notin \{\boldsymbol{x}, \boldsymbol{x} \oplus \mathbf{1}\})$. We then claimed the minimum sample complexity (above which one can make $P_e \to 0$, under which one cannot make $P_e \to 0$ no matter what we do and whatsoever) is:

$$S^* = \frac{n \ln n}{2}, \quad \text{i.e., } p^* = \frac{\ln n}{n}.$$

Outline In this section, we will prove that p^* is achievable:

$$p > \frac{\ln n}{n} \Longrightarrow P_e \to 0 \text{ as } n \to \infty. \tag{3.3}$$

It consists of three parts. First, we will employ the maximum likelihood (ML) decoding to derive the optimal decoder. We will then analyze the error probability under the ML decoding rule. Using a couple of bounding techniques, we will derive an upper bound of the error probability instead of attacking the exact probability

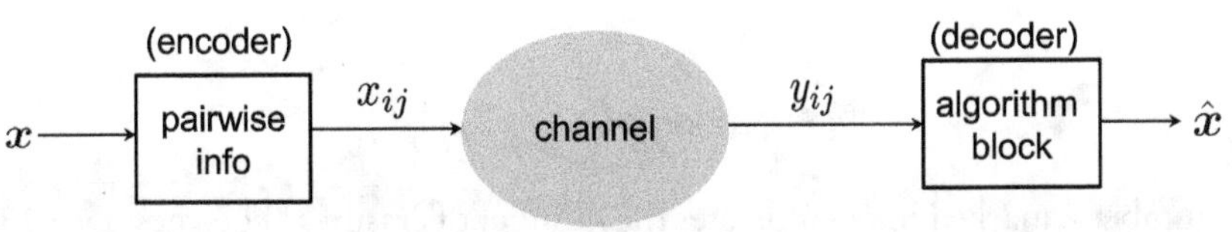

Figure 3.6. Translation of community detection into a communication problem.

directly. Lastly we will show that as long as $p > \frac{\ln n}{n}$, the upper bound approaches 0 as the number n of users tends to infinity, thereby proving the achievability.

Encoder The encoder converts $\boldsymbol{x}$ into $x_{ij} := x_i \oplus x_j$. See Fig. 3.6. Unlike the communication setting in which the encoder is subject to our design choice, it is not of our design, but it is given by the context. Let $\mathbf{X}$ be the output matrix of size n-by-n which contains x_{ij}'s as its entries. As in the communication setting, we call it a codeword (encoder output). Obviously $\mathbf{X}$ is symmetric as $x_{ij} = x_{ji}$. For instance, when $\boldsymbol{x} = (1000)$ and $n = 4$,

$$\mathbf{X}(\boldsymbol{x}) = \begin{bmatrix} & \underline{1} & 1 & 1 \\ & & 0 & 0 \\ & & & 0 \\ & & & \end{bmatrix}. \tag{3.4}$$

We omit diagonal components and symmetric counterparts as they can be trivially inferred. Let us call, a collection of codewords $\mathbf{X}(\boldsymbol{x})$'s, codebook.

The optimal decoder Let $\mathbf{Y} = [y_{ij}]$. The codebook is assumed to be known at the decoder. This assumption makes a trivial sense because the structure of pairwise measurements is revealed. The decoder employs an optimal decision rule: the MAP rule. In this setting, we have no idea on the statistics of $\boldsymbol{x}$. So we consider the worst-case scenario in which $\boldsymbol{x}$ is uniformly distributed. Notice that the randomness of $\boldsymbol{x}$, quantified as its entropy $H(\boldsymbol{x})$, is maximized when x_i's are i.i.d. each being uniformly distributed. In this setting, the MAP becomes equivalent to the ML decoder:

$$\hat{\boldsymbol{x}}_{\mathsf{ML}} = \arg\max_{\boldsymbol{x}} \mathbb{P}(\mathbf{Y}|\mathbf{X}(\boldsymbol{x})).$$

The calculation of $\mathbb{P}(\mathbf{Y}|\mathbf{X}(\boldsymbol{x}))$ is straightforward, being the same as the one in the communication setting. For instance, suppose $n = 4$, $\boldsymbol{x} = (0000)$, and y_{ij}'s are all zeros except $(y_{13}, y_{14}) = (\mathsf{e}, \mathsf{e})$:

$$\mathbf{Y}(\boldsymbol{x}) = \begin{bmatrix} & \underline{0} & \mathsf{e} & \mathsf{e} \\ & & 0 & 0 \\ & & & 0 \\ & & & \end{bmatrix}. \tag{3.5}$$

Then,

$$\mathbb{P}(\mathbf{Y}|\mathbf{X}(0000)) = (1-p)^2 p^4.$$

The number 2 marked in red indicates the number of erasures. This message (0000) is *compatible* since the corresponding likelihood is not zero. On the other hand, for

$\boldsymbol{x} = (1000)$,

$$\mathbb{P}(\mathbf{Y}|\mathbf{X}(1000)) = 0.$$

Note that $x_{12} = 1$ (underscored in (3.4)) is different from $y_{12} = 0$ (underscored in (3.5)), thus forcing the likelihood to be 0. This implies that the message $\boldsymbol{x} = (1000)$ can never be a solution, meaning that it is *incompatible* with $\mathbf{Y}$. Hence, the ML rule is summarized as follows.

1. Eliminate all the messages *incompatible* with $\mathbf{Y}$.
2. If there is only one survival, declare it as the correct message.

However, this procedure is not sufficient to describe the ML decoding rule. We may have a different erasure pattern that confuses the rule. To see this clearly, consider the following example. Suppose that y_{ij}'s are all zeros except $(y_{12}, y_{13}, y_{14}) = (\mathsf{e}, \mathsf{e}, \mathsf{e})$:

$$\mathbf{Y}(\boldsymbol{x}) = \begin{bmatrix} & \mathsf{e} & \mathsf{e} & \mathsf{e} \\ & & 0 & 0 \\ & & & 0 \\ & & & \end{bmatrix}. \tag{3.6}$$

Then,

$$\mathbb{P}(\mathbf{Y}|\mathbf{X}(0000)) = (1-p)^3 p^4;$$
$$\mathbb{P}(\mathbf{Y}|\mathbf{X}(0111)) = (1-p)^3 p^4.$$

The two patterns (0000, 0111) are compatible and the likelihood functions are equal. In this case, what we can do for the best is to flip a coin, choosing one out of the two in a random manner. This forms the last step of the ML decoding rule.

3. If there are multiple survivals, choose one randomly.

A setup for analysis of the error probability For the achievability proof (3.3), we analyze the probability of error when using the ML decoder. Starting with the definition of P_e and using the total probability law, we get:

$$\begin{aligned} P_e &:= \mathbb{P}\left(\hat{\boldsymbol{x}} \notin \{\boldsymbol{x}, \boldsymbol{x} \oplus \mathbf{1}\}\right) \\ &= \sum_{\boldsymbol{a}} \mathbb{P}(\boldsymbol{x} = \boldsymbol{a}) \mathbb{P}(\hat{\boldsymbol{x}} \notin \{\boldsymbol{a}, \boldsymbol{a} \oplus \mathbf{1}\} | \boldsymbol{x} = \boldsymbol{a}). \end{aligned}$$

For a fixed $\boldsymbol{a}$, $\mathbb{P}(\hat{\boldsymbol{x}} \notin \{\boldsymbol{a}, \boldsymbol{a} \oplus \mathbf{1}\} | \boldsymbol{x} = \boldsymbol{a})$ is a sole function of the likelihood which depends only on erasure patterns of the $\binom{n}{2}$ independent channels. Also, the erasure patterns are independent of the channel input affected by $\boldsymbol{a}$. Therefore, the probability is irrelevant of what the value of $\boldsymbol{a}$ is. Applying this to the above, we get:

$$\begin{aligned} P_e &= \mathbb{P}(\hat{\boldsymbol{x}} \notin \{\mathbf{0}, \mathbf{1}\} | \boldsymbol{x} = \mathbf{0}) \\ &\leq \sum_{\boldsymbol{a} \notin \{\mathbf{0}, \mathbf{1}\}} \mathbb{P}(\hat{\boldsymbol{x}} = \boldsymbol{a} | \boldsymbol{x} = \mathbf{0}) \end{aligned} \tag{3.7}$$

where the inequality comes from the union bound.

Further upper-bounding Consider $\mathbb{P}(\hat{\boldsymbol{x}} = \boldsymbol{a} | \boldsymbol{x} = \mathbf{0})$. To gain insights, consider an example where $n = 4$ and $\boldsymbol{a} = (1000)$. In this case, the error event implies that $\mathbf{X}(1000)$ must be *compatible.* A necessary condition for $\mathbf{X}(1000)$ being compatible under $\boldsymbol{x} = (0000)$ is: $(y_{12}, y_{13}, y_{14}) = (\mathsf{e}, \mathsf{e}, \mathsf{e})$. For all (i, j) entries whose values are different between the two codewords $\mathbf{X}(0000)$ and $\mathbf{X}(1000)$ (that we call *distinguishable positions*), erasures must occur; otherwise, (1000) cannot be compatible as its corresponding likelihood would be 0. Hence, we get:

$$\begin{aligned} &\mathbb{P}(\hat{\boldsymbol{x}} = (1000) | \boldsymbol{x} = \mathbf{0}) \\ &\quad \leq \mathbb{P}\left(\mathbf{X}(1000) \text{ compatible} | \boldsymbol{x} = \mathbf{0}\right) \\ &\quad \leq \mathbb{P}\left((y_{12}, y_{13}, y_{14}) = (\mathsf{e}, \mathsf{e}, \mathsf{e}) | \boldsymbol{x} = \mathbf{0}\right) \\ &\quad = (1 - p)^{3}. \end{aligned}$$

The number 3 marked in red indicates the number of erasures that must occur in those *distinguishable positions.*

A key to determine the above upper bound is the number of distinguishable positions between $\mathbf{X}(\boldsymbol{a})$ and $\mathbf{X}(\mathbf{0})$. One can easily verify that the number of distinguishable positions (w.r.t. $\mathbf{X}(\mathbf{0})$) depends on the number of 1's in $\boldsymbol{a}$. To see this, consider an example of $\mathbf{a} = (\underbrace{11 \cdots 1}_{k} \underbrace{00 \cdots 0}_{n-k})$. In this case,

$$\mathbf{X}(\boldsymbol{a}) = \begin{bmatrix} & 0 & 0 & 0 & 1 & 1 \\ & & 0 & 0 & 1 & 1 \\ & & & 0 & 1 & 1 \\ & & & & 0 & 0 \\ & & & & & 0 \\ & & & & & \end{bmatrix}.$$

where k denotes the number of 1's in $\boldsymbol{a}$. Each of the first k rows contains $(n - k)$ ones; hence, the total number of distinguishable positions w.r.t. $\mathbf{X}(\mathbf{0})$:

$$\# \text{ of distinguishable positions} = k(n - k).$$

For a succinct description of the summation term in (3.7), we classify the instance $\boldsymbol{a}$ depending on the number of 1's in $\boldsymbol{a}$. To this end, we introduce:

$$\mathcal{A}_k := \{\boldsymbol{a} \mid \|\boldsymbol{a}\|_1 = k\} \tag{3.8}$$

where $\|a\|_1 := |a_1| + |a_2| + \cdots + |a_n|$. Using this notation, we can then express (3.7) as:

$$\begin{aligned} P_e &\leq \sum_{k=1}^{n-1} \sum_{\boldsymbol{a} \in \mathcal{A}_k} (1-p)^{k(n-k)} \\ &= \sum_{k=1}^{n-1} |\mathcal{A}_k| (1-p)^{k(n-k)} \\ &= \sum_{k=1}^{n-1} \binom{n}{k} (1-p)^{k(n-k)} \qquad (3.9) \\ &= 2\sum_{k=1}^{\frac{n}{2}} \binom{n}{k} (1-p)^{k(n-k)} - \binom{n}{n} (1-p)^{n \cdot 0} \\ &\leq 2\sum_{k=1}^{\frac{n}{2}} \binom{n}{k} (1-p)^{k(n-k)} \end{aligned}$$

where the second last step follows from the fact that $\binom{n}{k}(1-p)^{k(n-k)}$ is symmetric around $k = n/2$.

The final step of the achievability proof Since we intend to prove the achievability when $p > \frac{\ln n}{n}$, focus on the regime: $\lambda > 1$ where λ is defined such that $p := \lambda \frac{\ln n}{n}$. Then, it suffices to show that for $\lambda > 1$,

$$\sum_{k=1}^{\frac{n}{2}} \binom{n}{k} (1-p)^{k(n-k)} \longrightarrow 0 \text{ as } n \to \infty. \tag{3.10}$$

In this setting of $p = \lambda \frac{\ln n}{n}$, p is arbitrarily close to 0 in the limit of n. This motivates us to employ the following upper bound on $1-p$ (which is very tight in the regime):

$$1 - p \leq e^{-p}. \tag{3.11}$$

Check the proof in Prob 7.1. When n is large, the following bound is good enough to prove the achievability:

$$\binom{n}{k} = \frac{n(n-1)(n-2)\cdots(n-k+1)}{k!} \leq \frac{n^k}{k!} \leq n^k \tag{3.12}$$

Applying the bounds (3.11) and (3.12) into (3.9), we obtain:

$$\begin{aligned} P_e &\leq 2\sum_{k=1}^{\frac{n}{2}} \binom{n}{k} (1-p)^{k(n-k)} \\ &\leq 2\sum_{k=1}^{\frac{n}{2}} n^k e^{-pk(n-k)} \\ &\overset{(a)}{=} 2\sum_{k=1}^{\frac{n}{2}} e^{k\ln n - \lambda k\left(1-\frac{k}{n}\right)\ln n} \\ &= 2\sum_{k=1}^{\frac{n}{2}} e^{-k(\lambda(1-\frac{k}{n})-1)\ln n} \end{aligned} \tag{3.13}$$

where (a) follows from $n^k = e^{k\ln n}$ and $p := \lambda\frac{\ln n}{n}$.

For the range of $1 \leq k \leq \frac{n}{2}$, $1-\frac{k}{n}$ (marked in blue in (3.13)) is minimized at $\frac{1}{2}$. Applying this to the above, we get:

$$P_e \leq 2\sum_{k=1}^{\frac{n}{2}} e^{-k(\frac{\lambda}{2}-1)\ln n} \tag{3.14}$$

Case I: $\lambda > 2$: If $\lambda > 2$, we can apply the well-known summation formula w.r.t. the geometric series where the common ratio is $e^{-(\frac{\lambda}{2}-1)\ln n}$, thus obtaining:

$$P_e \leq 2\sum_{k=1}^{\frac{n}{2}} \left(e^{-(\frac{\lambda}{2}-1)\ln n}\right)^k \leq 2\cdot\frac{e^{-(\frac{\lambda}{2}-1)\ln n}}{1-e^{-(\frac{\lambda}{2}-1)\ln n}} \longrightarrow 0 \text{ as } n \to \infty.$$

Hence, we can make $P_e \to 0$ for the case of $\lambda > 2$.

Case II: $1 < \lambda \leq 2$: Observe in the last step in (3.13) that

$$\lambda\left(1-\frac{k}{n}\right)-1 = 0 \quad\Longleftrightarrow\quad k = \left(1-\frac{1}{\lambda}\right)n,$$

and $\lambda(1-\frac{k}{n})-1$ is a decreasing function in k. See Fig. 3.7.

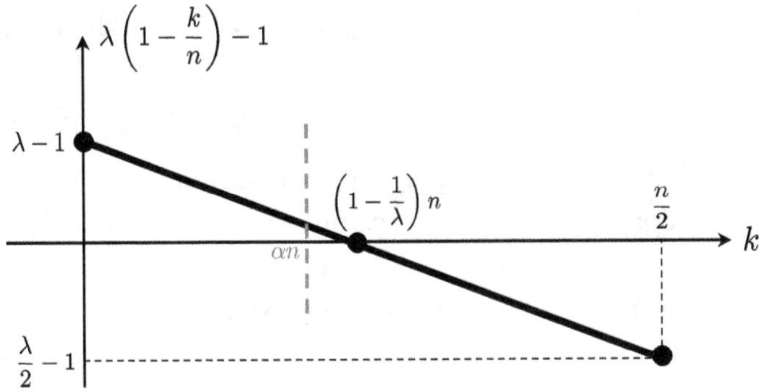

Figure 3.7. Behavior of $\lambda(1-\frac{k}{n})-1$ in k.

This together with a choice of $\alpha < 1-\frac{1}{\lambda}$ gives:

$$\lambda\left(1-\frac{k}{n}\right)-1>0 \quad \text{when } k \leq \alpha n.$$

Applying this to (3.13), we get:

$$\begin{aligned} P_e &\leq 2\sum_{k=1}^{\frac{n}{2}}\binom{n}{k}(1-p)^{k(n-k)} \\ &\leq 2\sum_{k=1}^{\alpha n} e^{-k(\lambda(1-\frac{k}{n})-1)\ln n} + \sum_{k=\alpha n+1}^{\frac{n}{2}}\binom{n}{k}(1-p)^{k(n-k)}. \end{aligned}$$

Again applying the summation formula of the geometric series to the first term in the last step of the above, the first term vanishes as $n\rightarrow\infty$. Hence, we obtain:

$$\begin{aligned} P_e &\lesssim \sum_{k=\alpha n+1}^{\frac{n}{2}}\binom{n}{k}(1-p)^{k(n-k)} \\ &\overset{(a)}{\leq} (1-p)^{\alpha(1-\alpha)n^2}\sum_{k=\alpha n+1}^{\frac{n}{2}}\binom{n}{k} \\ &\overset{(b)}{\leq} (1-p)^{\alpha(1-\alpha)n^2}\cdot 2^n \\ &\overset{(c)}{\leq} e^{-\lambda\alpha(1-\alpha)n\ln n}\cdot e^{n\ln 2}. \end{aligned}$$

where (a) is due to the fact that $k(n-k)\geq\alpha(1-\alpha)n^2$ (where the equality holds when $k=\alpha n+1$); (b) comes from the binomial theorem ($\sum_{k=0}^{n}\binom{n}{k}=2^n$); and

(c) comes from the fact that $1 - p \le e^{-p}$ and $p := \lambda \frac{\ln n}{n}$. The last term in the above tends to 0 as $n \to \infty$. This is because $\lambda\alpha(1-\alpha)n \ln n$ grows much faster than $n \ln 2$. This implies $P_e \to 0$, which completes the achievability proof (3.3).

Look ahead Using the ML decoding rule, we proved the achievability of the community detection limit (3.3). It turns out the converse holds: $p > \frac{\ln n}{n}$ for reliable detection, meaning that the condition of $p > \frac{\ln n}{n}$ is *necessary* for reliable detection. In the next section, we will prove the converse.

3.3 Social Networks: Converse Proof

Recap In the previous section, we proved the achievability of the limit on observation probability p for community detection:

$$p > \frac{\ln n}{n} \Longrightarrow P_e \to 0 \quad \text{as } n \to \infty.$$

Outline In this section, we will prove that the condition $p > \frac{\ln n}{n}$ is also necessary for reliable community detection, meaning the converse (the other way around) holds:

$$p < \frac{\ln n}{n} \Longrightarrow P_e \nrightarrow 0.$$

Proof strategy The converse proof relies on a lower bound of P_e which does not vanish under the condition of $p < \frac{\ln n}{n}$. In Part II, we learned about one important inequality that played a significant role in deriving such a lower bound. That is, Fano's inequality. However, Fano's inequality does not yield such a good lower bound in the context of community detection. Check in Prob 7.4. Hence, we will take a different approach.

The different approach builds upon another important concept that has been extensively employed in the graph theory literature. That is, *graph connectivity*. We say that a graph is *connected* if there exists a path (i.e., a sequence of connected edges) between any pair of two nodes. Otherwise, it is said to be *disconnected*. See Fig. 3.8 for a pictorial illustration.

The graph connectivity has a close relationship with an error event of community detection. Suppose that an edge in the graph indicates a situation where a pairwise measurement of the associated nodes is obtained. Then, the graph disconnectivity implies that there exist(s) isolated node(s) (node 1 in the example illustrated in the right side of Fig. 3.8). In this case, there is no way to decode $\boldsymbol{x}$ even up to a

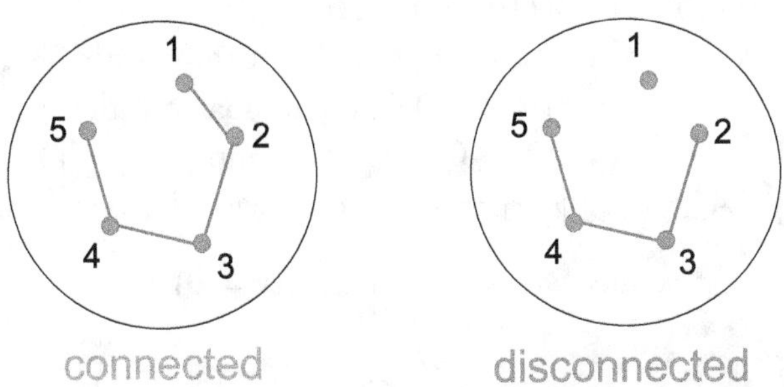

Figure 3.8. Graph connectivity: (Left) An example of a connected graph; (Right) An example of a disconnected graph.

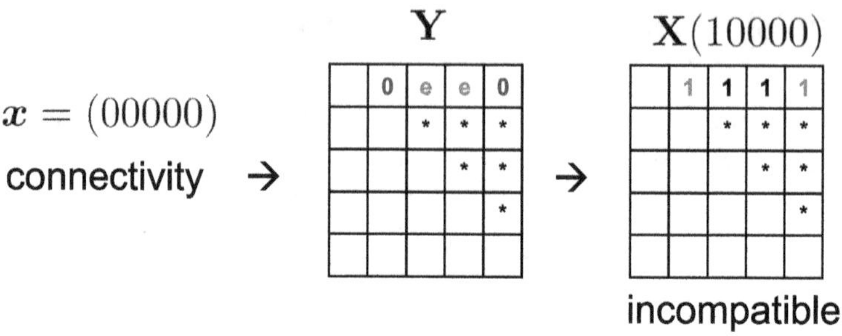

Figure 3.9. A necessary condition for graph connectivity.

global shift. Note in the example that there are four possible candidates for a solution. For instance, suppose $(y_{23}, y_{34}, y_{45}) = (0, 0, 0)$. Then, the four candidates are: (i) $\boldsymbol{x} = [0, 0, 0, 0, 0]$; (ii) $\boldsymbol{x} = [0, 1, 1, 1, 1]$; (iii) $\boldsymbol{x} = [1, 0, 0, 0, 0]$; and (iv) $\boldsymbol{x} = [1, 1, 1, 1, 1]$. Obviously this does not ensure successful community detection, hence $P_e \nrightarrow 0$. Therefore, it suffices to show that

$$p < \frac{\ln n}{n} \Longrightarrow \text{graph is disconnected, i.e., } \mathbb{P}(\text{connected}) \to 0.$$

An upper bound of connectivity probability $\mathbb{P}$(connected) Since we intend to show $\mathbb{P}(\text{connected}) \to 0$, it suffices to prove that its upper bound tends to 0 as $n \to \infty$. The graph connectivity has nothing to do with the value of $\boldsymbol{x}$. Hence, without loss of generality, one can assume that the ground truth $\boldsymbol{x} = \mathbf{0}$. This then gives:

$$\mathbb{P}(\text{connected}) = \mathbb{P}(\text{connected}|\boldsymbol{x} = \mathbf{0}).$$

The event of a graph being connected implies that node 1 is connected with at least one different node; otherwise, node 1 is isolated. This suggests that there exists at least one observation in the first row of the received signal matrix $\mathbf{Y}$. In the example of Fig. 3.9, (y_{12}, y_{15}) are observed as $(0, 0)$. This observation is compatible with the all-zero ground truth vector. On the other hand, the codeword w.r.t. the message $(10 \cdots 0)$ $(\mathbf{X}(10 \cdots 0))$ is *incompatible* with $\mathbf{Y}$ since the revealed components in $\mathbf{Y}$ contradict with corresponding components in $\mathbf{X}(10 \cdots 0)$. In the example of Fig. 3.9, $(y_{12}, y_{15}) = (0, 0)$ do not match with $(x_{12}, x_{15}) = (1, 1)$. Similarly $\mathbf{X}(010 \cdots 0), \ldots, \mathbf{X}(0 \cdots 01)$ are all incompatible with $\mathbf{Y}$. Using this and the following set $\mathcal{A}_1 := \{\boldsymbol{a} : \|\boldsymbol{a}\|_1 = 1\}$, we can rewrite $\mathbb{P}$(connected) as:

$$\begin{aligned}\mathbb{P}(\text{connected}) &= \mathbb{P}(\text{connected}|\boldsymbol{x} = \mathbf{0}) \\ &\leq \mathbb{P}\left(\bigcap_{\boldsymbol{a}\in\mathcal{A}_1} \{\mathbf{X}(\boldsymbol{a}) \text{ incomp.}\}\right).\end{aligned} \tag{3.15}$$

A key upper-bounding technique One key observation is that the computation of the above probability can be greatly simplified when the associated events are *independent*. Remember that $\mathbb{P}(A \cap B) = \mathbb{P}(A)\mathbb{P}(B)$ for independent events A and B. Also the more independent events are, the tighter bound we get. This motivates us to search for independent events (among n events) as many as possible.

To this end, we intend to see the functional relationship between $\mathbf{X}(\boldsymbol{a})$ and erasure patterns. Only the first row in $\mathbf{X}(10\cdots0)$ are distinct with those in $\mathbf{X}(\mathbf{0})$. This implies that whether $\mathbf{X}(10\cdots0)$ is *incompatible* depends solely on the pairwise measurements y_{ij}'s in those distinguishable positions. Similarly the event of $\mathbf{X}(010\cdots0)$ being incompatible depends on y_{ij}'s in the second row. Here the symmetric matrix property gives $x_{12} = x_{21}$, hence this may pose *dependency* between the two events.

However, the dependency can be removed for certain situations. Suppose the following event occurs: $y_{12} = \mathsf{e}$. Then, the two events share no overlapping positions, since the overlapping $(1, 2)$ entry is now erased. Hence, given $y_{12} = \mathsf{e}$, the two events become independent. Similarly, given $y_{ij} = \mathsf{e}$ for all $i, j \in \{1, 2, 3\}$,

$$\{\mathbf{X}(10\cdots0)\text{ incomp.}\}\perp\{\mathbf{X}(010\cdots0)\text{ incomp.}\}\perp\{\mathbf{X}(0010\cdots0)\text{ incomp.}\}.$$

This enables us to identify a general erasure pattern that makes mutiple events (say L events) independent:

$$\begin{aligned}
&y_{ij} = \mathsf{e} \quad \forall i, j \in \{1, 2, \ldots, L\} \Longrightarrow \\
&\{\mathbf{X}(10\cdots0)\text{ incomp.}\}\perp\{\mathbf{X}(010\cdots0)\text{ incomp.}\}\perp \\
&\cdots\perp\{\mathbf{X}(0\cdots \underbrace{1}_{L\text{th position}} \cdots0)\text{ incomp.}\}.
\end{aligned}$$

In view of graph, the number L refers to the number of nodes that are locally disconnected. In the example of Fig. 3.10, we have no observation for any pair of nodes 1,2,3,4,5. So $L \geq 5$ in this case.

This motivates us to find the maximum number of nodes that are locally disconnected. It reveals as many as independent events, thus leading to a tight upper bound on the connectivity probability. To this end, we consider the following situation. Suppose we consider the first m nodes in the graph. We will choose m such that L is maximized and m tends to ∞ as $n \to \infty$. Then, as per the WLLN, the number of edges in the subgraph consisting of the m nodes would be concentrated around

$$\binom{m}{2}p \quad \text{w.h.p.}$$

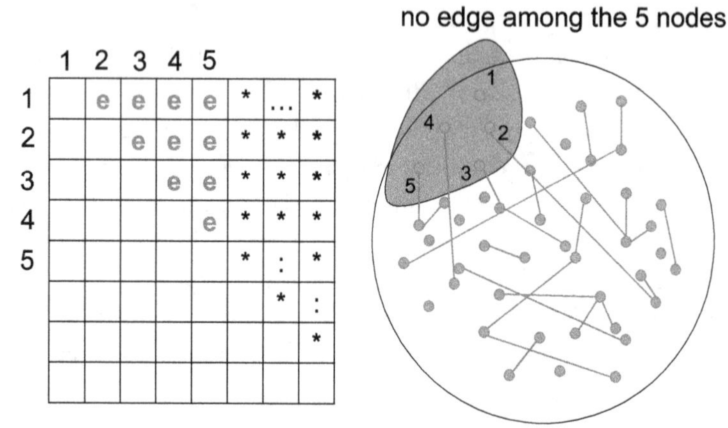

Figure 3.10. An example in which the number of nodes locally disconnected is greater than or equal to 5.

Using the fact that one edge is associated with two nodes, the number of locally disconnected nodes is at least $m - 2\binom{m}{2}p$:

$$L \geq \left\lfloor m - 2\binom{m}{2}p \right\rfloor \quad \text{w.h.p.}$$

Let $p = \lambda \frac{\ln n}{n}$ for some $\lambda < 1$. Then,

$$L \geq \left\lfloor m - \lambda m^2 \frac{\ln n}{n} \right\rfloor.$$

In an effort to maximize the above bound, we choose m such that the first term m and the second term $\lambda m^2 \frac{\ln n}{n}$ in the bound are of the same order. We make a particular choice for such m: $m = \lfloor \frac{n}{2 \ln n} \rfloor$, thus obtaining:

$$L \geq \left\lfloor \left(\frac{1}{2} - \frac{\lambda}{4}\right) \frac{\ln n}{n} \right\rfloor = \left\lfloor \frac{\alpha n}{\ln n} \right\rfloor$$

where $\alpha := \frac{1}{2} - \frac{\lambda}{4}$. We then reorder node indices such that the locally disconnected nodes are numbered as $1, 2, \ldots, \lfloor \frac{\alpha n}{\ln n} \rfloor$. Let $\mathcal{T}$ be such an event:

$$\mathcal{T} = \left\{ L \geq \left\lfloor \frac{\alpha n}{\ln n} \right\rfloor \right\}.$$

We call it a typical event as it happens w.h.p.:

$$\mathbb{P}(\mathcal{T}) \rightarrow 1 \quad \text{as } n \rightarrow \infty.$$

Given the typical event $\mathcal{T}$,

$$\begin{aligned}&\{\mathbf{X}(10\cdots 0)\text{ incomp.}\}\perp\{\mathbf{X}(010\cdots 0)\text{ incomp.}\}\perp\\&\cdots\perp\{\mathbf{X}(0\cdots \underbrace{1}_{\lfloor \frac{\alpha n}{\ln n}\rfloor\text{th position}}\cdots 0)\text{ incomp.}\}.\end{aligned} \tag{3.16}$$

Applying this into (3.15) together with the total probability law, we get:

$$\begin{aligned}\mathbb{P}(\text{connected}) &= \mathbb{P}(\text{connected}|\boldsymbol{x}=\mathbf{0})\\&\le \mathbb{P}\left(\bigcap_{\boldsymbol{a}\in\mathcal{A}_1}\{\mathbf{X}(\boldsymbol{a})\text{ incomp.}\}|\boldsymbol{x}=\mathbf{0}\right)\\&= \mathbb{P}\left(\left\{\bigcap_{\boldsymbol{a}\in\mathcal{A}_1}\{\mathbf{X}(\boldsymbol{a})\text{ incomp.}\}\right\},\mathcal{T}|\boldsymbol{x}=\mathbf{0}\right)\\&\quad+\mathbb{P}\left(\left\{\bigcap_{\boldsymbol{a}\in\mathcal{A}_1}\{\mathbf{X}(\boldsymbol{a})\text{ incomp.}\}\right\},\mathcal{T}^c|\boldsymbol{x}=\mathbf{0}\right)\\&\overset{(a)}{\approx} \mathbb{P}\left(\bigcap_{\boldsymbol{a}\in\mathcal{A}_1}\{\mathbf{X}(\boldsymbol{a})\text{ incomp.}\}|\boldsymbol{x}=\mathbf{0},\mathcal{T}\right)\\&\overset{(b)}{\le}\prod_{\boldsymbol{a}\in\mathcal{B}_1}\mathbb{P}\left(\mathbf{X}(\boldsymbol{a})\text{ incomp.}|\boldsymbol{x}=\mathbf{0},\mathcal{T}\right)\\&\overset{(c)}{=}\mathbb{P}\left(\mathbf{X}(10\cdots 0)\text{ incomp.}|\boldsymbol{x}=\mathbf{0},\mathcal{T}\right)^{\lfloor\frac{\alpha n}{\ln n}\rfloor}\end{aligned} \tag{3.17}$$

where (a) follows from $\mathbb{P}(\mathcal{T})\to 1, \mathbb{P}(\mathcal{T}^c)\to 0$ and (b) comes from (3.16) and the definition $\mathcal{B}_1 := \{\boldsymbol{b} : \|\boldsymbol{b}\|_1 = 1, b_i = 1, i = 1,2,\ldots,\lfloor\frac{\alpha n}{\ln n}\rfloor\}$; and (c) is by symmetry.

The final step The event of $\mathbf{X}(10\cdots 0)$ being incompatible given $(\boldsymbol{x}=\mathbf{0},\mathcal{T})$ implies that there exists at least one observation among the last $n-\lfloor\frac{\alpha n}{\ln n}\rfloor$ components in the first row of $\mathbf{Y}$. Hence,

$$\begin{aligned}&\mathbb{P}(\mathbf{X}(10\cdots 0)\text{ incomp.}|\boldsymbol{x}=\mathbf{0},\mathcal{T})\\&\quad\le 1-(1-p)^{n-\lfloor\frac{\alpha n}{\ln n}\rfloor}\end{aligned}$$

$$\leq 1-(1-p)^n$$

$$\stackrel{(a)}{\leq} e^{-(1-p)^n}$$

where (a) is due to the fact that $1-x \leq e^{-x}$ for $x > 0$. Applying this to (3.17), we get:

$$\begin{aligned}
\mathbb{P}(\text{connected}) &\leq \mathbb{P}(\mathbf{X}(10\cdots 0)\text{ incomp.}|\mathbf{x}=\mathbf{0},\mathcal{T})^{\lfloor\frac{\alpha n}{\ln n}\rfloor} \\
&\leq \exp\left(-(1-p)^n\left\lfloor\frac{\alpha n}{\ln n}\right\rfloor\right) \\
&\stackrel{(a)}{\approx} \exp\left(-e^{-pn}\frac{\alpha n}{\ln n}\right) \\
&= \exp\left(-\alpha e^{(1-\lambda)\ln n-\ln\ln n}\right)
\end{aligned}$$

where (a) comes from the fact that $(1-p)^n \approx e^{-pn}$ for sufficiently large n and small $p=\lambda\frac{\ln n}{n}$ (which is our case) and the fact that $\lfloor\frac{\alpha n}{\ln n}\rfloor \approx \frac{\alpha n}{\ln n}$ for large n. Therefore, if $\lambda < 1$, the upper bound goes to 0 as $n \to \infty$. This completes the proof.

Look ahead We have thus far proved the achievability and converse of the community detection limit. Remember that the achievable scheme is the ML decoding rule:

$$\hat{\boldsymbol{x}}_{\mathsf{ML}} = \arg\max_{\boldsymbol{x}} \mathbb{P}(\mathbf{Y}|\mathbf{X}(\boldsymbol{x})).$$

A practical issue arises with the implementation of the ML rule due to its prohibitive complexity. The number of likelihood computations required for the rule is 2^n and grows exponentially with the number of users n, which is typically very large in practice. As a result, the complexity becomes enormous. However, there is another algorithm that is much more efficient and provides nearly the same performance as the ML rule. In the following section, we will examine this efficient algorithm.

3.4 An Efficient Algorithm and Python Implementation

Recap In Section 3.2, we employed the ML decoding rule to prove the achievability of the limit $p^* = \frac{\ln n}{n}$ for community detection. Recall the ML decoding rule:

$$\hat{\boldsymbol{x}}_{\mathsf{ML}} = \arg\max_{\boldsymbol{x}} \mathbb{P}(\mathbf{Y}|\mathbf{X}(\boldsymbol{x}))$$

where $\boldsymbol{x} = [x_1, \ldots, x_n]^T$ indicates the community membership vector; $\mathbf{X}(\boldsymbol{x})$ denotes a codeword matrix taking $x_{ij} = x_i \oplus x_j$ as the (i,j) entry; and $\mathbf{Y}$ is an observation matrix with y_{ij}'s ($y_{ij} = x_{ij}$ w.p. p and e otherwise). One critical issue in the ML rule is that its complexity is significant. Since $\boldsymbol{x}$ takes one of the 2^n possible patterns, the ML rule requires the number 2^n of likelihood computations that grows exponentially with n. Hence, it is crucial to develop a computationally efficient algorithm that possibly yields the same performance as the ML rule. Indeed, efficient algorithms have been developed that achieve the optimal ML performance.

Outline In this section, we will examine one such efficient algorithm for community detection. Although the optimal algorithm is quite complex, we will focus on its simpler version that achieves sub-optimal performance while still including the main components of the algorithm. We will cover four main points in this section. First, we will introduce the adjacency matrix, which plays a fundamental role in the algorithm. Second, we will explain how the algorithm works, including the process of finding the principal eigenvector of the adjacency matrix. Third, we will explore an efficient method of computing the principal eigenvector, known as the power method. Finally, we will provide a Python implementation of the spectral algorithm.

Adjacency matrix Since the number n of users is often huge, pairwise measurements are big data although only part of them are observed. In data science, there is a useful entity that represents such big data in a succinct way. That is, the *adjacency matrix* (Chartrand, 1977). The adjacency matrix, say $\mathbf{A}$, is an equivalent representation of the pairwise data where each column and row represents users. Each entry, say a_{ij}, indicates whether users i and j are in the same community. Precisely, $a_{ij} = +1$ means that the two users are in the same community; $a_{ij} = -1$ indicates the opposite; and $a_{ij} = 0$ denotes no measurement:

$$a_{ij} = \begin{cases} 1 - 2y_{ij}, & \text{w.p. } p; \\ 0, & \text{otherwise } (y_{ij} = \mathsf{e}). \end{cases} \tag{3.18}$$

For instance, when $\boldsymbol{x} = [1, 0, 0, 1]^T$, we might have:

$$\mathbf{A} = \begin{bmatrix} +1 & -1 & 0 & 0 \\ -1 & +1 & +1 & -1 \\ 0 & +1 & +1 & -1 \\ 0 & -1 & -1 & +1 \end{bmatrix} \tag{3.19}$$

where we have erasures for (y_{13}, y_{14}).

In the full measurement setting ($p = 1$), one can make an observation that gives a significant insight into algorithms. When $p = 1$, we would obtain:

$$\mathbf{A} = \begin{bmatrix} +1 & -1 & -1 & +1 \\ -1 & +1 & +1 & -1 \\ -1 & +1 & +1 & -1 \\ +1 & -1 & -1 & +1 \end{bmatrix}. \tag{3.20}$$

Note that the rank of $\mathbf{A}$ is 1, i.e., each row is a linear combination of the other rows. Can we extract the community pattern from this rank-1 matrix? It turns out the answer is yes, and this forms the basis of the spectral algorithm that we will investigate in the sequel.

Spectral algorithm (Shen *et al.*, 2011) Using the fact that the rank of $\mathbf{A}$ (3.20) is 1, we can easily derive its eigenvector.

$$\mathbf{v} = \begin{bmatrix} +1 \\ -1 \\ -1 \\ +1 \end{bmatrix}. \tag{3.21}$$

Check that $\mathbf{Av} = 4\mathbf{v}$ indeed. Here the $+1$ (or -1) entries of $\mathbf{v}$ tell us the community memberships of the users. Therefore, in the ideal situation where every pair is sampled, the principal eigenvector (the sole eigenvector) recovers the communities. This approach, taking the adjacency matrix and computing its principal eigenvector, is called the *spectral algorithm*.

What about for the partial measurement case $p < 1$? In this case, we are not clear if the principal eigenvector is able to return the community memberships. Also, the components in the eigenvector may not necessarily take $+1$ (or -1). To address the second issue, we may take a thresholded eigenvector, say $\mathbf{v}_{\text{th}}$, where its entry takes the sign of v_i:

$$v_{\text{th},i} = \begin{cases} +1, & v_i > 0; \\ -1, & \text{otherwise}. \end{cases} \tag{3.22}$$

As long as p is big enough, $\mathbf{v}_{\text{th}}$ returns the ground truth of communities, as n tends to infinity. Due to the interest of this book, we will not analyze how big p is required for successful recovery of the spectral algorithm. Instead we will later provide empirical simulation via Python to demonstrate that for a large value of n, $\mathbf{v}_{\text{th}}$ indeed approaches the ground truth with an increase in p.

Power method (Golub and Van Loan, 2013) Another technical question arises when it comes to computing the principal eigenvector. What if the adjacency matrix is of a big size? Remember that the order of n is around 10^9 in Meta's social networks. A naive way of computing the eigenvector based on eigenvalue decomposition requires the complexity of around n^3. Hence, this way is prohibitive. Fortunately, there is one very efficient and useful way of computing the principal eigenvector. That is, the *power method*. The method is well-known and popular in the data science literature.

Prior to describing how it works in detail, let us make important observations that naturally lead to the method. Suppose that the adjacency matrix $\mathbf{A} \in \mathbb{R}^{n \times n}$ has m eigenvalues λ_i's and eigenvectors $\mathbf{v}_i$'s:

$$\mathbf{A} := \lambda_1 \mathbf{v}_1 \mathbf{v}_1^T + \lambda_2 \mathbf{v}_2 \mathbf{v}_2^T + \cdots + \lambda_m \mathbf{v}_m \mathbf{v}_m^T$$

where $\lambda_1 > \lambda_2 \geq \lambda_3 \geq \cdots \geq \lambda_m$ and $\mathbf{v}_i$'s are orthonormal: $\mathbf{v}_i^T \mathbf{v}_j = \mathbf{1}\{i = j\}$. By definition, $(\lambda_1, \mathbf{v}_1)$ are the principal eigenvalue and eigenvector respectively. Let $\mathbf{v} \in \mathbb{R}^n$ be an arbitrary non-zero vector such that $\mathbf{v}_1^T \mathbf{v} \neq 0$. Then, $\mathbf{Av}$ can be expressed as:

$$\begin{aligned} \mathbf{Av} &= \left(\sum_{i=1}^{m} \lambda_i \mathbf{v}_i \mathbf{v}_i^T \right) \mathbf{v} \\ &= \sum_{i=1}^{m} \lambda_i (\mathbf{v}_i^T \mathbf{v}) \mathbf{v}_i \end{aligned} \tag{3.23}$$

where the second equality is due to the fact that $\mathbf{v}_i^T \mathbf{v}$ is a scalar.

The first term $\lambda_1 (\mathbf{v}_1^T \mathbf{v}) \mathbf{v}_1$ in the above summation forms a major contribution, since λ_1 is the largest. This major effect becomes more dominant when we multiply the adjacency matrix to the resulting vector $\mathbf{Av}$. To see this, consider:

$$\begin{aligned}
\mathbf{A}^2\mathbf{v} &= \mathbf{A}(\mathbf{A}\mathbf{v}) \\
&= \left(\sum_{i=1}^{m} \lambda_i \mathbf{v}_i \mathbf{v}_i^T\right)\left(\sum_{j=1}^{m} \lambda_j (\mathbf{v}_j^T \mathbf{v})\mathbf{v}_j\right) \\
&\overset{(a)}{=} \sum_{i=1}^{m} \lambda_i^2 (\mathbf{v}_i^T \mathbf{v})(\mathbf{v}_i^T \mathbf{v}_i)\mathbf{v}_i + \sum_{i\neq j} \lambda_i \lambda_j (\mathbf{v}_j^T \mathbf{v})(\mathbf{v}_i^T \mathbf{v}_j)\mathbf{v}_j \\
&\overset{(b)}{=} \sum_{i=1}^{m} \lambda_i^2 (\mathbf{v}_i^T \mathbf{v})\mathbf{v}_i
\end{aligned} \tag{3.24}$$

where (a) comes from the fact that $\mathbf{v}_i^T\mathbf{v}_j$ is a scalar; and (b) $\mathbf{v}_i$'s are orthonormal vectors, i.e., $\mathbf{v}_i^T\mathbf{v}_i = 1$ and $\mathbf{v}_i^T\mathbf{v}_j = 0$ for $i \neq j$. The distinction in (3.24) relative to $\mathbf{Av}$ (3.23) is that we read λ_i^2 instead of λ_i. So the contribution from the principal component is more significant relative to the other components. Iterating this process (multiplying $\mathbf{A}$ to the resulting vector iteratively), we get:

$$\mathbf{A}^k\mathbf{v} = \sum_{i=1}^{m} \lambda_i^k (\mathbf{v}_i^T \mathbf{v})\mathbf{v}_i.$$

In the limit of k,

$$\frac{\mathbf{A}^k\mathbf{v}}{\lambda_1^k(\mathbf{v}_1^T\mathbf{v})} = \sum_{i=1}^{m}\left(\frac{\lambda_i}{\lambda_1}\right)^k \frac{\mathbf{v}_i^T\mathbf{v}}{\mathbf{v}_1^T\mathbf{v}}\mathbf{v}_i \longrightarrow \mathbf{v}_1 \quad \text{as } k \to \infty.$$

This implies that iterating the following process (multiplying $\mathbf{A}$ and then normalizing the resulting vector), the normalized vector converges to the principal eigenvector:

$$\frac{\mathbf{A}^k\mathbf{v}}{\sqrt{\|\mathbf{A}^k\mathbf{v}\|^2}} \longrightarrow \mathbf{v}_1 \quad \text{as } k \to \infty.$$

This observation leads to the power method:

1. Choose a random vector $\mathbf{v}$ and set $\mathbf{v}^{(0)} = \mathbf{v}$ and $t = 0$.
2. Compute $\mathbf{v}^{(t+1)} = \dfrac{\mathbf{A}\mathbf{v}^{(t)}}{\sqrt{\|\mathbf{A}\mathbf{v}^{(t)}\|^2}}$ and increase t by 1.
3. Iterate Step 2 until converged, e.g., $\|\mathbf{v}^{(t+1)} - \mathbf{v}^{(t)}\|^2 < \epsilon = 10^{-5}$.

The power method requires multiple (say k) matrix-vector multiplications, each having the complexity of n^2 multiplications. Hence, the complexity of the power method is still on the order of n^2, as long as the number k of iterations is not so large relative to n (this is often the case in practice). This is much smaller than the complexity n^3 of the eigenvalue decomposition, especially when n is very large. Due to this computational benefit, the power method is widely employed as an efficient algorithm for finding the principal eigenvector in many applications.

Python implementation of the spectral algorithm We implement the spectral algorithm via Python. We first generate the community memberships of n users.

```
from scipy.stats import bernoulli
import numpy as np

n = 8 # number of users
Bern = bernoulli(0.5)
# Generate n community memberships
x = Bern.rvs(n)
print(x)
```

```
[0 0 0 0 1 0 1 0]
```

We then construct random pairwise measurements.

```
# Construct the codebook
X = np.zeros((n,n))
for i in range(len(x)):
    for j in range(i,len(x)):
        # Compute xij = xi + xj (modulo 2)
        X[i,j] = (x[i]+x[j]) % 2
        # Symmetric component
        X[j,i] = X[i,j]
print(X)
```

```
[[0. 0. 0. 0. 1. 0. 1. 0.]
 [0. 0. 0. 0. 1. 0. 1. 0.]
 [0. 0. 0. 0. 1. 0. 1. 0.]
 [0. 0. 0. 0. 1. 0. 1. 0.]
 [1. 1. 1. 1. 0. 1. 0. 1.]
 [0. 0. 0. 0. 1. 0. 1. 0.]
 [1. 1. 1. 1. 0. 1. 0. 1.]
 [0. 0. 0. 0. 1. 0. 1. 0.]]
```

Next we compute the adjacency matrix.

```
# observation probability
p = 0.8
obs_bern = bernoulli(p)
```

```
# Construct an n-by-n mask matrix:
# entry = 1 (observed); 0 (otherwise)
mask_matrix = obs_bern.rvs((n,n))

# Construct the adjacency matrix
A = (1-2*X)*mask_matrix

print(1-2*X)
print(mask_matrix)
print(A)
```

```
[[ 1.   1.   1.   1.  -1.   1.  -1.   1.]
 [ 1.   1.   1.   1.  -1.   1.  -1.   1.]
 [ 1.   1.   1.   1.  -1.   1.  -1.   1.]
 [ 1.   1.   1.   1.  -1.   1.  -1.   1.]
 [-1.  -1.  -1.  -1.   1.  -1.   1.  -1.]
 [ 1.   1.   1.   1.  -1.   1.  -1.   1.]
 [-1.  -1.  -1.  -1.   1.  -1.   1.  -1.]
 [ 1.   1.   1.   1.  -1.   1.  -1.   1.]]
[[1 1 0 0 1 0 1 1]
 [1 1 1 0 0 1 1 1]
 [1 1 0 1 1 1 1 1]
 [1 1 1 1 0 1 1 1]
 [1 1 1 1 1 1 1 1]
 [1 1 1 1 1 0 1 1]
 [1 1 1 1 1 1 1 1]
 [1 0 1 1 1 1 1 1]]
[[ 1.   1.   0.   0.  -1.   0.  -1.   1.]
 [ 1.   1.   1.   0.  -0.   1.  -1.   1.]
 [ 1.   1.   0.   1.  -1.   1.  -1.   1.]
 [ 1.   1.   1.   1.  -0.   1.  -1.   1.]
 [-1.  -1.  -1.  -1.   1.  -1.   1.  -1.]
 [ 1.   1.   1.   1.  -1.   0.  -1.   1.]
 [-1.  -1.  -1.  -1.   1.  -1.   1.  -1.]
 [ 1.   0.   1.   1.  -1.   1.  -1.   1.]]
```

We run the power method to compute the principal eigenvector of **A**.

```
def power_method(A, eps=1e-5):
    # A computationally efficient algorithm
    # for finding the principal eigenvector
    # Choose a random vector
    v = np.random.randn(n)
    # normalization
    v = v/np.linalg.norm(v)

    prev_v = np.zeros(len(v))
```

```
    t = 0
    while np.linalg.norm(prev_v-v) > eps:
        prev_v = v
        v = np.array(np.dot(A,v)).reshape(-1)
        v = v/np.linalg.norm(v)
        t += 1
    print("Terminated after %s iterations"%t)
    return v

v1 = power_method(A)
print(v1)
print(np.sign(v1))
print(1-2*x)
```

```
Terminated after 8 iterations
[ 0.25389707  0.29847768  0.35720613  0.34957476 -0.40941936
  0.35720737 -0.40941936  0.36579104]
[ 1.  1.  1.  1. -1.  1. -1.  1.]
[ 1  1  1  1 -1  1 -1  1]
```

In the above experiment, the thresholded principal eigenvector np.sign(v1) coincides with the ground truth community vector 1-2*x.

Python: Performance of the spectral algorithm We will demonstrate via Python experiments that the principal eigenvector is getting closer to the ground truth of the community memberships as p increases. Consider a practical scenario in which n is large, say $n = 4000$. To measure the similarity between the principal eigenvector and the community vector, we employ a well-known correlation measure, called the Pearson correlation (Freedman *et al.*, 2007):

$$\rho_{X,Y} := \frac{\sigma_{X,Y}}{\sigma_X \sigma_Y} \tag{3.25}$$

where $\sigma_{X,Y} = \mathbb{E}[XY] - \mathbb{E}[X]\mathbb{E}[Y]$, $\sigma_X = \sqrt{\mathbb{E}[X^2] - (\mathbb{E}[X])^2}$, and $\sigma_Y = \sqrt{\mathbb{E}[Y^2] - (\mathbb{E}[Y])^2}$. To compute the Person correlation, we use a built-in function pearsonr defined in the scipy.stats module.

```
from scipy.stats import bernoulli
from scipy.stats import pearsonr
import numpy as np
import matplotlib.pyplot as plt

n = 4000 # number of users
Bern = bernoulli(0.5)
# Generate n community memberships
x = Bern.rvs(n)
```

```
# Construct the codebook
X = np.zeros((n,n))
for i in range(len(x)):
    for j in range(i,len(x)):
        # Compute xij = xi + xj (modulo 2)
        X[i,j] = (x[i]+x[j]) % 2
        # Symmetric component
        X[j,i] = X[i,j]

p = np.linspace(0.0003,0.0025,30)
limit = np.log(n)/n
p_norm = p/limit

def power_method(A, eps=1e-5):
    # A computationally efficient algorithm
    # for finding the principal eigenvector
    # Choose a random vector
    v = np.random.randn(n)
    # normalization
    v = v/np.linalg.norm(v)

    prev_v = np.zeros(len(v))
    t = 0
    while np.linalg.norm(prev_v-v) > eps:
        prev_v = v
        v = np.array(np.dot(A,v)).reshape(-1)
        v = v/np.linalg.norm(v)
        t += 1
    print("Terminated after %s iterations"%t)
    return v

corr = np.zeros_like(p)

for i,val in enumerate(p):
    obs_bern = bernoulli(val)
    # Construct an n-by-n mask matrix:
    # entry = 1 (observed); 0 (otherwise)
    mask_matrix = obs_bern.rvs((n,n))

    # Construct the adjacency matrix
    A = (1-2*X)*mask_matrix
    # Power method
    v1 = power_method(A)
    # Threshold the principal eigenvector
    v1 = np.sign(v1)
    # Compute the ground truth
```

```
    ground_truth = 1-2*x
    # Compute Pearson correlation
    corr[i] = np.abs(pearsonr(ground_truth,v1)[0])
    print(p_norm[i], corr[i])

plt.figure(figsize=(5,5), dpi=200)
plt.plot(p_norm, corr)
plt.title('Pearson correlation btw estimate and ground truth')
plt.grid(linestyle=':', linewidth=0.5)
plt.show()
```

Notice in Fig. 3.11 that the thresholded principal eigenvector is getting closer to the ground-truth community vector (or its flip version) with an increase in p, reflected in the high Pearson correlation for a large p. Especially when p is around the limit $p^* = \frac{\ln n}{n}$, the Pearson correlation is very close to 1, demonstrating that the spectral algorithm achieves almost the optimal performance promised by the ML decoding rule. This is sort of a heuristic argument. In order to give a precise argument, we should actually rely upon the empirical error rate (instead of the Pearson correlation) computed over sufficiently many random realizations of community vectors. For computational simplicity, we employ instead the Pearson correlation which can be reliably computed only with one random trial per each p.

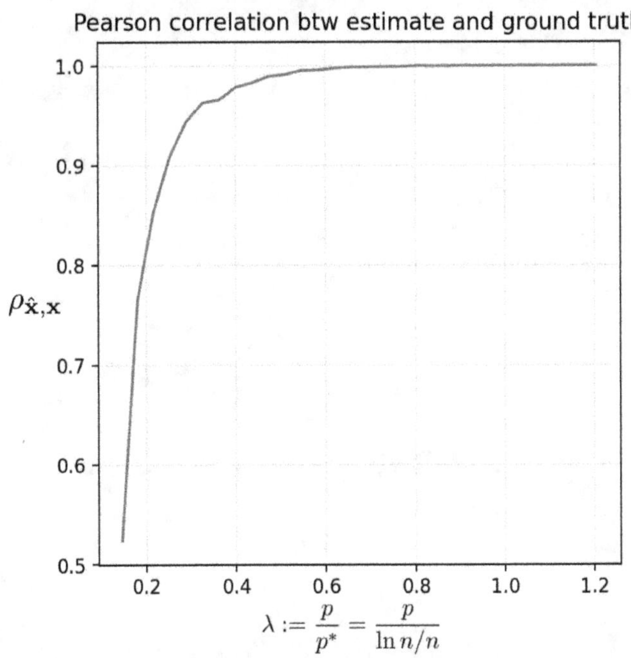

Figure 3.11. Pearson correlation between the thresholded principal eigenvector and the ground-truth community vector as a function of $\lambda := \frac{p}{p^*} = \frac{p}{\ln n/n}$.

Look ahead We have delved into one particular data science application of information theory, namely community detection, and highlighted the concept of phase transition, as well as confirming its occurrence through Python simulations. Moving forward, the next section will focus on a closely related application to community detection in the field of computational biology, known as *Haplotype phasing*. We will explore the nature of this problem and how it is connected to community detection.

Problem Set 7

Prob 7.1 (Basics on bounds and combinatorics)

(*a*) Let $p \geq 0$. Prove that

$$1 - p \leq e^{-p}.$$

Also specify the condition under which the equality holds.

(*b*) Show that for non-negative integers n and k ($k \leq n$):

$$\binom{n}{k} \leq e^{k \ln n}.$$

(*c*) Show that for integers $n \geq 0$:

$$\sum_{k=0}^{n} \binom{n}{k} = 2^n.$$

Prob 7.2 (The concept of reliable community detection) Suppose there are n users clustered into two communities. Let $x_i \in \{0, 1\}$ indicate a membership of community with regard to user $i \in \{1, 2, \ldots, n\}$. We are given part of the pairwise measurements:

$$y_{ij} = \begin{cases} x_i \oplus x_j, & \text{w.p. } p; \\ \mathsf{e}, & \text{w.p. } 1 - p, \end{cases}$$

for every pair $(i, j) \in \{(1, 2), (1, 3), \ldots, (1, n), (2, 3), \ldots, (n - 1, n)\}$ and $p \in [0, 1]$. Assume that y_{ij}'s are independent over (i, j). Given y_{ij}'s, one wishes to decode the community membership vector $\boldsymbol{x} := [x_1, x_2, \ldots, x_n]$ or its flipped counterpart $\boldsymbol{x} \oplus \mathbf{1} := [x_1 \oplus 1, x_2 \oplus 1, \ldots, x_n \oplus 1]$. Let $\hat{\boldsymbol{x}}$ be an estimate. Define the probability of error as:

$$P_e = \mathbb{P}(\hat{\boldsymbol{x}} \notin \{\boldsymbol{x}, \boldsymbol{x} \oplus \mathbf{1}\}).$$

(*a*) Let *sample complexity* be the number of pairwise measurements which are not erased. Show that

$$\frac{\text{sample complexity}}{\binom{n}{2}} \longrightarrow p \quad \text{as } n \to \infty.$$

(*b*) Consider the following optimization problem. Given p and n,

$$P_e^*(p, n) := \min_{\text{algorithm}} P_e.$$

State the definition of *reliable detection*. Also state the definition of *minimum sample complexity* using the concept of reliable detection.

(*c*) Consider a slightly different optimization problem. Given p,

$$P_e^*(p) := \min_{\text{algorithm},n} P_e.$$

The distinction here is that n is a design parameter. For $\epsilon > 0$, what are $P_e^*(\frac{\ln n}{n} + \epsilon)$ and $P_e^*(\frac{\ln n}{n} - \epsilon)$? Also explain why.

Prob 7.3 (An upper bound) Let $p = \lambda \frac{\ln n}{n}$ where $\lambda > 1$. Show that

$$\sum_{k=1}^{\frac{n}{2}} \binom{n}{k} (1-p)^{k(n-k)} \longrightarrow 0 \qquad \text{as } n \to \infty.$$

Hint: Use the bounds in Prob 7.1.

Prob 7.4 (Applying Fano's inequality to community detection) Consider an instance of community detection in which the goal is to figure out the community membership of each user between community 0 and community 1. Let $\boldsymbol{x} = [x_1, x_2, \ldots, x_n]$ be a collection of community memberships of n users: $x_i \in \{0, 1\}, 1 \leq i \leq n$. We are given part of pairwise measurements. With probability p, we observe $x_{ij} := x_i \oplus x_j$ independently over all pairs (i, j) where $i < j$:

$$y_{ij} = \begin{cases} x_{ij}, & \text{w.p. } p; \\ \mathsf{e}, & \text{otherwise.} \end{cases}$$

Let $\mathbf{X} = [x_{ij}] \in \mathbb{F}_2^{n \times n}$ and $\mathbf{Y} = [y_{ij}] \in \mathbb{F}_2^{n \times n}$.

(*a*) Suppose $n = 3$. Compute $H(\mathbf{X})$ and $H(\mathbf{Y})$.

(*b*) Consider the following upper bound:

$$H(\mathbf{Y}) \leq \sum_{i<j} H(y_{ij}).$$

Is this bound tight? If not, derive a tighter upper bound.

(*c*) Using Fano's inequality and data processing inequality, derive the following necessary condition for reliable detection:

$$p \geq \frac{2}{n-1}.$$

Does the result in part (*b*) lead to a tighter necessary condition? If so, derive the necessary condition.

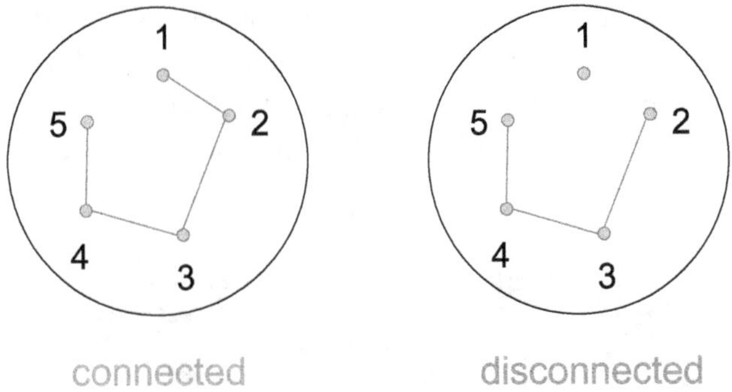

Figure 3.12. Illustration of graph connectivity and disconnectivity.

Prob 7.5 (Erdős-Rényi random graph) Consider a random graph $\mathcal{G}$ that two mathematicians (named Paul Erdős and Alfréd Rényi) introduced in (Erdős *et al.*, 1960). Hence, the graph is called the Erdős-Rényi graph. The graph contains n nodes and assumes that an edge appears w.p. $p \in [0, 1]$ for any pair of two nodes in an independent manner. We say that a graph is *connected* if for any two nodes, there exists a path (i.e., a sequence of edges) that connects one node to the other; otherwise, it is said to be *disconnected*. See Fig. 3.12 for examples.

(*a*) Show that

$$\mathbb{P}(\mathcal{G} \text{ is disconnected}) \leq \sum_{k=1}^{n-1} \binom{n}{k} (1-p)^{k(n-k)}.$$

Hint: Think about a necessary event for disconnectivity.

(*b*) Show that if $p > \frac{\ln n}{n}$, $\mathbb{P}(\mathcal{G} \text{ is disconnected}) \to 0$ as $n \to \infty$.
Hint: Recall the achievability proof for community detection that we did in Section 3.2.

(*c*) It has been shown that if $p < \frac{\ln n}{n}$, $\mathbb{P}(\mathcal{G} \text{ is disconnected}) \to 1$ as $n \to \infty$. This together with the result in part (*a*) implies that the sharp threshold on p for graph connectivity is the same as the one for community detection:

$$p^* = \frac{\ln n}{n}.$$

Relate this to the fundamental limit on observation probability in community detection, i.e., explain why the limits are same.

Prob 7.6 (The coupon collector problem) There are n different coupons. Suppose that for any cracker, the probability that the cracker contains a particular coupon among the n coupons is $\frac{1}{n}$, i.e., the n kinds of coupons are uniformly distributed over the entire crackers that are being sold.

(*a*) Suppose that Alice has $k(< n)$ distinct coupons. When Alice buys a new cracker, what is the probability that the new cracker contains a new coupon (i.e., being different from the k coupons that Alice possesses)? Let X_k be the number of crackers that Alice needs to buy to acquire a new coupon. Show that

$$\mathbb{P}(X_k = m) = \left(\frac{k}{n}\right)^{m-1} \frac{n-k}{n},$$

$$\mathbb{E}[X_k] = \frac{n}{n-k}.$$

(*b*) Suppose Bob has no coupon. Let $K := \sum_{k=0}^{n-1} X_k$ indicate the number of crackers that Bob needs to buy to collect all the coupons. Show that

$$\mathbb{E}[K] = n\left(1 + \frac{1}{2} + \frac{1}{3} + \cdots + \frac{1}{n}\right).$$

(*c*) Using the fact that $\sum_{k=1}^{n} \frac{1}{k} \approx \ln n$ in the limit of n (check Euler-Maclaurin formula in wikepidia), argue that in the limit of n,

$$\mathbb{E}[K] \approx n \ln n. \tag{3.26}$$

Note: This is order-wise the same as the minimal sample complexity $\frac{n \ln n}{2}$ required for reliable community detection.

(*d*) Using Python, plot $\mathbb{E}[K]$ and $n \ln n$ in the same figure for a proper range of n, say $1 \leq n \leq 10,000$.

Prob 7.7 (Converse for community detection) Suppose there are n users clustered into two communities. Let $x_i \in \{0, 1\}$ indicates a membership of community with regard to user $i \in \{1, 2, \ldots, n\}$. Let $\mathbf{X}$ be a codeword matrix whose (i,j)-th entry is $x_{ij} = x_i \oplus x_j$. Assume that we are given part of the comparison pairs:

$$y_{ij} = \begin{cases} x_{ij}, & \text{w.p. } p; \\ \mathsf{e}, & \text{o.w.} \end{cases}$$

for $(i,j) \in \{(1,2), (1,3), \cdots, (n-1, n)\}$ and $p \in [0, 1]$. We also assume that y_{ij}'s are independent over (i,j). Let $\mathbf{Y}$ be a received signal matrix whose (i,j)-th entry is y_{ij}. Given $\mathbf{Y}$, one wishes to decode the community membership vector $\boldsymbol{x} := [x_1, x_2, \ldots, x_n]$ or $\boldsymbol{x} \oplus \mathbf{1} := [x_1 \oplus 1, x_2 \oplus 1, \ldots, x_n \oplus 1]$. Let $\hat{\boldsymbol{x}}$ be an estimate. Define the probability of error as

$$P_e = \mathbb{P}(\hat{\boldsymbol{x}} \notin \{\boldsymbol{x}, \boldsymbol{x} \oplus \mathbf{1}\}).$$

(*a*) Show that

$$H(\boldsymbol{x}|\hat{\boldsymbol{x}}) \leq 1 + nP_e.$$

(*b*) Show that

$$I(\boldsymbol{x};\hat{\boldsymbol{x}}) \leq I(\mathbf{X};\mathbf{Y}).$$

(*c*) Assume x_i's are i.i.d. $\sim$ Bern$(\frac{1}{2})$. Using parts (*a*) and (*b*), derive a necessary condition on p under which P_e can be made arbitrarily close to 0 as $n \to \infty$.

Prob 7.8 (True or False?)

(*a*) Consider an instance of community detection with two communities. Let $\boldsymbol{x} := [x_1, x_2, \ldots, x_n]$ be the community membership vector in which $x_i \in \{0, 1\}$ and n denotes the total number of users. Suppose we are given part of the comparison pairs with observation probability p. In Section 3.1, we formulated an optimization problem which aims to minimize the probability of error defined as $P_e := \mathbb{P}\left(\hat{\boldsymbol{x}} \notin \{\boldsymbol{x}, \boldsymbol{x} \oplus \mathbf{1}\}\right)$. Given p and n, denote by $P_e^*(p, n)$ the minimum probability of error. In Section 3.2, we did not intend to derive the exact $P_e^*(p, n)$. Instead we developed a lower bound of $P_e^*(p, n)$ to demonstrate that for any $p > \frac{\ln n}{n}$, the probability of error can be made arbitrarily close to 0 as n tends to infinity.

(*b*) Consider an inference problem in which we wish to decode $X \in \mathcal{X}$ from $Y \in \mathcal{Y}$ where $\mathcal{X}$ and $\mathcal{Y}$ indicate the ranges of X and Y, respectively. Given $Y = y$, the optimal decoder is:

$$\hat{X} = \arg\max_{x \in \mathcal{X}} \mathbb{P}(Y = y | X = x).$$

3.5 DNA Sequencing: Fundamental Limits

Recap In the previous sections, we proved both the achievability and converse of the fundamental limit on the observation probability p needed to achieve reliable community detection:

$$p > \frac{\ln n}{n} \quad \Longleftrightarrow \quad P_e \to 0 \quad \text{as } n \to \infty.$$

Via Python simulation, we also observed phase transition on the limit $\frac{\ln n}{n}$. The estimated vector via the spectral algorithm indeed converges to the ground-truth community vector when p is close to $\frac{\ln n}{n}$.

Now we will move onto another data science application concerning phase transition. The application that we will focus on is w.r.t. computational biology. Specifically we will explore one of the important DNA sequencing problems, named *Haplotype phasing* (Browning and Browning, 2011; Das and Vikalo, 2015; Chen *et al.*, 2016a; Si *et al.*, 2014). Interestingly, it has a close connection with community detection.

Outline In this section, we will examine Haplotype phasing and explore its relationship to community detection. The section is divided into four parts. First, we will investigate two relevant keywords: DNA sequencing and Haplotype. We will then figure out what Haplotype phasing is. Next, drawing upon computational biology expertise, we will establish a link to community detection. Finally, we will examine the sharp threshold present in this problem, as in community detection.

Our 23 pairs of chromosome We will begin by discussing two important terms: (1) DNA sequence and (2) Haplotype. Our body consists of numerous cells, and each cell has a vital component known as the nucleus. The nucleus contains 23 pairs of chromosomes. In each pair, one comes from the mother (maternal chromosome), while the other comes from the father (paternal chromosome). Fig. 3.13 illustrates the 23 pairs of chromosomes. Each chromosome in a pair consists of a series of elements known as bases, and each base can take on one of four letters: A, C, T, or G. This series of bases is referred to as a DNA sequence, with a typical length of around 3 billion.

Looking inside a pair of chromosome, we see an interesting sequence pattern. The maternal sequence is almost identical to the paternal counterpart, $\sim$ 99.9% being identical. See Fig. 3.14. Differences occur only at certain positions. Such differences are called "Single Nucleotide Polymorphisms (SNPs)", being pronounced as "snips". It is well known that knowing the SNPs patterns is useful for personalized medicine. It can help predicting the probability of a certain cancer occurring. It

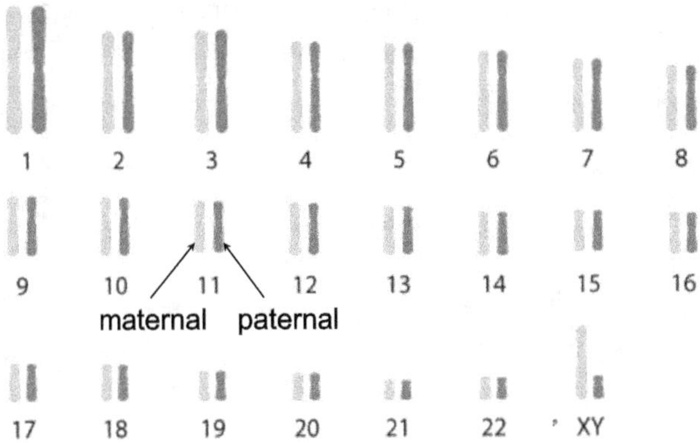

source: epigenetics tutorial of Umich BioSocial Collaborative

Figure 3.13. 23 pairs of chromosome.

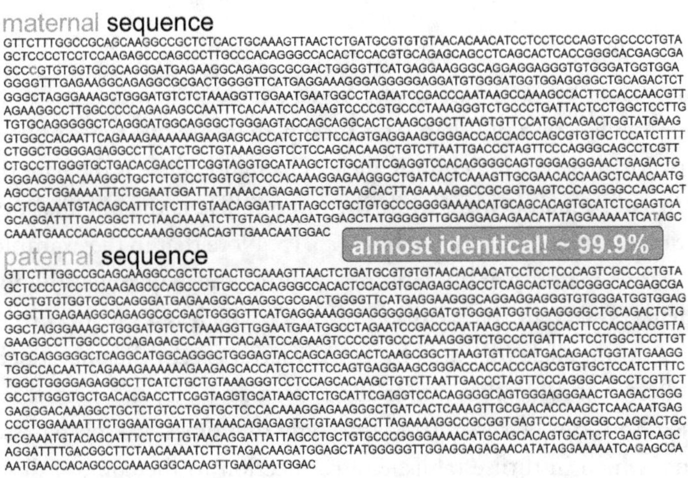

Figure 3.14. The maternal sequence is almost (99.9%) identical to the paternal sequence.

determines somatic mutations such as HIV. It also serves to understand phylogetic trees, exhibiting relationships between a variety of distinct species. The second keyword "Haplotype" refers to a pair of the two sequences of SNPs.

Haplotype phasing Haplotype phasing is the process of identifying the pair of two SNPs, which involves two sub-tasks: (1) identifying the locations of the SNPs, and (2) decoding the sequence pattern. The locations of SNPs are typically determined using "SNP calling" (Nielsen *et al.*, 2011). Therefore, Haplotype phasing usually refers to the second task: identifying the pattern of the SNPs. Each element in the pattern takes a value from the set of four letters $\{A, C, T, G\}$. You may be wondering how this is related to the community detection problem, where each component in the community membership vector is binary.

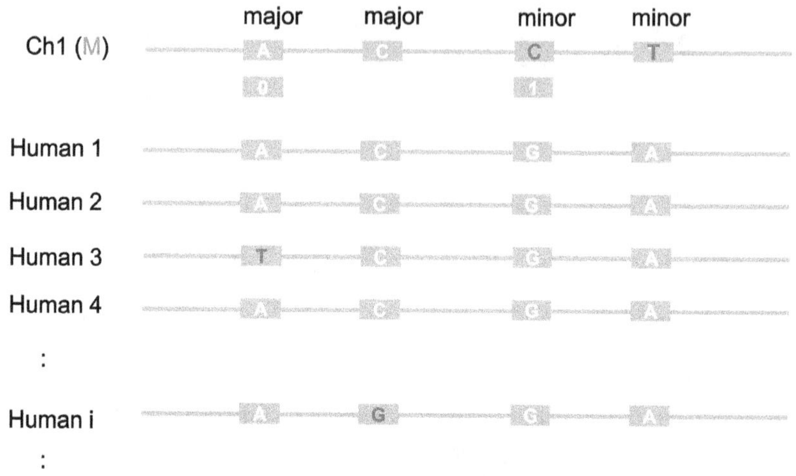

Figure 3.15. Major vs. minor allele.

Binary representation of major and minor allele In fact, a key property of the values taken by each base in the SNPs enables a concrete connection with the community detection problem. Specifically, there are two types of base components: (1) major allele; and (2) minor allele. The major allele is the base that occurs for a majority of human beings, while the minor allele is the one that occurs for a minor portion of humans, as shown in Fig. 3.15. Note that in the example, the first SNP reads A for a majority of humans, representing the major allele (denoted by 0), while Human 3's SNP in that position reads T, which is a rare occurrence and is classified as a minor allele (denoted by 1). Any letter except the one associated with the major allele is considered a minor allele, such as T, C, and G in this example. Because each SNP can be categorized into only two types, we can represent it as a binary value, which in turn establishes a connection to the community detection problem.

Two types of SNP positions Have we made the connection between the two problems? Not quite yet. Although there are similarities between Haplotype phasing and community detection, there is still a fundamental difference: Haplotype phasing involves decoding two sequences (vectors), whereas community detection involves decoding only one vector. However, it is possible to cast the Haplotype phasing problem into a problem of decoding only one vector (up to a global shift). To understand this, we need to consider another property of SNP positions: each position is of only two types – "heterozygous" or "homozygous". A heterozygous position refers to a position where the maternal base is a complement of the paternal base, while a homozygous position refers to a position where the maternal and paternal bases are the same. Using a standard method, one can determine the type of all SNP positions. Assuming that all SNP positions are heterozygous simplifies the problem and makes it identical to the community detection problem. Let $\boldsymbol{x}$ be the sequence of SNPs from a mother. Then, father's sequence would be its flipped version $\boldsymbol{x} \oplus \mathbf{1}$. The goal of Haplotype phasing is to decode $\boldsymbol{x}$ or $\boldsymbol{x} \oplus \mathbf{1}$.

Mate-pair read (Browning and Browning, 2011; Das and Vikalo, 2015) To establish the connection, we need to examine the type of information we have access to for Haplotype phasing. The information we can access is related to the current sequencing technology, which relies on a technique called "shotgun sequencing". This technique yields short fragments of the entire DNA sequence, known as "reads". The length of a typical read is between 100 and 500 bases, while SNPs consist of around 3 million bases, and the entire DNA sequence is around 3 billion bases in length. As a result, the average distance between SNPs is approximately 1000 bases, but the read length is much shorter than this distance. This implies that one read (spanning $100 \sim 200$ bases) usually contains only one SNP. This presents a challenge: we do not know which chromosome each read comes from (either maternal or paternal). To see this, let y_i denote the ith SNP contained in a read. Then, what we obtain is:

$$y_i = \begin{cases} x_i \text{ (mother's)}, & \text{w.p. } \frac{1}{2}; \\ x_i \oplus 1 \text{ (father's)}, & \text{w.p. } \frac{1}{2}. \end{cases}$$

Since the probabilities of getting x_i and $x_i \oplus 1$ are all equal to $\frac{1}{2}$, there is no way to figure out x_i from y_i.

To overcome this challenge, a more advanced sequencing technique has been developed which allows for simultaneous reading of two fragments, known as "mate-pair reads". As shown in Fig. 3.16, mate-pair reads have proved to be useful. One advantage of the sequencing technology is that both reads come from the same individual, which means that the information we obtain is:

$$(y_i, y_j) = \begin{cases} (x_i, x_j) \text{ (mother's)}, & \text{w.p. } \frac{1}{2}; \\ (x_i \oplus 1, x_j \oplus 1) \text{ (father's)}, & \text{w.p. } \frac{1}{2}. \end{cases}$$

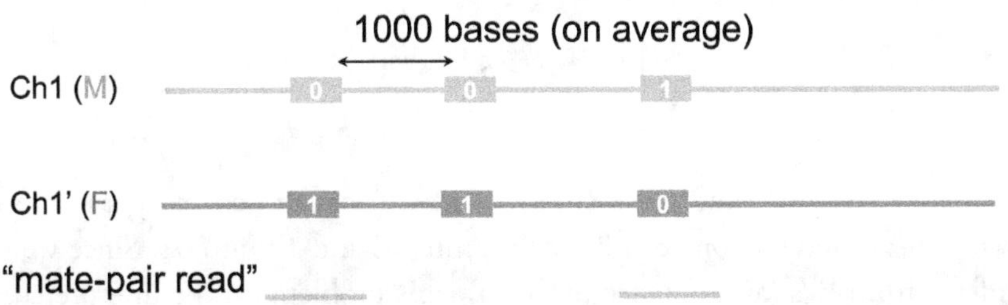

Figure 3.16. Mate-pair read.

Connection to community detection What we know for sure from (y_i, y_j) is their *parity*: $y_i \oplus y_j = x_i \oplus x_j$. Note that this is exactly the pairwise measurement given for community detection. In reality, however, what we get is a *noisy* version of (y_i, y_j):

$$(y_i, y_j) = \begin{cases} (x_i \oplus z_i, x_j \oplus z_j) \text{ (mother's)}, & \text{w.p. } \frac{1}{2}; \\ (x_i \oplus 1 \oplus z_i, x_j \oplus 1 \oplus z_j) \text{ (father's)}, & \text{w.p. } \frac{1}{2} \end{cases}$$

where z_i indicates an additive noise induced at the ith SNP. As per extensive experiments, it is found that z_i's can be modeled as i.i.d., each being according to say $\mathsf{Bern}(q)$ (meaning that the noise statistics are identical and independent across all SNPs). The parity is

$$y_{ij} := y_i \oplus y_j = x_i \oplus x_j \oplus z_i \oplus z_j.$$

Let $z_{ij} := z_i \oplus z_j$. Then, its statistics would be $z_{ij} \sim \mathsf{Bern}(2q(1-q))$. Why? Using the total probability law, we get:

$$\begin{aligned} \mathbb{P}(z_{ij} = 1) &= \mathbb{P}(z_i = 1)\mathbb{P}(z_j = 0|z_i = 1) + \mathbb{P}(z_i = 0)\mathbb{P}(z_j = 1|z_i = 0) \\ &= q(1-q) + (1-q)q \\ &= 2q(1-q) \end{aligned}$$

where the second equality is due to the independence of z_i and z_j. Denoting $\theta = 2q(1-q)$, we see that the measurement y_{ij} is a noisy version of x_{ij}. One may wonder if looking at the parity only (instead of individual measurements (y_i, y_j)) suffices to decode $\boldsymbol{x}$. In other words, is y_{ij} is a sufficient statistic? It is indeed the case. Check in Prob 8.5(c). This suggests that we do not lose any information loss although we consider only parities.

Translation into a communication problem The connection as above was made in (Chen *et al.*, 2016a). The authors in (Chen *et al.*, 2016a) applied the connection into a partial and random measurement model where the parity is observed with probability p, independently from others:

$$y_{ij} = \begin{cases} x_i \oplus x_j \oplus z_{ij}, & \text{w.p. } p; \\ \mathsf{e}, & \text{w.p. } 1-p \end{cases}$$

where $z_{ij} \sim \mathsf{Bern}(\theta)$ and $\theta \in (0, \frac{1}{2})$. Without loss of generality, assume that $0 \le \theta < \frac{1}{2}$; otherwise, one can flip all 0's into 1's and 1's into 0's. Since we wish to infer $\boldsymbol{x}$ from y_{ij}'s (an inference problem), this problem can be interpreted as a communication problem illustrated in Fig. 3.17.

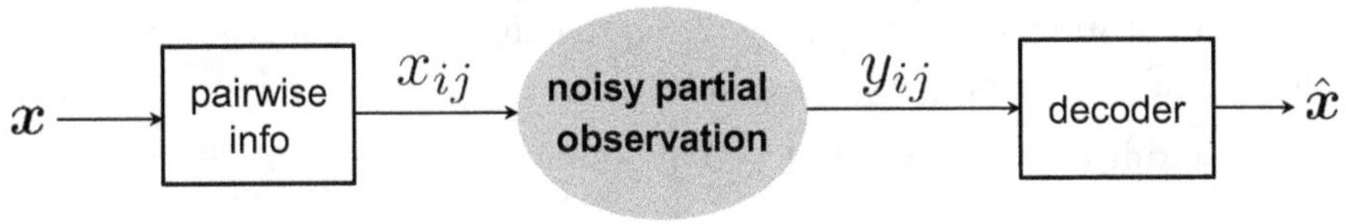

Figure 3.17. Translation of Haplotype phasing into a communication problem under a noisy channel with partial observations.

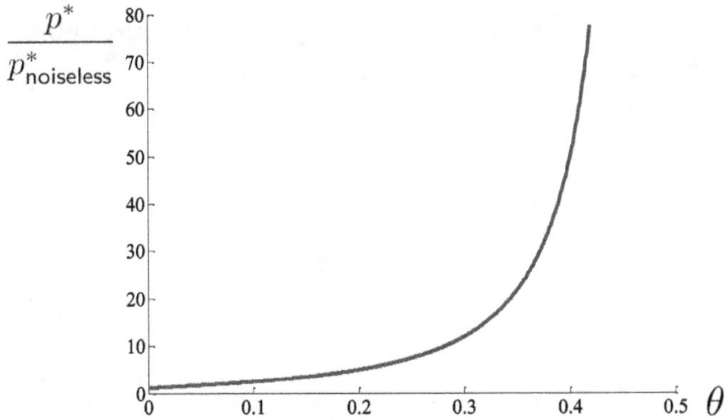

Figure 3.18. The effect of noise upon the limit.

The fundamental limit (Chen *et al.*, 2016a) The above model subsumes the noiseless scenario $\theta = 0$ as a special case. One can easily expect that the larger θ, the larger the limit p^*. It is shown that the fundamental tradeoff behaves like:

$$p^* = \frac{\ln n}{n} \cdot \frac{1}{1 - e^{-\mathsf{KL}(0.5\|\theta)}} \tag{3.27}$$

where $\mathsf{KL}(\cdot\|\cdot)$ denotes the Kullback-Leibler (KL) divergence defined w.r.t. a natural logarithm:

$$\begin{aligned}\mathsf{KL}(0.5\|\theta) &= 0.5 \ln \frac{0.5}{\theta} + 0.5 \ln \frac{0.5}{1-\theta} \\ &= 0.5 \ln \frac{1}{4\theta(1-\theta)}.\end{aligned}$$

Plugging this into (3.27), we get:

$$p^* = \frac{\ln n}{n} \cdot \frac{1}{1 - \sqrt{4\theta(1-\theta)}}. \tag{3.28}$$

Indeed the limit is an increasing function of θ. It grows exponentially with θ. See Fig. 3.18.

The only distinction in the noisy setting relative to the noiseless counterpart is that we have a factor $\frac{1}{1-e^{-\mathsf{KL}(0.5\|\theta)}}$ in (3.27), reflecting the noise effect. In the noiseless setting $\theta = 0$, $\mathsf{KL}(0.5\|\theta) = \infty$, so it reduces to the limit $\frac{\ln n}{n}$.

Look ahead We related community detection to one of applications in computational biology: Haplotype phasing. We showed that Haplotype phasing is a noisy version of the community detection problem, and claimed that phase transition occurs on observation probability like: $p^* = \frac{\ln n}{n} \cdot \frac{1}{1-e^{-\mathsf{KL}(0.5\|\theta)}}$. In the next section, we will prove the achievability of the claimed limit.

3.6 DNA Sequencing: Achievability Proof

Recap In the previous section, we made a connection between Haplotype phasing and community detection. We showed that Haplotype phasing is a noisy version of community detection, wherein the goal is to decode $\boldsymbol{x} = [x_1, \ldots, x_n]$ from y_{ij}'s:

$$y_{ij} = \begin{cases} x_i \oplus x_j \oplus z_{ij}, & \text{w.p. } p; \\ \mathsf{e}, & \text{w.p. } 1-p \end{cases}$$

where $z_{ij} \sim \mathsf{Bern}(\theta)$ and $\theta \in (0, \frac{1}{2})$. We then claimed that as in community detection, phase transition occurs on observation probability:

$$p^* = \frac{\ln n}{n} \cdot \frac{1}{1 - e^{-\mathsf{KL}(0.5\|\theta)}}$$

where $\mathsf{KL}(0.5\|\theta)$ denotes the KL divergence between $\mathsf{Bern}(0.5)$ and $\mathsf{Bern}(\theta)$ defined w.r.t. the natural logarithm:

$$\mathsf{KL}(0.5\|\theta) := 0.5 \ln \frac{0.5}{\theta} + 0.5 \ln \frac{0.5}{1-\theta} = 0.5 \log \frac{1}{4\theta(1-\theta)}.$$

Outline In this section, we will demonstrate that the limit is achievable, and we will do so in three steps. First, we will derive the optimal ML decoder. Next, we will analyze the error probability. Although the overall procedure of the proof is similar to that in the noiseless case, there are a few key differences that require the use of important bounding techniques, which we will detail. Finally, by applying these bounding techniques to the error probability, we will prove the achievability.

The optimal ML decoder As in the noiseless case, we employ the same deterministic encoder which yields a codeword matrix $\mathbf{X}(\boldsymbol{x})$ with the (i,j) entry $x_{ij} = x_i \oplus x_j$. A distinction arises in the decoder side. The optimal decoder takes the ML decision rule. But the ML decoder is not the one based solely on the concept of compatibility vs. incompatibility which formed the basis of the noiseless case. To see this, consider the following example in which $n = 4$, the ground-truth vector $\boldsymbol{x} = \mathbf{0}$, an observation matrix $\mathbf{Y}$ reads:

$$\mathbf{Y} = \begin{bmatrix} & 1 & \mathsf{e} & \mathsf{e} \\ & & 0 & 0 \\ & & & 1 \\ & & & \end{bmatrix}. \tag{3.29}$$

A key observation is that not all the observed components match the corresponding x_{ij}'s. In this example, noises are added to the $(1,2)$ and $(3,4)$ entries, yielding $(y_{12}, y_{34}) = (1,1)$. This makes the calculation of $\mathbb{P}(\mathbf{Y}|\mathbf{X}(\boldsymbol{x}))$ a bit more involved, relative to the noiseless case. The likelihood function is not solely determined by erasure patterns but is also influenced by flipping error patterns, which creates a distinction between the noiseless and noisy cases. In the absence of noise, the likelihood takes on either a value of 0 or a specific non-zero value. However, in the presence of noise, the likelihood can take on multiple non-zero values, depending on the flipping error patterns. To illustrate this, consider the following examples. Given **Y** in (3.29), the likelihoods are:

$$\mathbb{P}(\mathbf{Y}|\mathbf{X}(0000)) = (1-p)^2 p^4 \cdot \theta^2 (1-\theta)^2;$$
$$\mathbb{P}(\mathbf{Y}|\mathbf{X}(1000)) = (1-p)^2 p^4 \cdot \theta^1 (1-\theta)^3;$$
$$\mathbb{P}(\mathbf{Y}|\mathbf{X}(0100)) = (1-p)^2 p^4 \cdot \theta^3 (1-\theta)^1$$

where

$$\mathbf{X}(1000) = \begin{bmatrix} & 1 & 1 & 1 \\ & & 0 & 0 \\ & & & 0 \\ & & & \end{bmatrix}, \quad \mathbf{X}(0100) = \begin{bmatrix} & 1 & 0 & 0 \\ & & 1 & 1 \\ & & & 0 \\ & & & \end{bmatrix}.$$

In all likelihoods, the first product term $(1-p)^2 p^4$ is common. So the second term associated with a flipping error pattern will decide the ML solution. The numbers marked in red indicate the numbers of flips. Since the flipping probability θ is assumed to be less than $\frac{1}{2}$, the smallest number of flips would maximize the likehlihood, thus yielding:

$$\hat{\boldsymbol{x}}_{\mathsf{ML}} = \arg\min_{\boldsymbol{x}} d(\mathbf{X}(\boldsymbol{x}), \mathbf{Y}) \tag{3.30}$$

where $d(\cdot,\cdot)$ denotes the Hamming distance: the number of distinct bits between the two arguments (Hamming, 1950).

A setup for the analysis of the error probability For the achievability proof, we analyze the probability of error. Taking the same procedures as in the noiseless case, we obtain:

$$\begin{aligned} P_e &:= \mathbb{P}(\hat{\boldsymbol{x}}_{\mathsf{ML}} \notin \{\boldsymbol{x}, \boldsymbol{x} \oplus \mathbf{1}\}) \\ &= \mathbb{P}(\hat{\boldsymbol{x}}_{\mathsf{ML}} \notin \{\mathbf{0}, \mathbf{1}\} | \boldsymbol{x} = \mathbf{0}) \\ &\le \sum_{\boldsymbol{a} \notin \{\mathbf{0},\mathbf{1}\}} \mathbb{P}(\hat{\boldsymbol{x}}_{\mathsf{ML}} = \boldsymbol{a} | \boldsymbol{x} = \mathbf{0}) \\ &= \sum_{k=1}^{n-1} \sum_{\boldsymbol{a}_k \in \mathcal{A}_k} \mathbb{P}(\hat{\boldsymbol{x}}_{\mathsf{ML}} = \boldsymbol{a}_k | \boldsymbol{x} = \mathbf{0}) \\ &= \sum_{k=1}^{n-1} \binom{n}{k} \mathbb{P}(\hat{\boldsymbol{x}}_{\mathsf{ML}} = \boldsymbol{a}_k | \boldsymbol{x} = \mathbf{0}) \end{aligned} \tag{3.31}$$

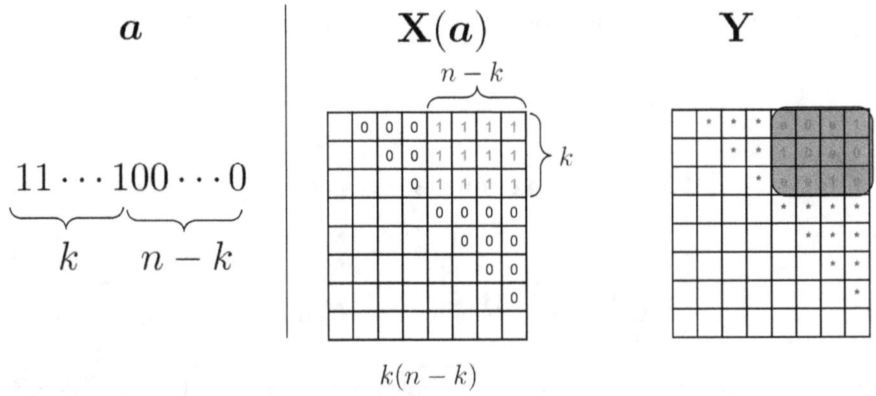

Figure 3.19. Illustration of the distinguishable positions colored in purple. Unlike the noiseless case, all that matters in a decision is the Hamming distance. So the error event implies that the number of 1's is greater than or equal to the number of 0's in the $k(n-k)$ distinguishable positions.

where the first equality is by symmetry (each error probability does not depend on the input vector pattern); and the inequality comes from the union bound. In the second last equality, we take an equivalent yet more insightful expression, by introducing a set $\mathcal{A}_k := \{\boldsymbol{a} : \|\boldsymbol{a}\|_1 = k\}$.

A bound on $\mathbb{P}(\hat{x}_{\mathsf{ML}} = a_k | x = 0)$ Focus on $\mathbb{P}(\hat{\boldsymbol{x}}_{\mathsf{ML}} = \boldsymbol{a}_k | \boldsymbol{x} = \mathbf{0})$. Unlike the noiseless case, deriving its upper bound is not that straightforward. To see this, consider a case in which $\boldsymbol{a}_k = (1\cdots10\cdots0) \in \mathcal{A}_k$. Notice that $\mathbf{X}(\boldsymbol{a}_k)$ takes 1's only in the last $n-k$ positions of the first k rows (that we called *distinguishable positions* in light of $\mathbf{X}(\mathbf{0})$). See the middle in Fig. 3.19.

In the noiseless case, the error event $\{\hat{\boldsymbol{x}}_{\mathsf{ML}} = \boldsymbol{a}_k | \boldsymbol{x} = \mathbf{0}\}$ must imply that such positions are *all erased*, since otherwise $\mathbf{X}(\boldsymbol{a}_k)$ is *incompatible*. In the noisy case, on the other hand, the error event does not necessarily imply that the distinguishable positions are all erased, since $\mathbf{X}(\boldsymbol{a}_k)$ could still be a candidate for the solution even if a few observations are made in such positions. As indicated in (3.30), what matters in a decision is the Hamming distance. As long as $d(\mathbf{X}(\boldsymbol{a}_k), \mathbf{Y}) \le d(\mathbf{X}(\mathbf{0}), \mathbf{Y})$, the codeword $\mathbf{X}(\boldsymbol{a}_k)$ can be chosen as a solution, no matter what the number of erasures is in those positions. Hence, the error event only suggests that the number of 1's is greater than or equal to the number of 0's in the $k(n-k)$ distinguishable positions. Let M be the number of observations made in the distinguishable positions. Then, we get: for $\boldsymbol{a} \in \mathcal{A}_k$,

$$
\begin{aligned}
&\mathbb{P}(\hat{\boldsymbol{x}}_{\mathsf{ML}} = \boldsymbol{a}_k | \boldsymbol{x} = \mathbf{0}) \\
&\le \mathbb{P}(d(\mathbf{X}(\boldsymbol{a}_k), \mathbf{Y}) \le d(\mathbf{X}(\mathbf{0}), \mathbf{Y}) | \boldsymbol{x} = \mathbf{0})
\end{aligned}
$$

$$= \mathbb{P}(\#\ 1\text{'s} \geq \#\ 0\text{'s in the distinguishable positions}|\boldsymbol{x} = \mathbf{0})$$

$$\stackrel{(a)}{=} \sum_{\ell=0}^{k(n-k)} \mathbb{P}(M = \ell|\boldsymbol{x} = \mathbf{0}) \times \mathbb{P}(\#\ 1\text{'s} \geq \#\ 0\text{'s in distinguishable positions}|\boldsymbol{x} = \mathbf{0}, M = \ell) \tag{3.32}$$

where (a) is due to the total probability law. Here $\mathbb{P}(M = \ell|\boldsymbol{x} = \mathbf{0}) = \binom{k(n-k)}{\ell}p^{\ell}(1-p)^{k(n-k)-\ell}$.

Consider $\mathbb{P}(\#\ 1\text{'s} \geq \#\ 0\text{'s in distinguishable positions}|\boldsymbol{x} = \mathbf{0}, M = \ell)$. Let Z_i be the measured value at the ith observed entry in the $k(n-k)$ distinguishable positions. Then, Z_i's are i.i.d. $\sim \mathsf{Bern}(\theta)$ where $i \in \{1, 2, \ldots, \ell\}$. Using this, we get:

$$\mathbb{P}(\#\ 1\text{'s} \geq \#\ 0\text{'s in distinguishable positions}|\boldsymbol{x} = \mathbf{0}, M = \ell) = \mathbb{P}\left(\frac{Z_1 + Z_2 + \cdots + Z_\ell}{\ell} \geq 0.5\right). \tag{3.33}$$

Since Z_i's are i.i.d. $\sim \mathsf{Bern}(\theta)$, one may expect that the empirical mean of Z_i's would be concentrated around the true mean θ as ℓ increases. It is indeed the case and it can be proved via the WLLN. Also by our assumption, $\theta < 0.5$. Hence, the probability $\mathbb{P}(\frac{Z_1+Z_2+\cdots+Z_\ell}{\ell} \geq 0.5)$ would converge to zero, as ℓ tends to infinity. What we are interested in here is how fast the probability converges to zero. There is a very well-known concentration bound which characterizes a convergence behavior of the probability. That is, the *Chernoff bound* (Bertsekas and Tsitsiklis, 2008; Gallager, 2013), formally stated below:

$$\mathbb{P}\left(\frac{Z_1 + Z_2 + \cdots + Z_\ell}{\ell} \geq 0.5\right) \leq e^{-\ell \mathsf{KL}(0.5\|\theta)}. \tag{3.34}$$

Check Prob 4.4 for the proof.

Applying this into (3.32), we get:

$$\mathbb{P}(\hat{\boldsymbol{x}}_{\mathsf{ML}} = \boldsymbol{a}_k|\boldsymbol{x} = \mathbf{0}) \leq \sum_{\ell=0}^{k(n-k)} \mathbb{P}(M = \ell|\boldsymbol{x} = \mathbf{0}) \times \mathbb{P}(\#\ 1\text{'s} \geq \#\ 0\text{'s in distinguishable positions}|\boldsymbol{x} = \mathbf{0}, M = \ell)$$

$$\begin{aligned}
&\overset{(a)}{\leq} \sum_{\ell=0}^{k(n-k)} \binom{k(n-k)}{\ell} p^{\ell}(1-p)^{k(n-k)-\ell} e^{-\ell \mathsf{KL}(0.5\|\theta)} \\
&= (1-p)^{k(n-k)} \sum_{\ell=0}^{k(n-k)} \binom{k(n-k)}{\ell} \left(\frac{pe^{-\mathsf{KL}(0.5\|\theta)}}{1-p} \right)^{\ell} \\
&\overset{(b)}{=} (1-p)^{k(n-k)} \left(1 + \frac{pe^{-\mathsf{KL}(0.5\|\theta)}}{1-p} \right)^{k(n-k)} \\
&= (1 - p(1 - e^{-\mathsf{KL}(0.5\|\theta)}))^{k(n-k)}
\end{aligned}$$

where (a) comes from $\mathbb{P}(M = \ell | \boldsymbol{x} = \mathbf{0}) = \binom{k(n-k)}{\ell} p^{\ell}(1-p)^{k(n-k)-\ell}$ and the Chernoff bound (3.34); and (b) is due to the binomial theorem: $\sum_{k=0}^{n} \binom{n}{k} x^{n-k} y^k = (x+y)^n$.

The final step of the achievability proof Putting the above to (3.31), we get:

$$\begin{aligned}
P_e &\leq \sum_{k=1}^{n-1} \binom{n}{k} \mathbb{P}(\hat{\boldsymbol{x}}_{\mathsf{ML}} = \boldsymbol{a}_k | \boldsymbol{x} = \mathbf{0}) \\
&\leq \sum_{k=1}^{n-1} \binom{n}{k} (1 - p(1 - e^{-\mathsf{KL}(0.5\|\theta)}))^{k(n-k)}.
\end{aligned}$$

Remember what we proved in the noiseless case (check the precise statement in Prob 7.3):

$$\sum_{k=1}^{n-1} \binom{n}{k} (1-q)^{k(n-k)} \longrightarrow 0 \quad \text{if } q > \frac{\ln n}{n}.$$

Hence, by replacing q with $p(1 - e^{-\mathsf{KL}(0.5\|\theta)})$ in the above, one can make P_e arbitrarily close to 0, provided that

$$p(1 - e^{-\mathsf{KL}(0.5\|\theta)}) > \frac{\ln n}{n} \quad \Longleftrightarrow \quad p > \frac{\ln n}{n} \cdot \frac{1}{1 - e^{-\mathsf{KL}(0.5\|\theta)}}.$$

This completes the achievability proof.

Look ahead We have proved the achievability of the limit in the noisy observation model $p^* = \frac{\ln n}{n} \cdot \frac{1}{1-e^{-\mathsf{KL}(0.5\|\theta)}}$. In the next section, we will prove the converse.

3.7 DNA Sequencing: Converse Proof

Recap We proved the achievability of the limit on observation probability p in a noisy community detection problem:

$$p > \frac{\ln n}{n} \cdot \frac{1}{1 - e^{-\mathsf{KL}(0.5\|\theta)}} \Longrightarrow P_e \to 0 \quad \text{as } n \to \infty.$$

Outline In this section, we will prove the converse:

$$p < \frac{\ln n}{n} \cdot \frac{1}{1 - e^{-\mathsf{K}(0.5\|\theta)}} \Longrightarrow P_e \nrightarrow 0.$$

Proof strategy In the noiseless case, the converse proof is based on graph connectivity:

$$\text{graph is connected} \Longleftrightarrow P_e \to 0.$$

Hence, we focused on checking whether $\mathbb{P}$(connected) converges to 1 depending on conditions of observation probability. In the noisy case, however, checking graph connectivity is not sufficient because the event of graph being connected does not necessarily imply reliable detection:

$$\text{graph is connected} \overset{\text{X}}{\Longrightarrow} P_e \to 0.$$

So we will start from scratch. Starting with the definition of the probability of error, we get:

$$\begin{aligned}
P_e &:= 1 - \mathbb{P}(\text{success}) \\
&\overset{(a)}{=} 1 - \mathbb{P}\left(\{\hat{\boldsymbol{x}} = \mathbf{0}\} \cup \{\hat{\boldsymbol{x}} = \mathbf{1}\} | \boldsymbol{x} = \mathbf{0}\right) \\
&\overset{(b)}{=} 1 - \mathbb{P}(\hat{\boldsymbol{x}} = \mathbf{0} | \boldsymbol{x} = \mathbf{0}) - \mathbb{P}(\hat{\boldsymbol{x}} = \mathbf{1} | \boldsymbol{x} = \mathbf{0}) \\
&\overset{(c)}{=} 1 - 2\mathbb{P}(\hat{\boldsymbol{x}} = \mathbf{0} | \boldsymbol{x} = \mathbf{0})
\end{aligned}$$

where (a) is by symmetry; (b) follows from the fact that the two events are disjoint; and (c) is due to the fact that $\{\hat{\boldsymbol{x}} = \mathbf{0}\}$ and $\{\hat{\boldsymbol{x}} = \mathbf{1}\}$ are equally likely (there is no way to disambiguate $\boldsymbol{x}$ and $\boldsymbol{x} \oplus \mathbf{1}$ from pairwise comparisons; the only way that we can do in this case is to flip a fair coin). Hence, it suffices to show that

$$p < \frac{\ln n}{n} \cdot \frac{1}{1 - e^{-D^*}} \quad \Longrightarrow \quad \mathbb{P}(\hat{\boldsymbol{x}} = \mathbf{0} | \boldsymbol{x} = \mathbf{0}) \to 0 \quad \text{as } n \to \infty. \qquad (3.35)$$

Here we define $D^* := \mathsf{KL}(0.5\|\theta)$ for notational simplicity.

An upper bound on $\mathbb{P}(\hat{x} = 0|x = 0)$ In the converse proof, one cannot make any assumption on the decoder type. This is because we wish to come up with a necessary condition that holds under any arbitrary decoder. However, there is one exceptional case where one can make an assumption. That is the case in which the *optimal* decoder is employed. Notice that a necessary condition w.r.t. the optimal decoder also holds for any other decoder. Let $P_e(\text{opt})$ and $P_e(\text{a decoder})$ be error probabilities w.r.t. the optimal decoder and a particular decoder, respectively. Then, by the definition of optimality:

$$P_e(\text{a decoder}) \geq P_e(\text{opt}).$$

We see that $P_e(\text{opt}) \nrightarrow 0$ implies $P_e(\text{a decoder}) \nrightarrow 0$. Hence, it suffices to show that $P_e \nrightarrow 0$ under the optimal decoder. This allows us to assume the use of the optimal decoder:

$$\hat{\boldsymbol{x}} = \arg\min_{\boldsymbol{x}} d(\mathbf{X}(\boldsymbol{x}), \mathbf{Y})$$

where $d(\cdot, \cdot)$ indicates the Hamming distance. In the noisy observation model, the optimal decoder minimizes the Hamming distance; see (3.30). For notational simplicity, define $d(\boldsymbol{x}) := d(\mathbf{X}(\boldsymbol{x}), \mathbf{Y})$.

The interested event $\{\hat{\boldsymbol{x}} = \mathbf{0}\}$ implies that $\{d(\boldsymbol{a}) \geq d(\mathbf{0})\}$, $\forall \boldsymbol{a} \neq \mathbf{0}$. This together with the fact that $\mathbb{P}(A \cap B) \leq \mathbb{P}(A)$ for any two events (A, B) gives:

$$\begin{aligned}\mathbb{P}(\hat{\boldsymbol{x}} = \mathbf{0}|\boldsymbol{x} = \mathbf{0}) &\leq \mathbb{P}\left(\bigcap_{\boldsymbol{a}\neq\mathbf{0}}\{d(\boldsymbol{a}) \geq d(\mathbf{0})\}|\boldsymbol{x} = \mathbf{0}\right) \\ &\leq \mathbb{P}\left(\bigcap_{\boldsymbol{a}\in\mathcal{A}_1}\{d(\boldsymbol{a}) \geq d(\mathbf{0})\}|\boldsymbol{x} = \mathbf{0}\right)\end{aligned} \tag{3.36}$$

where $\mathcal{A}_1 = \{\boldsymbol{a} : \|\boldsymbol{a}\|_1 = 1, a_i \in \{0, 1\}\}$.

Remember the two key observations that we made in the noiseless case: (1) the calculation of the above probability can be greatly simplified when the associated events are *independent*; and (2) the more independent events are, the tighter bound we get. This again motivates us to search for independent events (among n events) as many as possible.

As in the noiseless case, the number of independent events is the same as the number of locally disconnected nodes. To see this, refer to Fig. 3.20.

Like the noiseless case, consider an event $\mathcal{T}$ in which there are at least $\lfloor \frac{\alpha n}{\ln n} \rfloor$ nodes which are locally disconnected, i.e., $y_{ij} = \mathsf{e}$ for $i, j \in \{1, 2, \ldots, \lfloor \frac{\alpha n}{\ln n} \rfloor\}$:

$$\mathcal{T} := \left\{L \geq \left\lfloor \frac{\alpha n}{\ln n} \right\rfloor\right\}$$

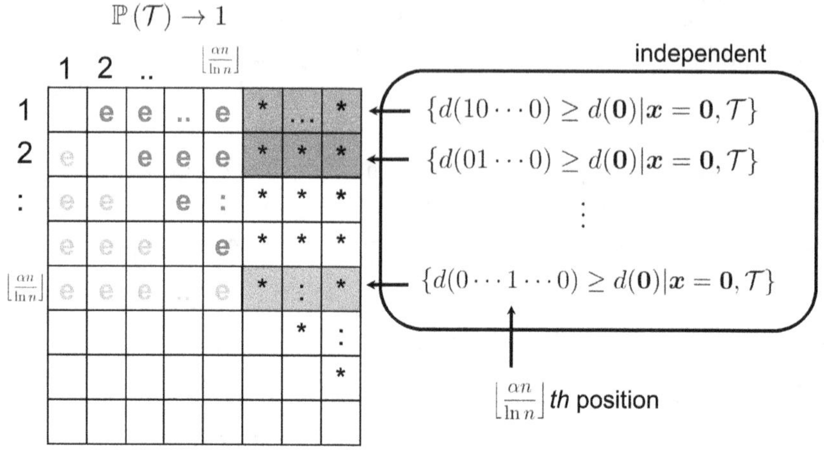

Figure 3.20. Independent events vs. locally disconnected nodes.

where L indicates the number of locally disconnected nodes. We assume that the locally disconnected nodes are numbered as $1, 2, \ldots, \lfloor \frac{\alpha n}{\ln n} \rfloor$.

Given $\mathcal{T}$, the event $\{d(10\cdots0) \geq d(\mathbf{0})|\boldsymbol{x} = \mathbf{0}\}$ is a sole function of y_{1j}'s (marked in light blue in the figure) where $j \in \{\lfloor \frac{\alpha n}{\ln n} \rfloor + 1, \ldots, n\}$. Notice that $(1, j)$'s indicate *distinguishable positions* where the corresponding codeword entries differ w.r.t. the messages $(10\cdots0)$ and $\mathbf{0}$. Hence, the difference between Hamming distances, $d(10\cdots0) - d(\mathbf{0})$, depends only on the positions. Similarly the event $\{d(010\cdots0) \geq d(\mathbf{0})|\boldsymbol{x} = \mathbf{0}\}$ is a sole function of y_{2j}'s (marked in purple). Hence, the two events are *independent* since y_{1j}'s are y_{2j}'s are disjoint. Extending this argument to other events, given $(\mathcal{T}, \boldsymbol{x} = \mathbf{0})$, we obtain:

$$\begin{aligned} &\{d(10\cdots0) \geq d(\mathbf{0})\} \perp \{d(010\cdots0) \geq d(\mathbf{0})\} \perp \cdots \\ &\perp \{d(0\cdots \underbrace{1}_{\lfloor \frac{\alpha n}{\ln n} \rfloor \text{th position}} \cdots 0) \geq d(\mathbf{0})\}. \end{aligned} \tag{3.37}$$

As was shown in Section 3.3, the event $\mathcal{T}$ occurs w.h.p. for $\alpha = \frac{1}{2} - \frac{\lambda}{4}$ where λ is the prefractor that appears in $p = \lambda \frac{\ln n}{n}$: $\mathbb{P}(\mathcal{T}) \to 1$ as $n \to \infty$. Applying this to (3.36), we get:

$$\begin{aligned} \mathbb{P}(\hat{\boldsymbol{x}} = \mathbf{0}|\boldsymbol{x} = \mathbf{0}) &\leq \mathbb{P}\left(\bigcap_{\boldsymbol{a} \in \mathcal{A}_1} \{d(\boldsymbol{a}) \geq d(\mathbf{0})\} | \boldsymbol{x} = \mathbf{0} \right) \\ &= \mathbb{P}\left(\left\{ \bigcap_{\boldsymbol{a} \in \mathcal{A}_1} \{d(\boldsymbol{a}) \geq d(\mathbf{0})\} \right\}, \mathcal{T} | \boldsymbol{x} = \mathbf{0} \right) \end{aligned}$$

$$
\begin{aligned}
&+\mathbb{P}\left(\left\{\bigcap_{a\in\mathcal{A}_1}\{d(a)\geq d(\mathbf{0})\}\right\},\mathcal{T}^c|x=\mathbf{0}\right)\\
&\overset{(a)}{\approx}\mathbb{P}\left(\left\{\bigcap_{a\in\mathcal{A}_1}\{d(a)\geq d(\mathbf{0})\}\right\}|x=\mathbf{0},\mathcal{T}\right)\\
&\overset{(b)}{\leq}\prod_{a\in\mathcal{B}_1}\mathbb{P}(d(a)\geq d(\mathbf{0})|x=\mathbf{0},\mathcal{T})\\
&\overset{(c)}{=}\mathbb{P}(d(10\cdots 0)\geq d(\mathbf{0})|x=\mathbf{0},\mathcal{T})^{\lfloor\frac{\alpha n}{\ln n}\rfloor}\\
&=\{1-\mathbb{P}(d(10\cdots 0)< d(\mathbf{0})|x=\mathbf{0},\mathcal{T})\}^{\lfloor\frac{\alpha n}{\ln n}\rfloor}
\end{aligned}
\tag{3.38}
$$

where (*a*) follows from $\mathbb{P}(\mathcal{T})\to 1$ and $\mathbb{P}(\mathcal{T}^c)\to 0$; (*b*) comes from (3.37) and the definition $\mathcal{B}_1 := \{b : \|b\|_1 = 1, b_i \in \{1,2,\ldots,\lfloor\frac{\alpha n}{\ln n}\rfloor\}\}$; and (*c*) is due to symmetry.

A lower bound on $\mathbb{P}(d(10\cdots 0)<d(0)|x=0,\mathcal{T})$ The last equation in (3.38) motivates us to explore a lower bound on $\mathbb{P}(d(10\cdots 0)<d(\mathbf{0})|x=\mathbf{0},\mathcal{T})$, as it yields an upper bound on $\mathbb{P}(\hat{x}=\mathbf{0}|x=\mathbf{0})$.

Given $\mathcal{T}$, only the last $\tilde{n} := n-\lfloor\frac{\alpha n}{\ln n}\rfloor$ positions in the first row of $\mathbf{Y}$ are *distinguishable* between messages $\mathbf{0}$ and $(10\cdots 0)$. The event $\{d(10\cdots 0)<d(\mathbf{0})\}$ depends solely on those positions, which in turn suggests that the number of 1's is greater than the number of 0's in the distinguishable positions. See Fig. 3.21.

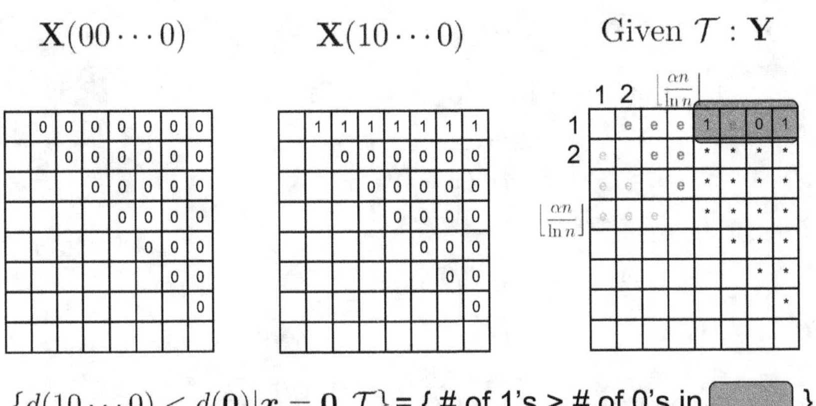

Figure 3.21. An equivalent condition.

Let M be the number of observations in the distinguishable positions. Then, we get:

$$
\begin{aligned}
&\mathbb{P}(d(10\cdots 0) < d(\mathbf{0})|\boldsymbol{x} = \mathbf{0}, \mathcal{T}) \\
&\qquad = \mathbb{P}(\#\text{ 1's} > \#\text{ 0's in the distinguishable positions}|\boldsymbol{x} = \mathbf{0}, \mathcal{T}) \\
&\qquad \overset{(a)}{=} \sum_{\ell=0}^{\tilde{n}} \mathbb{P}(M = \ell|\boldsymbol{x} = \mathbf{0}, \mathcal{T}) \\
&\qquad\qquad \times \mathbb{P}\left(\#\text{ 1's} > \#\text{ 0's in distinguishable positions}|\boldsymbol{x} = \mathbf{0}, \mathcal{T}, M = \ell\right) \\
&\qquad \overset{(b)}{=} \sum_{\ell=0}^{\tilde{n}} \binom{\tilde{n}}{\ell} p^{\ell}(1-p)^{\tilde{n}-\ell} \mathbb{P}\left(Z_1 + Z_2 + \cdots + Z_\ell > \frac{\ell}{2}\right).
\end{aligned}
\tag{3.39}
$$

where (a) is due to the total probability law; and (b) comes from $\mathbb{P}(M = \ell|\boldsymbol{x} = \mathbf{0}, \mathcal{T}) = \binom{\tilde{n}}{\ell} p^{\ell}(1-p)^{\tilde{n}-\ell}$. We define Z_i as the measured value at the ith observed entry in the $\tilde{n}$ distinguishable positions. Then, Z_i i.i.d. $\sim$ Bern(θ) where $i \in \{1, \ldots, \ell\}$.

The Chernoff bound (see Prob 4.4) provides an upper bound on $\mathbb{P}(Z_1 + Z_2 + \cdots + Z_\ell > \frac{\ell}{2})$. What we are interested in is a lower bound though. It turns out the interested probability has the same order of the upper bound $e^{-\ell D^*}$. More concretely, given $\epsilon > 0$, there exists n_0 such that for $\ell \geq n_0$:

$$
\mathbb{P}\left(Z_1 + Z_2 + \cdots + Z_\ell > \frac{\ell}{2}\right) \geq e^{-(1+\epsilon)\ell D^*}.
$$

Here $n_0 = \ln n$ is one such choice under which the above holds. Check this in Prob 8.2. This together with (3.39) gives:

$$
\begin{aligned}
&\mathbb{P}(d(10\cdots 0) < d(\mathbf{0})|\boldsymbol{x} = \mathbf{0}, \mathcal{T}) \\
&\qquad \geq \sum_{\ell=\ln n}^{\tilde{n}} \binom{\tilde{n}}{\ell} p^{\ell}(1-p)^{\tilde{n}-\ell} e^{-(1+\epsilon)\ell D^*} \\
&\qquad \overset{(a)}{\geq} (1-\delta_n) \sum_{\ell=0}^{\tilde{n}} \binom{\tilde{n}}{\ell} p^{\ell}(1-p)^{\tilde{n}-\ell} e^{-(1+\epsilon)\ell D^*} \\
&\qquad \overset{(b)}{\approx} (1-p)^{\tilde{n}} \sum_{\ell=0}^{\tilde{n}} \binom{\tilde{n}}{\ell} \left(\frac{pe^{-(1+\epsilon)D^*}}{1-p}\right)^{\ell}
\end{aligned}
$$

$$
\begin{aligned}
&\stackrel{(c)}{=} (1-p)^{\tilde{n}}\left(1+\frac{pe^{-(1+\epsilon)D^*}}{1-p}\right)^{\tilde{n}} \\
&= (1-p(1-e^{-(1+\epsilon)D^*}))^{\tilde{n}} \\
&\geq (1-p(1-e^{-(1+\epsilon)D^*}))^{n}. \qquad (3.40)
\end{aligned}
$$

The step (a) is due to the fact that the summation up to $\ln n$ is negligible relative to the entire sum. Precisely speaking, there exists $\delta_n \to 0$ such that the step (a) holds. Check this in Prob 8.2. The step (b) comes from $\delta_n \to 0$ and (c) is due to the binomial theorem.

The final step We are ready to complete the proof. Putting (3.40) to (3.38), we get:

$$
\begin{aligned}
&\mathbb{P}(\hat{\boldsymbol{x}} = \mathbf{0} | \boldsymbol{x} = \mathbf{0}) \\
&\lesssim \{1 - \mathbb{P}(d(10\cdots 0) < d(\mathbf{0}) | \boldsymbol{x} = \mathbf{0}, \mathcal{T})\}^{\lfloor \frac{\alpha n}{\ln n} \rfloor} \\
&\lesssim \{1 - (1-p(1-e^{-(1+\epsilon)D^*}))^{n}\}^{\frac{\alpha n}{\ln n}} \\
&\stackrel{(a)}{\leq} \exp\left\{-(1-p(1-e^{-(1+\epsilon)D^*}))^{n}\frac{\alpha n}{\ln n}\right\} \\
&\stackrel{(b)}{\approx} \exp\left\{-e^{-np(1-e^{-(1+\epsilon)D^*})}\frac{\alpha n}{\ln n}\right\} \\
&= \exp\left\{-\alpha e^{(1-\lambda(1-e^{-(1+\epsilon)D^*}))\ln n - \ln\ln n}\right\}
\end{aligned}
$$

where (a) is due to the fact that $1-x \leq e^{-x}$ for $x > 0$; and (b) comes from the fact that $(1-p(1-e^{-(1+\epsilon)D^*}))^n \approx e^{-p(1-e^{-(1+\epsilon)D^*})n}$ for sufficiently large n and small $p = \lambda\frac{\ln n}{n}$ (which is our case). Therefore, if $\lambda < \frac{1}{1-e^{-(1+\epsilon)D^*}}$, the upper bound goes to 0 as $n \to \infty$. This completes the proof (3.35).

Look ahead We have proven the limit on p for the noisy community detection:

$$
p > \frac{\ln n}{n}\cdot\frac{1}{1-e^{-(1+\epsilon)D^*}} \quad\Longleftrightarrow\quad P_e \to 0.
$$

Our achievability is based on the ML decoder which suffers from high computational complexity. In the next section, we will explore efficient algorithms that possibly yield the optimal performance as the ML rule.

3.8 DNA Sequencing: Algorithm and Python Implementation

Recap We have proved the achievability and converse of the limit in the noisy observation model (inspired by Haplotype phasing):

$$p^* = \frac{\ln n}{n} \cdot \frac{1}{1 - e^{-\mathsf{KL}(0.5\|\theta)}} = \frac{\ln n}{n} \cdot \frac{1}{1 - \sqrt{4\theta(1-\theta)}}$$

where θ is the flipping error rate. Since the optimal ML decoding rule comes with a challenge in computational complexity (as in the noiseless case), it is important to develop computationally efficient algorithms that achieve the optimal ML performance yet with much lower complexities. Even in the noisy setting, such efficient algorithms are already developed.

Outline In this section, we will investigate two efficient algorithms for Haplotype phasing in the presence of noise. The first method is the same as the one we used in the noiseless scenario, which is the spectral algorithm. The second algorithm is a slightly more complex version that involves obtaining an initial estimate through the spectral algorithm and refining it with an additional operation to improve performance (Chen *et al.*, 2016a). This second algorithm is not only still efficient but also optimal. However, as the focus of this book is not on optimality proofs, we will concentrate on explaining how the algorithms work and providing their Python implementations. This section consists of four parts. First, we will review the spectral algorithm. Second, we will apply it to the noisy observation setting and evaluate its performance through Python simulation. Third, we will describe how the second algorithm works. Finally, we will implement the second algorithm with the additional operation in Python to demonstrate that it outperforms the first spectral algorithm.

Review of the spectral algorithm The spectral algorithm is based on the adjacency matrix $\mathbf{A} \in \mathbb{R}^{n \times n}$ where each entry a_{ij} indicates whether users i and j are in the same community:

$$a_{ij} = \begin{cases} 1 - 2y_{ij}, & \text{w.p. } p; \\ 0, & \text{w.p. } 1 - p \ (y_{ij} = \mathsf{e}) \end{cases} \tag{3.41}$$

where $y_{ij} = x_i \oplus x_j \oplus z_{ij}$ and z_{ij}'s are i.i.d. $\sim \mathsf{Bern}(\theta)$. Here $a_{ij} = 0$ denotes no measurement. In the noiseless case, $a_{ij} = +1$ means that the two users are in the same community; and $a_{ij} = -1$ indicates the opposite.

Inspired by the fact that the rank of $\mathbf{A}$ is 1 under the ideal situation (the noiseless full measurement setting) and the principal eigenvector matches the ground-truth community vector (up to a global shift), the spectral algorithm computes the principal eigenvector to declare its thresholded version as an estimate. For computational efficiency, it employs the power method for computing the principal eigenvector (instead of eigenvalue decomposition). Here is the summary of how it works.

1. Construct the adjacency matrix $\mathbf{A}$ as per (3.41).
2. Choose a random vector $\mathbf{v}$ and set $\mathbf{v}^{(0)} = \mathbf{v}$ and $t = 0$.
3. $\mathbf{v}^{(t+1)} = \dfrac{\mathbf{A}\mathbf{v}^{(t)}}{\sqrt{\|\mathbf{A}\mathbf{v}^{(t)}\|^2}}$ and increase t by 1.
4. Iterate Step 3 until converged, e.g., $\|\mathbf{v}^{(t+1)} - \mathbf{v}^{(t)}\|^2 < \epsilon = 10^{-5}$.

Python implementation of the spectral algorithm We implement the spectral algorithm via Python. The code below is almost the same as in the noiseless setting. The only distinction is that y_{ij} is a noisy version of $x_i \oplus x_j$. We consider a setting where the flipping error rate $\theta = 0.1$ and $n = 4000$.

```
from scipy.stats import bernoulli
from scipy.stats import pearsonr
import numpy as np
import matplotlib.pyplot as plt

n = 4000 # number of users
Bern = bernoulli(0.5)
# Generate n community memberships
x = Bern.rvs(n)

# Construct the codebook
X = np.zeros((n,n))
for i in range(len(x)):
    for j in range(i,len(x)):
        # Compute xij = xi + xj (modulo 2)
        X[i,j] = (x[i]+x[j]) % 2
        # Symmetric component
        X[j,i] = X[i,j]

# Construct an observation matrix
theta = 0.1 # noise flipping error rate
noise_Bern = bernoulli(theta)
noise_matrix = noise_Bern.rvs((n,n))
Y = (X + noise_matrix)%2

p = np.linspace(0.001,0.0065,30)
limit = 1/(1-np.sqrt(4*theta*(1-theta)))*np.log(n)/n
p_norm = p/limit
```

```
def power_method(A, eps=1e-5):
    # A computationally efficient algorithm
    # for finding the principal eigenvector
    # Choose a random vector
    v = np.random.randn(n)
    # normalization
    v = v/np.linalg.norm(v)

    prev_v = np.zeros(len(v))
    t = 0
    while np.linalg.norm(prev_v-v) > eps:
        prev_v = v
        v = np.array(np.dot(A,v)).reshape(-1)
        v = v/np.linalg.norm(v)
        t += 1
    print("Terminated after %s iterations"%t)
    return v

corr = np.zeros_like(p)

for i,val in enumerate(p):
    obs_bern = bernoulli(val)
    # Construct an n-by-n mask matrix:
    # entry = 1 (observed); 0 (otherwise)
    mask_matrix = obs_bern.rvs((n,n))

    # Construct the adjacency matrix
    A = (1-2*Y)*mask_matrix
    # Power method
    v1 = power_method(A)
    # Threshold the principal eigenvector
    v1 = np.sign(v1)
    # Compute the ground truth
    ground_truth = 1-2*x
    # Compute Pearson correlation
    corr[i] = np.abs(pearsonr(ground_truth,v1)[0])
    print(p_norm[i], corr[i])

plt.figure(figsize=(5,5), dpi=200)
plt.plot(p_norm, corr)
plt.title('Pearson correlation btw estimate and ground truth')
plt.grid(linestyle=':', linewidth=0.5)
plt.show()
```

As depicted in Fig. 3.22, the Pearson correlation approaches 1 as p approaches the limit p^*. As previously noted, there exists an alternative method that surpasses the

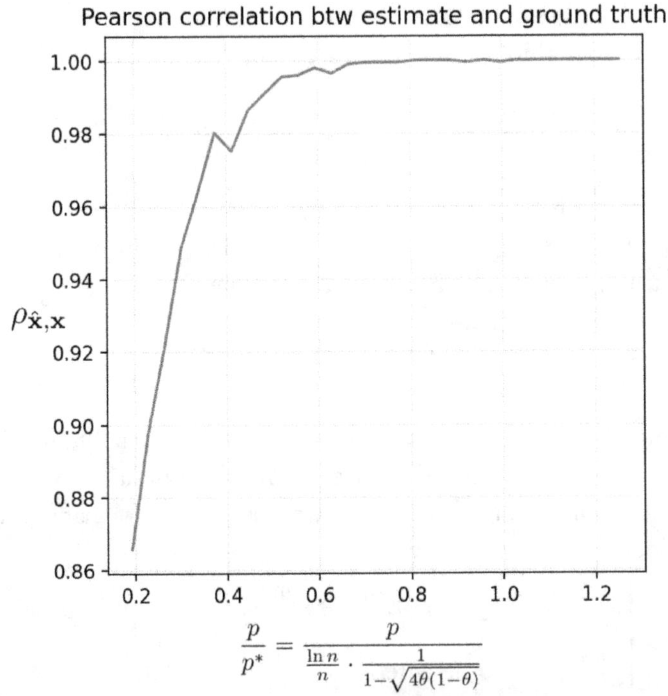

Figure 3.22. Pearson correlation between the ground-truth community vector and the estimate obtained from the spectral algorithm: $n = 4000$ and $\theta = 0.1$.

performance of the spectral algorithm. Let us now examine this algorithm in further detail.

Additional step: Local refinement The other algorithm takes two steps: (i) running the spectral algorithm; and then (2) performing an additional step called *local refinement*. The role of the second step is to detect any errors assuming that a majority of the components in the estimate vector are correct. The idea of local refinement is to use coordinate-wise maximum likelihood estimator (MLE). Here is how it works. Suppose we pick up an user, say user i. We then compute the coordinate-wise likelihood w.r.t. the membership value of user i. However, computing the coordinate-wise likelihood presents a challenge as it requires knowledge of the ground-truth memberships of other users, which are not revealed. As a surrogate, we employ an initial *estimate*, say $\boldsymbol{x}^{(0)} = [x_1^{(0)}, \ldots, x_n^{(0)}]$, obtained in the earlier step. Specifically, we compute:

$$x_i^{\mathsf{MLE}} = \arg \max_{a \in \{x^{(0)}, x^{(0)} \oplus 1\}} \mathbb{P}(\mathbf{Y} | x_1^{(0)}, \ldots, x_{i-1}^{(0)}, a, x_{i+1}^{(0)}, \ldots, x_n^{(0)}).$$

To find x_i^{MLE}, we only need to compare the two likelihood functions w.r.t. $a = x^{(0)}$ and $a = x^{(0)} \oplus 1$. It boils down to comparing the following two:

$$\sum_{j:(i,j)\in\Omega} y_{ij} \oplus x_i^{(0)} \oplus x_j^{(0)} \quad \text{vs} \quad \sum_{j:(i,j)\in\Omega} y_{ij} \oplus x_i^{(0)} \oplus 1 \oplus x_j^{(0)} \tag{3.42}$$

where Ω indicates the set of (i,j) pairs such that $y_{ij} \neq \mathsf{e}$. In order to gain an insights into the above two terms, consider an idealistic setting where $\boldsymbol{x}^{(0)}$ matches the ground truth. In this case, the two terms become:

$$\sum_{j:(i,j)\in\Omega} z_{ij} \quad \text{vs} \quad \sum_{j:(i,j)\in\Omega} z_{ij} \oplus 1. \tag{3.43}$$

Since the noise flipping error rate θ is assumed to be less than 0.5, the first term is likely to be smaller than the second. Hence, it would be reasonable to take the candidate that yields a smaller value among the two. This is exactly what the coordinate-wise MLE does:

$$x_i^{\mathsf{MLE}} = \begin{cases} x_i^{(0)}, & \sum_j y_{ij} \oplus x_i^{(0)} \oplus x_j^{(0)} < \sum_j y_{ij} \oplus x_i^{(0)} \oplus 1 \oplus x_j^{(0)}; \\ x_i^{(0)} \oplus 1, & \text{otherwise.} \end{cases}$$

We follow the same procedure for all other users, performing the local refinement step iteratively and step-by-step. This forms one iteration to yield $\boldsymbol{x}^{(1)} = [x_1^{\mathsf{MLE}}, \ldots, x_n^{\mathsf{MLE}}]$. It has been shown in (Chen *et al.*, 2016a) that with multiple iterations (around the order of $\ln n$ iterations), the coordinate-wise MLE converges to the ground-truth community vector. We will not provide a proof of this fact, but instead, we will present simulation results that demonstrate the improved performance offered by local refinement. Below is the code for local refinement:

```
# initial estimate obtained from the spectral algorithm
x0 = (1- v1)//2
xt = x0

ITER = 3
for t in range(1,ITER):
    xt1 = xt
    for i in range(len(xt)):
        # Likelihood w.r.t. x_i^(t)
        L1 = (Y[i,:] + xt[i] + xt)%2
        L1 = L1*mask_matrix[i,:]
        # Likelihood w.r.t. x_i^(t)+1
        L2 = (Y[i,:] + xt[i] + 1 + xt)%2
        L2 = L2*mask_matrix[i,:]
        xt1[i] = xt[i]*(sum(L1) <= sum(L2)) \
```

```
            + ((xt[i]+1)%2)*(sum(L1)>sum(L2))
    xt = xt1
```

Performances of the spectral algorithm vs local refinement We compare the Pearson correlations of the spectral algorithm and the two-step approach with local refinement, for a setting where $n = 4000$ and $\theta = 0.1$. We set the number of iterations in the local refinement step as 9, since it is close to the suggested number ($\ln 4000 = 8.294$). Here is a code for simulation.

```
from scipy.stats import bernoulli
from scipy.stats import pearsonr
import numpy as np
import matplotlib.pyplot as plt

n = 4000 # number of users
Bern = bernoulli(0.5)
# Generate n community memberships
x = Bern.rvs(n)

# Construct the codebook
X = np.zeros((n,n))
for i in range(len(x)):
    for j in range(i,len(x)):
        # Compute xij = xi + xj (modulo 2)
        X[i,j] = (x[i]+x[j]) % 2
        # Symmetric component
        X[j,i] = X[i,j]

# Construct an observation matrix
theta = 0.1 # noise flipping error rate
noise_Bern = bernoulli(theta)
noise_matrix = noise_Bern.rvs((n,n))
Y = (X + noise_matrix)%2

p = np.linspace(0.001,0.0065,30)
limit = 1/(1-np.sqrt(4*theta*(1-theta)))*np.log(n)/n
p_norm = p/limit


corr1 = np.zeros_like(p)
corr2 = np.zeros_like(p)

for i, val in enumerate(p):
    obs_bern = bernoulli(val)
    # Construct an n-by-n mask matrix:
```

```
    # entry = 1 (observed); 0 (otherwise)
    mask_matrix = obs_bern.rvs((n,n))

    ####################################
    ######## Spectral algorithm ######
    ####################################
    # Construct the adjacency matrix
    A = (1-2*Y)*mask_matrix
    # Power method
    v1 = power_method(A)
    # Threshold the principal eigenvector
    v1 = np.sign(v1)

    ####################################
    ######## Local refinement ########
    ####################################
    # initial estimate (from the spectral algorithm)
    x0 = (1- v1)//2
    xt = x0
    # number of iterations
    ITER = 9
    for t in range(1,ITER):
        xt1 = xt
        # coordinate-wise MLE
        for k in range(len(xt)):
            # Likelihood w.r.t. x_k^{(t)}
            L1 = (Y[k,:] + xt[k] + xt)%2
            L1 = L1*mask_matrix[k,:]
            # likelihood w.r.t. x_k^{(t)}+1
            L2 = (Y[k,:] + xt[k] + 1 + xt)%2
            L2 = L2*mask_matrix[k,:]
            xt1[k] = xt[k]*(sum(L1) <= sum(L2)) \
                  + ((xt[k]+1)%2)*(sum(L1)>sum(L2))
        xt = xt1

    # Compute the ground truth
    ground_truth = 1-2*x
    # Compute Pearson correlation
    corr1[i] = np.abs(pearsonr(ground_truth,v1)[0])
    corr2[i] = np.abs(pearsonr(ground_truth,1-2*xt)[0])
    print(p_norm[i], corr1[i], corr2[i])

plt.figure(figsize=(5,5), dpi=200)
plt.plot(p_norm, corr1, label='spectral algorithm' )
plt.plot(p_norm, corr2, label='local refinement')
plt.title('Pearson correlation btw estimate and ground truth')
```

```
plt.legend()
plt.grid(linestyle=':', linewidth=0.5)
plt.show()
```

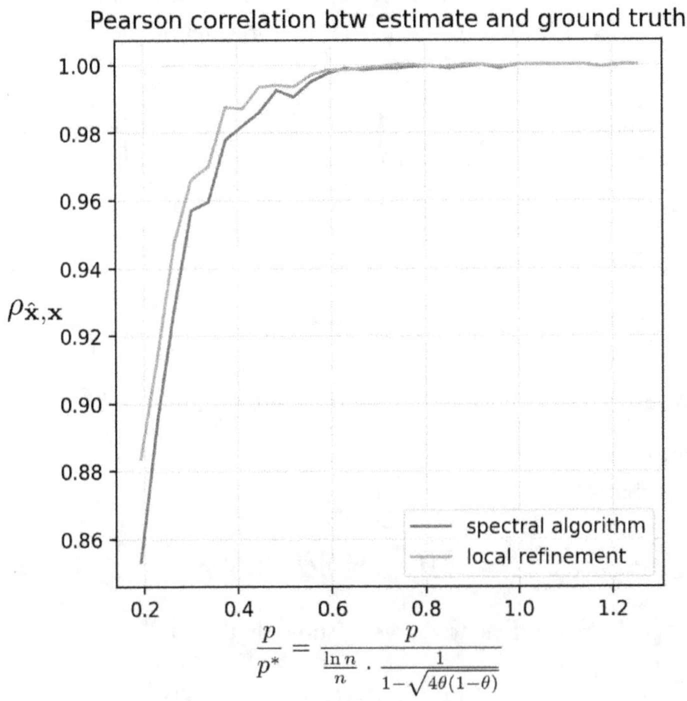

Figure 3.23. Pearson correlation performances of the spectral algorithm and local refinement: $n = 4000$ and $\theta = 0.1$.

Fig. 3.23 demonstrates that the inclusion of the local refinement step leads to an improvement in performance.

Look ahead We have investigated two data science applications that exhibit phase transitions. The subsequent section will delve into another data science application that involves phase transition.

Problem Set 8

Prob 8.1 (Reverse Chernoff bound) Suppose we observe n i.i.d. discrete random variables $Y^n := (Y_1, \ldots, Y_n)$. Consider two hypotheses $H_0 : Y_i \sim \mathbb{P}_0(y)$; and $H_1 : Y_i \sim \mathbb{P}_1(y)$ for $y \in \mathcal{Y}$ and $i \in \{1, \ldots, n\}$. Define the Chernoff information D^* as:

$$D^* = -\min_{0 \le \lambda \le 1} \ln \left\{ \sum_{y \in \mathcal{Y}} \mathbb{P}_0(y)^{\lambda} \mathbb{P}_1(y)^{1-\lambda} \right\}.$$

Let $\mathcal{L}_k$ be the likelihood function w.r.t. H_k:

$$\mathcal{L}_k = \mathbb{P}(Y^n | H_k).$$

Assume that $\mathbb{P}_0 \sim \mathsf{Bern}(\theta)$ and $\mathbb{P}_1 \sim \mathsf{Bern}(1-\theta)$ for a fixed $\theta \in (0, \frac{1}{2})$.

(*a*) Compute D^*.

(*b*) Show that

$$\mathbb{P}(\mathcal{L}_1 \ge \mathcal{L}_0 | H_0) \le e^{-nD^*}.$$

(*c*) Fix $\epsilon > 0$. For sufficiently large n, show that

$$\mathbb{P}(\mathcal{L}_1 \ge \mathcal{L}_0 | H_0) \ge e^{-(1+\epsilon)nD^*}.$$

Prob 8.2 (Useful bounds) Let $p = \lambda \frac{\ln n}{n}$ for some positive constant λ.

(*a*) Show that

$$\sum_{\ell=0}^{\ln n - 1} \binom{n}{\ell} p^{\ell} (1-p)^{n-\ell} \le (1-p)^n \ln n \left(\frac{np}{1-p} \right)^{\ln n}.$$

(*b*) Show that

$$\sum_{\ell=0}^{n} \binom{n}{\ell} p^{\ell} (1-p)^{n-\ell} = (1-p)^n \left(1 + \frac{p}{1-p} \right)^n.$$

(*c*) Show that there exists $\epsilon_n > 0$ such that $\epsilon_n \to 0$ as $n \to \infty$ and

$$\sum_{\ell=\ln n}^{n} \binom{n}{\ell} p^{\ell} (1-p)^{n-\ell} \ge (1-\epsilon_n)(1-p)^n \left(1 + \frac{p}{1-p} \right)^n.$$

Prob 8.3 (Generalized Chernoff bound) Suppose we observe n i.i.d. discrete random variables $Y^n := (Y_1, \ldots, Y_n)$. Consider two hypotheses $H_0 : Y_i \sim \mathbb{P}_0(y)$; and $H_1 : Y_i \sim \mathbb{P}_1(y)$ for $y \in \mathcal{Y}$. Define the Chernoff information as:

$$D^* := -\min_{0 \le \lambda \le 1} \ln \left\{ \sum_{y \in \mathcal{Y}} \mathbb{P}_0(y)^{\lambda} \mathbb{P}_1(y)^{1-\lambda} \right\}.$$

Let $\mathcal{L}_k$ be the likelihood function w.r.t. H_k: $\mathcal{L}_k = \mathbb{P}(Y^n | H_k)$.

(*a*) Show that for $a, b > 0$ and $\lambda \in [0, 1]$,

$$\min\{a, b\} \le a^{\lambda} b^{1-\lambda}.$$

(*b*) Let $A := \{y^n : \mathbb{P}(y^n | H_1) \ge \mathbb{P}(y^n | H_0)\}$. Show that

$$\mathbb{P}(\mathcal{L}_1 \ge \mathcal{L}_0 | H_0) = \sum_{y^n \in A} \prod_{i=1}^{n} \mathbb{P}_0(y_i).$$

(*c*) Using parts (*a*) and (*b*), show that

$$\mathbb{P}(\mathcal{L}_1 \ge \mathcal{L}_0 | H_0) \le e^{-nD^*}. \tag{3.44}$$

Prob 8.4 (A generalized model for community detection) Suppose there are n users clustered into two communities. Let $x_i \in \{0, 1\}$ indicate a community membership with regard to user $i \in \{1, 2, \ldots, n\}$. Assume that we are given part of the comparison pairs:

$$y_{ij} = \begin{cases} \sim \mathbb{P}(y_{ij} | x_{ij}), & \text{w.p. } p; \\ \mathsf{e}, & \text{w.p. } 1 - p \end{cases}$$

for every pair $(i, j) \in \{(1, 2), (1, 3), \cdots, (n-1, n)\}$ and $p \in [0, 1]$. Whenever an observation is made, y_{ij} is generated as per:

$$\mathbb{P}(y_{ij} | x_{ij}) = \begin{cases} \mathbb{P}_0(y_{ij}), & x_{ij} = 0; \\ \mathbb{P}_1(y_{ij}), & x_{ij} = 1. \end{cases}$$

We also assume that y_{ij}'s are independent over (i, j). Given y_{ij}'s, one wishes to decode the community membership vector $\boldsymbol{x} := [x_1, x_2, \ldots, x_n]$ or $\boldsymbol{x} \oplus \mathbf{1} := [x_1 \oplus 1, x_2 \oplus 1, \ldots, x_n \oplus 1]$. Let $\hat{\boldsymbol{x}}$ be a decoded vector. Define the probability of error as $P_e = \mathbb{P}(\hat{\boldsymbol{x}} \notin \{\boldsymbol{x}, \boldsymbol{x} \oplus \mathbf{1}\})$. Suppose we employ the ML decoding rule:

$$\hat{\boldsymbol{x}}_{\mathsf{ML}} = \arg\max_{\boldsymbol{x}} \mathcal{L}(\boldsymbol{x})$$

where $\mathcal{L}(\boldsymbol{x}) := \mathbb{P}(\mathbf{Y} | \mathbf{X}(\boldsymbol{x}))$ indicates the likelihood function w.r.t. $\boldsymbol{x}$.

(*a*) Show that

$$P_e \leq \sum_{k=1}^{n-1} \binom{n}{k} \mathbb{P}(\hat{\boldsymbol{x}}_{\mathsf{ML}} = \boldsymbol{a}_k | \boldsymbol{x} = \boldsymbol{0})$$

where $\boldsymbol{a}_k = [\underbrace{1, 1, \ldots, 1}_{k}, \underbrace{0, 0, \ldots, 0}_{n-k}]$.

(*b*) Let M be the number of observations made in the distinguishable positions between $\mathbf{X}(\boldsymbol{a}_k)$ and $\mathbf{X}(\mathbf{0})$. Show that

$$\mathbb{P}(\hat{\boldsymbol{x}}_{\mathsf{ML}} = \boldsymbol{a}_k | \boldsymbol{x} = \boldsymbol{0}) \leq \sum_{\ell=0}^{k(n-k)} \binom{k(n-k)}{\ell} p^{\ell} (1-p)^{k(n-k)-\ell}$$
$$\times \mathbb{P}(\mathcal{L}(\boldsymbol{a}_k) \geq \mathcal{L}(\mathbf{0}) | \boldsymbol{x} = \mathbf{0}, M = \ell).$$

(*c*) Using the above and part (*c*) in Prob 8.3, show that

$$P_e \leq \sum_{k=1}^{n-1} \binom{n}{k} (1 - p(1 - e^{-D^*}))^{k(n-k)}$$

where D^* denotes the Chernoff information:

$$D^* := - \min_{0 \leq \lambda \leq 1} \ln \left\{ \sum_{y \in \mathcal{Y}} \mathbb{P}_0(y)^{\lambda} \mathbb{P}_1(y)^{1-\lambda} \right\}.$$

(*d*) Using the above and Prob 7.3, show that if $p > \frac{\ln n}{n} \cdot \frac{1}{1-e^{-D^*}}$, P_e can be made arbitrarily close to 0 as $n \to \infty$.

Prob 8.5 (True or False?)

(*a*) Suppose that $X_1 \sim \mathsf{Bern}(p), X_2 \sim \mathsf{Bern}(\frac{1}{2})$ and $S = X_1 \oplus X_2$. Then, S follows $\mathsf{Bern}(\frac{1}{2})$ for any $p \in [0, 1]$.

(*b*) Let $\mathbf{X} = [X_1, X_2, \ldots, X_n]^T$ be an i.i.d. random vector, each being according to $\mathsf{Bern}(\frac{1}{2})$. Let $\mathbf{A}$ be an n-by-n full-rank matrix with $A_{ij} \in \{0, 1\}$ entries. Let $\mathbf{Y} = \mathbf{AX}$, i.e., $Y_i = \sum_{j=1}^{n} A_{ij} X_j$. Here the summation is modulo-2 addition. Then, Y_i's are i.i.d.

(*c*) Let X_1 and X_2 be independent random variables, each being according to $\text{Bern}(\frac{1}{2})$. Suppose we observe

$$(Y_1, Y_2) = \begin{cases} (X_1 \oplus Z_1, X_2 \oplus Z_2), & \text{w.p. } 1-\alpha; \\ (X_1 \oplus 1 \oplus Z_1, X_2 \oplus 1 \oplus Z_2), & \text{w.p. } \alpha \end{cases}$$

where Z_1 and Z_2 are independent random variables $\sim \text{Bern}(q)$, being also independent of (X_1, X_2). Then, $Y_1 \oplus Y_2$ is a sufficient statistic w.r.t. (X_1, X_2).

3.9 Top-K Ranking: Fundamental Limits

Recap Throughout the previous sections, we have examined two data science applications that incorporate information theory, wherein a precise threshold on the amount of information required to perform a specific task exists. Moreover, we have discovered that various bounding techniques covered in Parts I and II play a role in characterizing the fundamental limits of the explored problems, such as community detection and Haplotype phasing.

Outline In the upcoming sections, we will explore another data science application where phase transition occurs, specifically in the context of rank aggregation. Our focus will be on a particular type of ranking problem, known as top-K ranking, which aims to address the computational challenge posed by the huge number of items to be ranked in the big data era. We will first provide an overview of the ranking problem and highlight the aforementioned challenge. Then, we will introduce top-K ranking and explain how it differs from traditional ranking methods. Subsequently, we will present a benchmark model that can be used to evaluate the performance of ranking algorithms. Finally, we will investigate the fundamental limits of top-K ranking and show how phase transition occurs in terms of the amount of information required to achieve reliable ranking.

Ranking Ranking refers to the process of arranging items in order of significance. One example of ranking is a competition where several candidates participate, and the judges have to rank them in order of quality to determine the winner(s). In the simplest scenario, if each candidate has a score that indicates their quality, and those scores are known, then ranking is a straightforward task of sorting the candidates according to their scores. One can employ one of the well-known sorting algorithms (like quick sorting, bubble sorting, merge sorting, etc[1]) to obtain a ranking. Numerous sorting algorithms are available, which can sort a large number of items efficiently with a relatively small number of comparisons, typically scaling as $n \ln n$. However, the challenge arises when we do not know the individual scores of the candidates, which was assumed in the previous scenario. In reality, it is not that easy to assess the value of individuals. How can we assign an absolute value to people or items? It is nearly impossible. Therefore, in practical scenarios, individual scores are often not available. On the other hand, pairwise comparisons are relatively easy to obtain. For a given pair of candidates, it may be possible to determine which one is better than the other without having knowledge of their absolute qualities.

1. These are very well known in the computer science literature. For those who are not familiar with, please refer to an introductory book on data structure and/or wikepedia.

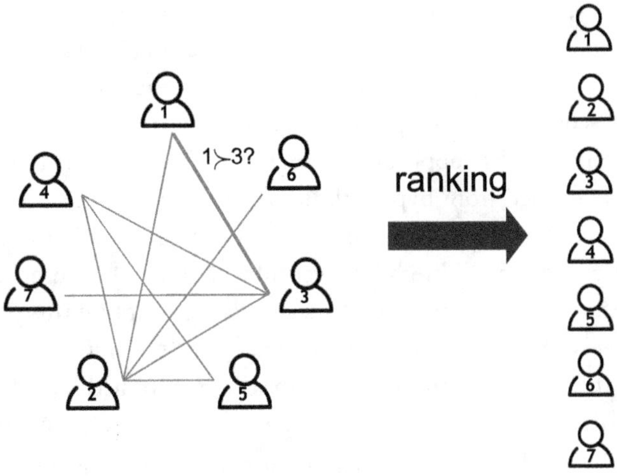

Figure 3.24. Small-scale ranking based on pairwise comparisons.

In situations where only pairwise comparisons are available, obtaining a ranking becomes a challenging task. One approach to tackle this problem is to aggregate all the comparison information to determine the order of all the candidates. By analyzing each comparison, one can determine which candidate is preferred over the other, and subsequently obtain a ranking. For instance, in Fig. 3.24, where the number of candidates is 7, this approach would require at most 21 ($= \binom{7}{2}$) comparisons. Thus, it is evident that ranking is straightforward for small-scale problems.

Large-scale ranking However, as the scale of the ranking problem increases, it becomes increasingly challenging. A prime example is web search, which is handled by search engines such as Google and Bing. The number of items (in this case, websites) to be ranked is enormous, with Google alone managing billions (on the order of 10^9) of websites. When a user enters a query, the search engine must provide a list of relevant websites, but the sheer volume of related websites poses a significant challenge.

Challenge in large-scale ranking To illustrate the challenge, let us consider the naive ranking approach discussed earlier. If we were to use this approach to rank the websites in the web search example, we would require an enormous number of comparisons given that the number of websites is in the billions. Specifically, we would need a total of $\frac{n^2}{2} (\approx \binom{n}{2})$ comparisons, which is an extremely large number, around 10^{18} for the web search example. This means that ranking is no longer a trivial task, and it begs the question of whether this number of comparisons is truly necessary or if there exists a more efficient ranking algorithm.

Unfortunately, two conventional assumptions impose a fundamental lower bound of around n^2 comparisons, which cannot be beaten. The first assumption is that we require a complete ordering of the items, while the second assumption is that pairwise comparisons are given probabilistically. This second assumption is relevant in many practical applications, such as web search where pairwise comparisons can be formed from hyperlink information. Since we cannot control the existence of such information, it is reasonable to assume that it is given probabilistically. Another example is Twitter's follower information, where user A follows user B, and this information is also given by a context. To capture this passively given information, we can assume that pairwise samples are given in a random manner.

Assuming the two conventional assumptions mentioned earlier, obtaining a ranking becomes a huge challenge because it requires almost all comparisons, which is around $\frac{n^2}{2}$. This is due to the fact that in order to identify the order between two consecutive candidates, a direct comparison between any two adjacent items is necessary. Therefore, all comparisons are required with probability 1, making the total number of required comparisons too large to handle, especially when dealing with a large scale ranking such as web search, where the number of items (websites) can be in the billions. This is why Google's ranking algorithm, PageRank, is an offline algorithm that pre-computes ranking results for popular queries and stores them in a table. However, this approach is not real-time and may result in outdated or missing results.

Why does the challenge arise? The challenge arises because we are focused on the ordering of all items, which requires a fundamental lower bound of $\frac{n^2}{2}$ comparisons.

Top-K ranking To tackle this challenge, a straightforward yet effective approach can be adopted based on the following observation: in many practical scenarios, the focus is only on identifying a small number of significant items, such as the top-K ranked items, among a large number of alternatives. A classic example is web search, where only the top 20 or 30 relevant websites are of interest, rather than the entire list of relevant websites.

This observation motivates us to shift our focus to top-K ranking, where the goal is to identify the top-K ranked items rather than obtaining a complete ordering. Clearly, this approach can significantly reduce the number of required comparisons for ranking. This gives rise to natural information-theoretic questions.

- What is the fundamental limit on the number of pairwise comparisons required for top-K ranking?
- Is there a computationally efficient algorithm that can achieve the limit?

There exists the fundamental limit which is far below $\sim n^2/2$. Also, there is an efficient algorithm that can achieve the limit. We will explore them in depth.

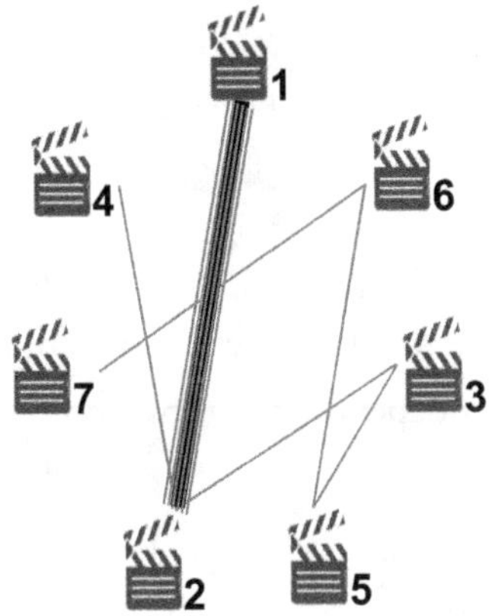

Figure 3.25. A comparison graph in the BTL model.

A benchmark model The information-theoretic limit is determined by a mathematical model that measures the accuracy of pairwise comparisons. The Bradley-Terry-Luce (BTL) model, which has been widely used as a benchmark model for evaluating ranking algorithms, is a prominent example of such a model (Bradley and Terry, 1952). It is commonly believed that an algorithm that performs well under the BTL model will perform well in practice, although it is not always the case.

The BTL model is based on two key assumptions. The first assumption is that there are ground truth scores, $\boldsymbol{x} := [x_1, x_2, \ldots, x_n]$, that determine the ranking of the items, with higher scores corresponding to higher rankings. The second assumption is related to the quality of information available. When the number n of items is very large, it is not feasible to observe all possible pairs of items, and we can only access a subset of them. To account for this, we introduce a comparison graph that shows which pairs have been observed (see Fig. 3.25). In the comparison graph, an edge indicates that a comparison has been made between the two items it connects. For a given pair of items i and j that have been observed, we are provided with pairwise comparison information that indicates whether item i is preferred over item j:

$$y_{ij} = \mathbf{1}\{\text{item } i \succ \text{item } j\}, \qquad (i,j) \in \mathcal{E}$$

where $\mathbf{1}\{\cdot\}$ denotes an indicator function and $\mathcal{E}$ indicates the edge set. This model assumes that the winning rate is proportional to the relative score of the two associated items. Hence, the probability of item i winning over item j is $\frac{x_i}{x_i+x_j}$, which in turn yields:

$$y_{ij} \sim \text{Bern}\left(\frac{x_i}{x_i + x_j}\right). \tag{3.45}$$

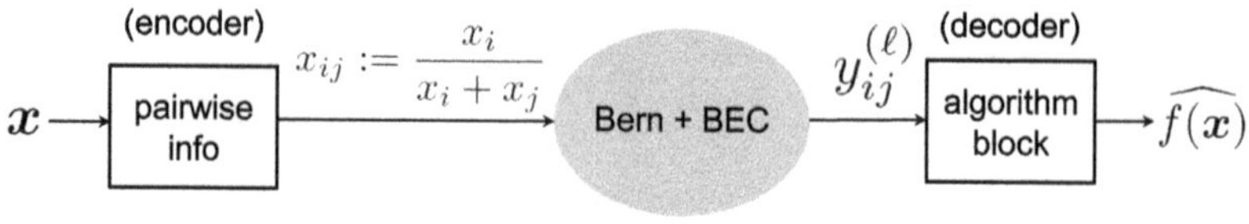

Figure 3.26. Translation of top-K ranking into a communication problem.

Notice that y_{ij} is a noisy data. In an effort to combat the noise effect, this model allows for repeated independent comparisons, say L repetitions: $y_{ij}^{(\ell)}$ i.i.d. $\sim \text{Bern}(\frac{x_i}{x_i+x_j})$ over $\ell \in \{1, 2, \ldots, L\}$.

Translation to a communication problem Interpreting $\boldsymbol{x}$ as a message that we wish to infer given $\{y_{ij}^{(\ell)}\}$, we can view the ranking problem as an inference problem. Note that $y_{ij}^{(\ell)}$'s are statistically related to $\boldsymbol{x}$ (see (3.45)). Hence, one can translate the problem into a communication problem.

We start with $\boldsymbol{x}$ and this is fed into an encoder. See Fig. 3.26. The encoder is a pairwise information function. Since the observation quality hinges upon the relative score of two items involved, one can set the function so as to yield:

$$x_{ij} := \frac{x_i}{x_i + x_j}.$$

Since we can access to only part of pairwise information in a random manner and also what we observe is the binary information, we can abstract the measurement process as an erasure channel:

$$y_{ij} = \begin{cases} \text{Bern}(x_{ij}), & \text{w.p. } p; \\ \text{e}, & \text{w.p. } 1 - p \end{cases}$$

where p indicates the observation probability. Assume that whenever an observation is made (i.e., $(i, j) \in \mathcal{E}$), L independent copies are given: $y_{ij}^{(\ell)}$'s i.i.d. over $\ell \in \{1, 2, \ldots, L\}$.

Given $\{y_{ij}^{(\ell)}\}$'s, the goal is to identify a set of top-K ranked items. Since it is a function of $\boldsymbol{x}$, we denote this by $f(\boldsymbol{x})$.[2] We consider a simple setting which aims at decoding the top-K set. One may want to consider another practically-relevant setting which targets the order of the top-K items as well.

An optimization problem As we did in the previous instances, we define two performance metrics. One is the quantity that we are interested in characterizing

2. It is in contrast to the previous instances (communication, community detection and Haplotype phasing) which aim at decoding $\boldsymbol{x}$. In general inference problems, what we wish to decode is a function of $\boldsymbol{x}$, so the case herein belongs to the general setting.

the limit on: the number of pairwise comparisons, sample complexity. Due to the WLLN, it is concentrated around $\binom{n}{2}pL$ in the limit of n. The second is the error probability defined as $P_e := \mathbb{P}(\hat{f}(\boldsymbol{x}) \neq f(\boldsymbol{x}))$. With these two metrics, we can formulate an optimization problem as we did earlier. Given (p, L, n):

$$P_e^*(p, L, n) := \min_{\text{algorithm}} P_e.$$

Similar to the communication and community detection problems, it is not that easy to derive the probability of error. It has been wide open. To make some progress, we focus on the asymptotic regime in which n is pretty large. This motivates us to ponder upon the following optimization. Given (p, L),

$$P_e^*(p, L) := \min_{\text{algorithm}, n} P_e.$$

The distinction is that n plays as a control variable that we optimize over. It turns out for some (p, L), we can make P_e arbitrarily close to zero as $n \longrightarrow \infty$. We say that such (p, L) is *achievable*. We are interested in characterizing the minimal achievable region $\mathcal{R}$ of (p, L).

Minimal achievable region of (p, L) The minimal achievable region depends highly on one key metric. The key metric is the separation score between the boundary items (the Kth and $(K+1)$th ranked items):

$$\Delta_K := \frac{x_K - x_{K+1}}{x_K}.$$

Here Δ_K indicates the normalized separation score.

Our intuition says that the larger separation score, the easier to rank, thus yielding the smaller achievable (p, L). This is indeed the case. More concretely, it has been shown that for some positive constants $0 < c_2 < c_1$ (Chen and Suh, 2015):

$$pL > c_1 \frac{\ln n}{n\Delta_K^2} \quad \Longrightarrow \quad P_e \to 0,$$

$$pL < c_2 \frac{\ln n}{n\Delta_K^2} \quad \Longrightarrow \quad P_e \nrightarrow 0.$$

We focus on a feasible regime in which $p > \frac{\ln n}{n}$. Otherwise, the comparison graph is *disconnected* (why?), thus ranking becomes impossible. See Fig. 3.27 for an illustration of the minimal achievable region.

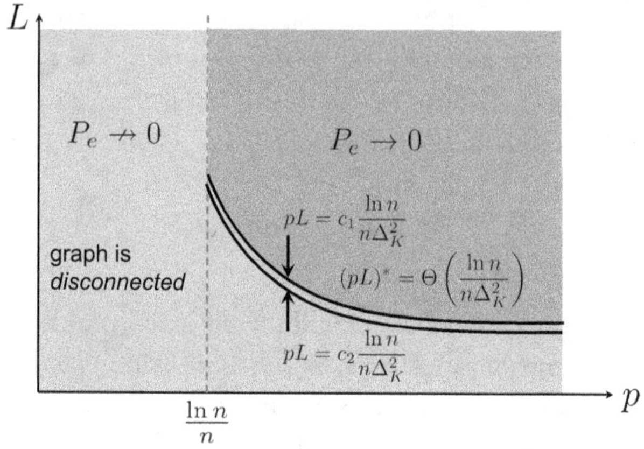

Figure 3.27. Minimal achievable region of (p, L).

Using the above result, one can characterize the order-wise tight minimal sample complexity:

$$S^* = \Theta\left(\frac{n \ln n}{\Delta_K^2}\right).$$

The standard notation $y = \Theta(x)$ means that there exist positive constants c_1 and c_2 such that $c_2 x \leq y \leq c_1 x$. This result makes an intuitive sense. The larger Δ_K^2, the easier to rank, thus reducing sample complexity. Also comparing to

$$S^*_{\text{total-order}} = O(n^2),$$

the result is promising. With top-K ranking, we can reduce the sample complexity from $\sim n^2$ to $\sim n \ln n$. Here the standard notation $y = O(x)$ means that there exists a positive constant c such that $y \leq cx$.

Look ahead In the next section, we will study an efficient algorithm that achieves the limit: for some positive constant c,

$$pL > c\frac{\ln n}{n\Delta_K^2} \quad \Longrightarrow \quad P_e \rightarrow 0.$$

3.10 Top-K Ranking: An Efficient Algorithm

Recap In the previous section, we delved into a ranking problem. Our focus was on top-K ranking, where the objective is to retrieve the top-K ranked items, driven by numerous practical applications. Using the well-established BTL model (Bradley and Terry, 1952) as a benchmark, we transformed it into a communication problem, where the ground-truth scores were represented by the input $\boldsymbol{x}$, and the target function $f(\boldsymbol{x})$ was the set of top-K items. Under the BTL model, every pair of items is observed randomly and uniformly with probability p, and each observed pair has L independent copies, each following $\mathsf{Bern}(\frac{x_i}{x_i+x_j})$ for items i and j. We asserted that the minimum number of pairwise comparisons required for reliable top-K ranking, i.e., the minimum sample complexity, is:

$$\Theta\left(\frac{n\ln n}{\Delta_K^2}\right) \tag{3.46}$$

where $\Delta_K := \frac{x_K - x_{K+1}}{x_K}$ denotes the normalized score separation between the Kth and $(K+1)$th items, reflecting a difficulty level of separating the two boundary items.

Outline In this section, our focus will be on a computationally efficient algorithm that achieves the aforementioned limit (3.46). This section consists of four parts. Firstly, we will introduce a performance metric that captures the ranking performance associated with the BTL model. Next, we will discuss one well-known algorithm that aims to maximize this ranking performance, which is a modified version of Google's PageRank. After that, we will draw attention to a challenge faced by this variant. Finally, we will examine a more advanced algorithm that overcomes the challenge and achieves the minimum sample complexity stated in (3.46).

How to estimate scores? In the BTL model, scores correspond to a ranking such that higher scores imply higher rankings. Therefore, we adopt a two-step approach consisting of score estimation followed by ranking based on the estimate. The question then becomes how to estimate the scores. To answer this, we first need to identify a suitable metric that quantifies the quality of an estimate. One commonly used performance metric is the probability of error, which is defined as:

$$P_e := \mathbb{P}(\hat{\boldsymbol{x}} \neq \boldsymbol{x}).$$

This is not a proper metric in our problem setting though. Why? P_e is always 1 no matter what the decoder is. It cannot distinguish good decoders from bad ones. This is because $\boldsymbol{x}$ is a continuous value, which leads the success event to have measure 0.

In the context of *estimation* in which one wishes to infer a *continuous* quantity, there is a well-known metric: Mean Square Error (MSE) defined as

$$\frac{1}{n}\|\hat{\boldsymbol{x}}-\boldsymbol{x}\|^2 := \frac{1}{n}\sum_{i=1}^{n}(\hat{x}_i - x_i)^2.$$

We wish to develop algorithms that minimize the MSE.

A slight variant of PageRank There is one popular algorithm that minimizes the MSE in the context of ranking problems (Negahban *et al.*, 2012). That is, a slight variant of PageRank, the backbone of Google's web search engine. We employ the variant.

Let us explain how the algorithm works. Here are a few key observations that give an inspiration to the algorithm. Remember that when a pair of (i,j) is observed, we are given L independent copies: $y_{ij}^{(1)}, y_{ij}^{(2)}, \ldots, y_{ij}^{(L)}$. From this, what we can compute is its empirical mean:

$$y_{ij} := \frac{1}{L}\sum_{\ell=1}^{L} y_{ij}^{(\ell)}, \qquad (i,j) \in \mathcal{E} \tag{3.47}$$

where $\mathcal{E}$ denotes the edge set of the comparison graph. One key observation is: by the WLLN, the empirical mean converges to its true mean:

$$y_{ij} := \frac{1}{L}\sum_{\ell=1}^{L} y_{ij}^{(\ell)} \overset{\text{in prob.}}{\longrightarrow} \mathbb{E}[y_{ij}^{(\ell)}] = \frac{x_i}{x_i + x_j}. \tag{3.48}$$

Also observe that

$$x_i \cdot \frac{x_j}{x_i + x_j} = x_j \cdot \frac{x_i}{x_i + x_j}. \tag{3.49}$$

This formula (3.49) reminds us of the *detailed balance equation* in a Markov chain:

$$\pi_i p_{ji} = \pi_j p_{ij}$$

where $\pi_i := \mathbb{P}(\text{state} = i)$ (stationary distribution) and $p_{ji} := \mathbb{P}(\text{state}_{\text{next}} = j|\text{state}_{\text{current}} = i)$ (transition probability). We can view x_i as π_i and $\frac{x_j}{x_i+x_j}$ as p_{ji}.

This motivates us to construct a Markov chain in which transition probability from j to i takes y_{ij} which converges to $\frac{x_i}{x_i+x_j}$ as $L \to \infty$. One caveat is that $\sum_i y_{ij}$ can exceed 1 while $\sum_i p_{ij}$ must be 1. To resolve this, we normalize y_{ij} by the maiximum out-degree (maximum number of out-going edges), denoted by $d_{\max}$:

$$p_{ij} = \frac{y_{ij}}{d_{\max}}.$$

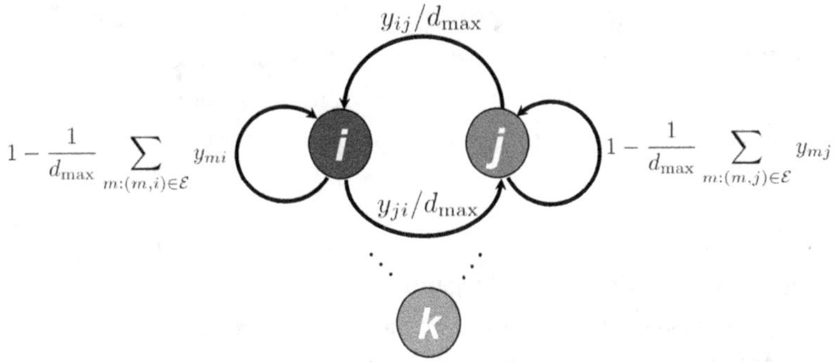

Figure 3.28. A Markov chain inspired by observations of (3.48) and (3.49).

Also we add self-transition to ensure $\sum_i p_{ij} = 1$:

$$p_{jj} = 1 - \frac{1}{d_{\max}} \sum_{m:(m,j)\in\mathcal{E}} y_{mj}.$$

Now by (3.48), one can see that

$$\pi_i \frac{x_j}{x_i + x_j} = \pi_j \frac{x_i}{x_i + x_j}.$$

This together with (3.49) concludes that the stationary distribution $\boldsymbol{\pi}$ converges to $\boldsymbol{x}$ up to some scaling as $L \to \infty$. This series of observations leads to the following natural idea:

$$\text{Take } \hat{\boldsymbol{x}} = \boldsymbol{\pi}.$$

Then, how to compute $\boldsymbol{\pi}$? To gain some insight, consider the following equation (called the *global balance equation* in the Markov chain literature):

$$\sum_j p_{ij}\pi_j = \pi_i.$$

Alternative matrix representation of this is:

$$\mathbf{P}\pi = \pi$$

where $\mathbf{P}$ denotes the transition probability matrix:

$$\mathbf{P} := \begin{bmatrix} p_{11} & p_{12} & \cdots & p_{1n} \\ p_{21} & p_{22} & \cdots & p_{2n} \\ \vdots & \vdots & \ddots & \vdots \\ p_{n1} & p_{n2} & \cdots & p_{nn} \end{bmatrix}.$$

From this, we see that π is an *eigenvector* of $\mathbf{P}$. It is well-known in the Markov chain literature that the stationary distribution π is the first eigenvector (with the largest eigenvalue) of the transition probability matrix. Remember that one very efficient and useful way of computing the first eigenvector is the *power method*. See Section 3.4.

Additional step: Local refinement The PageRank variant taking $\hat{\boldsymbol{x}} = \pi$ exhibits a great MSE performance (Negahban *et al.*, 2012). Is this algorithm enough then? We are interested in a ranking, not the scores $\boldsymbol{x}$ themselves. So one may ask: Does the MMSE solution implies high ranking accuracy? It turns out it is not necessarily the case. We can readily see this from the following scenario in which many of the estimates are very close to the ground truth scores while only a very few of the estimates (let's call them *outliers*) are far from the ground truth. In this case, the MSE can be very small while ranking accuracy is not that good due to the outliers. See such an example in Fig. 3.29. Notice that many of the estimates are close to the ground-truth scores, leading to a small MSE; however, its ranking result ($\widehat{\text{top-3}}$) is distinct from the ground-truth top-3. From this observation, we see that *coordinate-wise* errors are required to be small enough in order to ensure high ranking accuracy.

Remember the advanced algorithm for Haplotype phasing in Section 3.8. The advanced algorithm employs an additional step (called *local refinement*) in an effort to detect any coordinate-wise error. This motivates us to apply the same method herein. The role of the second step in this problem setup is to detect outliers and then control corresponding errors in a point-wise manner. To implement this, we use coordinate-wise maximum likelihood estimator (MLE) which tries to minimize coordinate-wise MSE. Here is how it works. Pick up an item, say item i. We then compute the coordinate-wise MLE w.r.t. item i. Similar to Haplotype phasing,

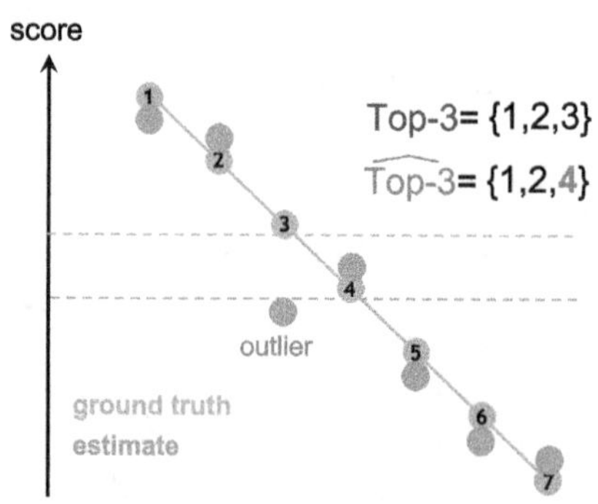

Figure 3.29. An example in which the MSE of an estimate is small while its ranking result ($\widehat{\text{top-3}}$) is distinct from the ground-truth top-3.

computing the coordinate-wise maximum likelihood estimate (MLE) poses a challenge. It necessitates having knowledge of the ground-truth scores of other items that are not accessible. Therefore, we adopt the same approach as before and utilize an estimate obtained in the previous step. Specifically, we compute:

$$x_i^{\mathsf{MLE}} = \arg\max_a \mathcal{L}(\hat{x}_1, \ldots, \hat{x}_{i-1}, a, \hat{x}_{i+1}, \ldots, \hat{x}_n)$$

where $\hat{x}_j$ denotes the estimate obtained in the previous stage; and $\mathcal{L}(\cdot)$ indicates the likelihood function:

$$\begin{aligned}
&\mathcal{L}(\hat{x}_1, \ldots, \hat{x}_{i-1}, a, \hat{x}_{i+1}, \ldots, \hat{x}_n) \\
&\quad = \prod_{j:(i,j)\in\mathcal{E}} \prod_{\ell=1}^{L} \mathbb{P}\left(y_{ij}^{(\ell)} | \hat{x}_1, \ldots, \hat{x}_{i-1}, a, \hat{x}_{i+1}, \ldots, \hat{x}_n\right) \\
&\quad = \prod_{j:(i,j)\in\mathcal{E}} \prod_{\ell=1}^{L} \left(\frac{a}{a+\hat{x}_j}\right)^{y_{ij}^{(\ell)}} \left(\frac{\hat{x}_j}{a+\hat{x}_j}\right)^{1-y_{ij}^{(\ell)}} \\
&\quad = \prod_{j:(i,j)\in\mathcal{E}} \left(\frac{a}{a+\hat{x}_j}\right)^{Ly_{ij}} \left(\frac{\hat{x}_j}{a+\hat{x}_j}\right)^{L(1-y_{ij})}
\end{aligned} \tag{3.50}$$

where the last equality is due to the definition of $y_{ij} := \frac{1}{L}\sum_{\ell=1}^{L} y_{ij}^{(\ell)}$ (see (3.47)).

We then compute the gap between the coordinate-wise MLE and its earlier estimate:

$$d := |x_i^{\mathsf{MLE}} - \hat{x}_i|.$$

We declare $\hat{x}_i$ to be an outlier if d exceeds a certain threshold which is carefully chosen, and then replace $\hat{x}_i$ with x_i^{MLE} in an effort to control the corresponding error. Otherwise, we keep $\hat{x}_i$ as it is. We repeat the same procedure for other items, one-by-one and step-by-step. This forms one iteration. We do multiple iterations. The formula of the threshold is suggested in detail from (Chen and Suh, 2015):

$$\delta^{(t)} = c\left\{\sqrt{\frac{\ln n}{npL}} + \frac{1}{2^t}\left(\sqrt{\frac{\ln n}{pL}} - \sqrt{\frac{\ln n}{npL}}\right)\right\} \tag{3.51}$$

where c indicates some positive constant and t refers to the iteration index: $0 \leq t \leq T_{\mathsf{iter}}$. The rationale behind this choice is two folded. First, in the initial stage $t = 0$, $\sqrt{\frac{\ln n}{pL}}$ can be shown to be order-wise greater than the maximum gap between

$\hat{x}_i$ and x_i^{MLE}:

$$\max_{1\le i\le n} |x_i^{\mathsf{MLE}} - \hat{x}_i| \le c_1 \sqrt{\frac{\ln n}{pL}} \tag{3.52}$$

for some positive constant c_1. Second, at the end of the last stage, $\sqrt{\frac{\ln n}{npL}}$ can be proven to be an upper bound:

$$\max_{1\le i\le n} |x_i^{\mathsf{MLE}} - x_i^{(T_{\mathsf{iter}}-1)}| \le c_2 \sqrt{\frac{\ln n}{npL}} \tag{3.53}$$

for some positive constant c_2. Here $x_i^{(t)}$ refers to the coordinate-wise MLE of the ith item at the t-th iteration. Due to the interest of this book, we will omit the proof of the bounds.

Minimum sample complexity It has been shown in (Chen and Suh, 2015) that the PageRank variant together with local refinement yields: if the number of iterations is around $\ln n$, then for some positive constant c,

$$\binom{n}{2} pL > c \cdot \frac{n \ln n}{\Delta_K^2} \quad \Longrightarrow \quad P_e \to 0.$$

This completes the achievability proof. In fact, the analysis for the above is not that simple. It requires a variety of techniques including: (i) non-trivial linear-algebra tricks; (ii) Chernoff-like bounds; (iii) a bunch of inequalities such as Cauchy-Schwarz, triangle, Pinsker's, etc. We omit details. We will also omit the other direction proof: for some other constant $c' > 0$,

$$P_e \to 0 \quad \Longrightarrow \quad \binom{n}{2} pL > c' \cdot \frac{n \ln n}{\Delta_K^2}.$$

To know more know, refer to (Chen and Suh, 2015).

Look ahead We have explored the third application of data science, namely top-K ranking, and examined the minimum sample complexity required for reliable top-K ranking. Additionally, we investigated an efficient algorithm that achieves the limit with a constant factor gap. In the subsequent section, we will implement this efficient algorithm using Python.

3.11 Top-*K* Ranking: Python Implementation

Recap In the previous sections, we explored the problem of top-K ranking, which is another inference problem concerning phase transition. We examined an efficient algorithm that achieves the minimal sample complexity, up to a constant factor gap.

$$S := \binom{n}{2} pL > c \frac{n \ln n}{\Delta_K^2} \quad \Longrightarrow \quad P_e \to \infty \tag{3.54}$$

where c is some positive constant and $\Delta_K := \frac{x_K - x_{K+1}}{x_K}$ denotes the normalized score separation between the Kth and $(K+1)$th items.

Outline In this section, we will delve into the algorithm by implementing it in Python. This section consists of four parts. First, we will provide a pseudocode of the algorithm that outlines all the detailed procedures. Second, we will focus on implementing the first stage of the algorithm, which is the PageRank variant. This variant finds the first eigenvector of the transition probability matrix in the relevant Markov chain. We will also evaluate the performance of the algorithm by calculating the mean squared error (MSE) between the ground truth score vector $\boldsymbol{x}$ and its estimate $\hat{\boldsymbol{x}}$. Third, we will incorporate the second stage of the algorithm (local refinement) to implement the advanced version. Finally, we will compare the MSE performance of the two algorithms.

Pseudocode of the efficient algorithm The efficient algorithm consists of two stages: (i) obtaining an initial estimate of $\boldsymbol{x}$ via the spectral algorithm (finding the first eigenvector of the transition probability matrix of a Markov chain); and (ii) performing local refinement of the initial estimate via the coordinate-wise MLE. Details of the first stage are given as below.

1. Given $y_{ij}^{(\ell)}$'s and $\mathcal{E}$, compute y_{ij} for all $(i,j) \in \mathcal{E}$: $y_{ij} = \frac{1}{L}\sum_{\ell=1}^{L} y_{ij}^{(\ell)}$.
2. Compute the transition matrix $\mathbf{P} = [p_{ij}]$: $p_{ij} = \frac{y_{ij}}{d_{\max}}$, $(i,j) \in \mathcal{E}$. Here $d_{\max} = \max_{1 \le j \le n} |\{(i,j) : (i,j) \in \mathcal{E}\ \forall i\}|$. For the (j,j) entry,

$$p_{jj} = 1 - \frac{1}{d_{\max}} \sum_{m:(m,j)\in\mathcal{E}} y_{mj}.$$

3. Compute the stationary distribution $\boldsymbol{\pi}$ to obtain $\boldsymbol{x}^{(0)} = n\boldsymbol{\pi}$.

In Step 3, we multiply $\boldsymbol{\pi}$ by n for a proper scaling. This way, an estimated score $x_i^{(0)}$ does not scale with n. The second stage for local refinement is described as follows.

4. For $t \in \{0, \ldots, T_{\text{iter}} - 1\}$, iterate the following: for each $1 \leq i \leq n$,

$$x_i^{\text{MLE}} = \arg\max_a \ln \mathcal{L}(x_1^{(t)}, \ldots, x_{i-1}^{(t)}, a, x_{i+1}^{(t)}, \ldots, x_n^{(t)})$$

$$x_i^{(t+1)} = \begin{cases} x_i^{\text{MLE}}, & |x_i^{\text{MLE}} - x_i^{(t)}| > \delta^{(t)}; \\ x^{(t)}, & |x_i^{\text{MLE}} - x_i^{(t)}| \leq \delta^{(t)}. \end{cases}$$

Here $\ln \mathcal{L}(\cdot)$ indicates the log-likelihood function:

$$\begin{aligned} &\ln \mathcal{L}(x_1^{(t)}, \ldots, x_{i-1}^{(t)}, a, x_{i+1}^{(t)}, \ldots, x_n^{(t)}) \\ &= \sum_{j:(i,j)\in\mathcal{E}} \left\{ L y_{ij} \ln\left(\frac{a}{a + x_j^{(t)}}\right) + L(1 - y_{ij}) \ln\left(\frac{x_j^{(t)}}{a + x_j^{(t)}}\right)\right\} \\ &= L \sum_{j:(i,j)\in\mathcal{E}} (1 - y_{ij}) \ln x_j^{(t)} + L \sum_{j:(i,j)\in\mathcal{E}} \left\{ y_{ij} \ln a - \ln(a + x_j^{(t)})\right\}, \end{aligned}$$

and $\delta^{(t)}$ denotes the threshold for comparison:

$$\delta^{(t)} = c\left\{\sqrt{\frac{\ln n}{npL}} + \frac{1}{2^t}\left(\sqrt{\frac{\ln n}{pL}} - \sqrt{\frac{\ln n}{npL}}\right)\right\}$$

for some positive constant c.

Python implementation of the PageRank variant We implement the PageRank variant (corresponding to Steps 1,2 and 3 in the above). For illustrative purpose, we first consider a simple setting where interested parameters are small numbers.

```
import numpy as np
n=4
K=2 # top-K ranking
p=2*np.log(n)/n # for graph connectivity
L=100
```

We set $p = \frac{2 \ln n}{n}$ to ensure graph connectivity ($p > \frac{\ln n}{n}$). The parameter L is chosen as 100. We generate the ground truth score vector $\boldsymbol{x}$. To emphasize the separation score between the Kth and $(K+1)$th scores, we set $x_K = 0.8$ and $x_{K+1} = 0.8 - \text{gap}$ such that Δ_K is fixed as $\frac{x_K - x_{K+1}}{x_K} = \frac{\text{gap}}{0.8}$. We choose gap $= 0.1$ to have $\Delta_K = 0.125$. On the other hand, we generate $(x_1, \ldots, x_{K-1})$ and $(x_{K+2}, \ldots, x_n)$, uniformly distributed from $[0.8, 1]$ and $[0.8 - \text{gap}, 0.5]$, respectively.

```
# Generate the ground truth score vector
def scoreVector(gap,n,K):
    # Generate top-K items btw 0.8 and 1
```

```
    xt=np.random.uniform(0.8,1,K)
    xt=sorted(xt,reverse=True)
    xt[K-1]=0.8
    # Generate n-K bottom items btw 0.5 and 0.8-gap
    xb=np.random.uniform(0.5,0.8-gap,n-K)
    xb=sorted(xb,reverse=True)
    xb[0]=0.8-gap
    x=np.concatenate((xt,xb))
    # score scaling
    x=x/sum(x)*n
    # Compute DeltaK
    DeltaK=(x[K-1]-x[K])/x[K-1]
    # sample complexity
    return x,DeltaK
```

```
gap=0.1
x,DeltaK=scoreVector(gap,n,K)
S=n*(n-1)/2*p*L
# claimed limit
limit=n*np.log(n)/(DeltaK**2)
print(x)
print(DeltaK)
print(S)
print(limit)
```

```
[1.1848639  1.04924926 0.9180931  0.84779374]
0.12500000000000003
415.88830833596717
354.8913564466918
```

In this setup, the sample complexity is above the claimed limit.

Next, we construct a comparison graph.

```
from scipy.stats import bernoulli

def genGraph(n,p):
    # Generate a comparison graph
    Bern = bernoulli(p)
    G = Bern.rvs((n,n))
    # make G symmetric
    for i in range(n):
        for j in range(i,n): G[j,i]=G[i,j]
    for i in range(n): G[i,i]=0
    # compute dmax=max_j |{(i,j):(i,j) in G forall i}|
   dmax=max(sum(G))
   return G,dmax
```

```
G,dmax=genGraph(n,p)
print(G)
print(dmax)
```

```
[[0 1 1 1]
 [1 0 1 1]
 [1 1 0 1]
 [1 1 1 0]]
3
```

Using the comparison graph, we generate observations $y_{ij}^{(\ell)}$'s and then compute y_{ij} and p_{ij} as per Steps 1 and 2 in the pseudocode.

```
from numpy.random import binomial

def transitionMatrix(x,n,L,G,dmax):
    # initialization
    Y=np.zeros((n,n))
    P=np.zeros((n,n))
    # construct transition matrix
    for i in range(n):
        for j in range(i,n):
            if G[i,j]==1:
                Y[i,j]=binomial(L,x[i]/(x[i]+x[j]))/L
                Y[j,i]=1-Y[i,j]
                P[i,j]=Y[i,j]/dmax
                P[j,i]=Y[j,i]/dmax
    # add self-transition
    for i in range(n): P[i,i]=1-sum(P[:,i])
    return P,Y
```

```
P,Y=transitionMatrix(x,n,L,G,dmax)
print(Y)
print(P)
```

```
[[0.   0.51 0.58 0.58]
 [0.49 0.   0.55 0.63]
 [0.42 0.45 0.   0.51]
 [0.42 0.37 0.49 0.  ]]
[[0.55666667 0.17       0.19333333 0.19333333]
 [0.16333333 0.55666667 0.18333333 0.21      ]
 [0.14       0.15       0.46       0.17      ]
 [0.14       0.12333333 0.16333333 0.42666667]]
```

In order to find the first eigenvector of **P**, we employ the power method (see Section 3.4). Below we copy the Python code of the power method.

```
def power_method(A, eps=1e-5):
    # A computionally efficient algorithm
```

```
    # for finding the principal eigenvector
    # Choose a random vector
    v = np.random.randn(n)
    # normalization
    v = v/np.linalg.norm(v)

    prev_v = np.zeros(len(v))
    t = 0
    while np.linalg.norm(prev_v-v) > eps:
        prev_v = v
        v = np.array(np.dot(A,v)).reshape(-1)
        v = v/np.linalg.norm(v)
        t += 1
    return v
```

```
x0=power_method(P)
# scaling
x0=x0/sum(x0)*n
print(x0)
print(x)
```

```
[1.17192056 1.1662718  0.87537578 0.78643185]
[1.1848639  1.04924926 0.9180931  0.84779374]
```

We see that $\hat{\boldsymbol{x}}$ has the same order as that of $\boldsymbol{x}$.

Now we perform an extensive experiment for a large value of n, say $n = 1000$, and for a range of p spanning the claimed limit $p^* = \frac{1}{\binom{n}{2}L} \cdot \frac{n \ln n}{\Delta_K^2}$.

```
import numpy as np
from scipy.stats import bernoulli
from numpy.random import binomial
import matplotlib.pyplot as plt

n=1000
K=5 # top-K ranking
L=20
p_range=np.linspace(0.02,0.12,30)
#ln(n)/n ~0.007
#nln(n)/(DeltaK**2)/(n*(n-1)*L/2)~0.0089
gap=0.1
# generate the ground truth score vector
x,DeltaK=scoreVector(gap,n,K)
MSE = np.zeros_like(p_range)
ITER=20
for idx,p in enumerate(p_range):
   for k in range(ITER):
       # Generate a comparison graph
```

```
        G,dmax=genGraph(n,p)
        # Transition probability matrix
        P,Y=transitionMatrix(x,n,L,G,dmax)
        # Compute the stationary distribution
        x0=power_method(P)
        # score scaling
        x0=x0/sum(x0)*n
        # Compute MSE between x and x0
        MSE[idx]+= \
        sum(np.square(x-x0))/sum(np.square(x))/ITER

plimit=n*np.log(n)/(DeltaK**2)/(n*(n-1)*L/2)
p_norm=p_range/plimit

plt.figure(figsize=(5,5), dpi=100)
plt.plot(p_norm,MSE,label='PageRank variant')
plt.yscale('log')
plt.title('Normalized MSE')
plt.legend()
plt.grid(linestyle=':', linewidth=0.5)
plt.show()
```

Fig. 3.30 demonstrates the MSE performance of the PageRank variant as a function of p. The range of p is set so as to exceed the graph connectivity threshold $\frac{\ln n}{n}$ as well as to span the claimed limit $0.4 \leq \frac{p}{p^*} \leq 2.7$. Notice that the MSE decreases with an increase in n.

Additional stage: Local refinement We investigate a more advanced algorithm that employs local refinement addigionally (Step 4 in the above pseudocode). To implement this, we need to solve the optimization problem taking the log-likelihood as an objective function:

$$x_i^{\mathsf{MLE}} = \arg\max_a \ln \mathcal{L}_i(a).$$

where we use a simpler notation for the log-likelihood:

$$\mathcal{L}_i(a) := \mathcal{L}(x_1^{(t)}, \ldots, x_{i-1}^{(t)}, a, x_{i+1}^{(t)}, \ldots, x_n^{(t)}).$$

One key observation is that the objective function is concave in the optimization variable a. Check this in Prob 9.4. Hence, it is a *convex* optimization problem. As mentioned in Section 1.5 and Prob 2.4, one can solve convex optimization via the Lagrange multiplier method. Sometimes, the method yields the closed form solution. But it is not always the case.

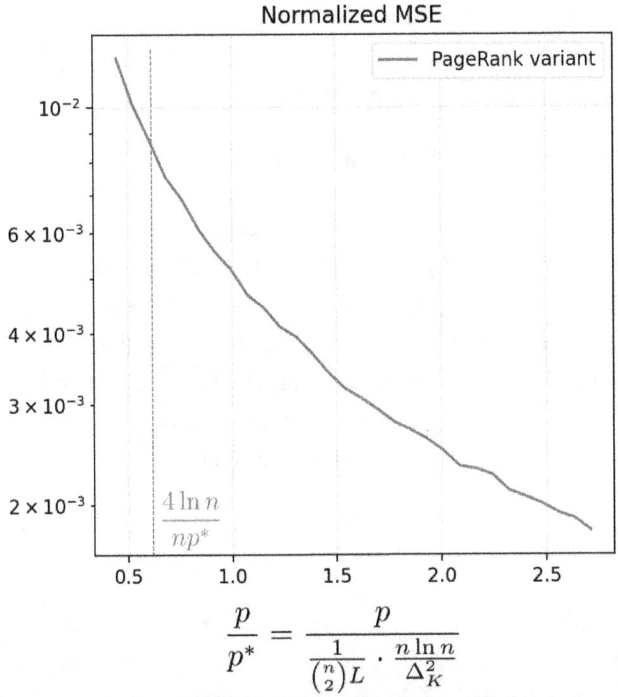

Figure 3.30. The MSE performance of the PageRank variant as a function of observation probability p: $n = 1000$, $L = 20$, $K = 5$ and $\Delta_K = 0.125$.

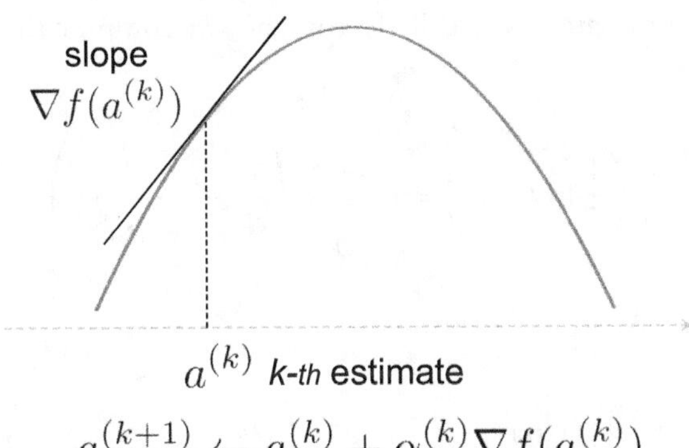

Figure 3.31. How gradient ascent works.

Gradient ascent Our problem has no closed-form solution. Hence, we employ an algorithm that allows us obtain the solution *numerically*. One prominent algorithm that yields the numerical solution is *gradient ascent*. Simply put, it is an algorithm that finds the unique stationary point when the interested function is concave. It is called *gradient descent* when the objective function is convex. Here is how the algorithm works. See Fig. 3.31. It is an iterative algorithm. Suppose that at the kth iteration, we have an estimate of a^*, say $a^{(k)}$. We then compute the gradient of the function evaluated at the estimate: $\nabla f(a^{(k)})$. Next we update the estimate along the same direction w.r.t. the gradient:

$$a^{(k+1)} \longleftarrow a^{(k)} + \alpha^{(k)} \nabla f(a^{(k)}) \tag{3.55}$$

where $\alpha^{(k)} > 0$ indicates the learning rate (or called a step size) that usually decays like $\alpha^{(k)} = \frac{1}{2^k}$. If you think about it, this update rule makes an intuitive sense. Suppose $a^{(k)}$ is placed left relative to the optimal point a^*, as in the two-dimensional case[3] illustrated in Fig. 3.31. Then, we should move $a^{(k)}$ to the right so that it becomes closer to a^*. The update rule actually does this, as we add by $\alpha^{(k)} \nabla f(a^{(k)})$. Notice that $\nabla f(a^{(k)})$ points to the right direction given that $a^{(k)}$ is placed left relative to a^*. We repeat this procedure until it converges. It turns out: as $k \rightarrow \infty$, it converges:

$$a^{(k)} \longrightarrow a^*, \tag{3.56}$$

as long as the learning rate is chosen properly, like the one decaying exponentially. We will not touch upon the proof of this convergence. In fact, the proof is difficult. There is a big field in statistics which intends to prove the convergence of a variety of algorithms (if the convergence holds).

Python implementation of local refinement We apply gradient ascent to implement the coordinate-wise MLE. To this end, we compute the gradient of the objective function:

$$\frac{d}{da}\left(\frac{1}{L}\ln \mathcal{L}_i(a)\right) = \sum_{j:(i,j)\in\mathcal{E}} \left(\frac{y_{ij}}{a} - \frac{1}{a + x_j^{(t)}}\right).$$

Then, the update rule for $a^{(k)}$ reads:

$$a^{(k+1)} \longleftarrow a^{(k)} + \alpha^{(k)} \sum_{j:(i,j)\in\mathcal{E}} \left(\frac{y_{ij}}{a^{(k)}} - \frac{1}{a^{(k)} + x_j^{(t)}}\right). \tag{3.57}$$

See below for a code implementation of the coordinate-wise MLE. We first set up parameters and run the PageRank variant to obtain an initial estimate.

3. In a higher-dimensional case, it is difficult to visualize how $a^{(k)}$ is placed. Hence, we focus on the two-dimensional case. It turns out that gradient ascent works even for high-dimensional settings although it is not 100% intuitive.

```
import numpy as np

n=4
K=2 # top-K ranking
p=2*np.log(n)/n # for graph connectivity
L=100
gap=0.1

# generate the ground truth score vector
x,DeltaK=scoreVector(gap,n,K)
# Generate a comparison graph
G,dmax=genGraph(n,p)
# Transition probability matrix
P,Y=transitionMatrix(x,n,L,G,dmax)
# Compute the stationary distribution
x0=power_method(P)
# score scaling
x0=x0/sum(x0)*n
print(x0)
```

[1.22440733 0.90509149 1.06241023 0.80809095]

```
def coordinateMLE(x0,G,Y,xmin,xmax):
    n=len(G[0])
    alpha=0.01
    nITER=10

    # initialization for xMLE
    xMLE=np.zeros_like(x0)

    for i in range(len(G[:,0])):
        # compute the ith coordinate MLE
        a=x0[i] # initialization
        for k in range(nITER):
            # compute the gradient of ln(L_i(a))/L
            grad=0 # initialization
            for j in range(len(G[i])):
                if G[i,j]==1: grad+=Y[i,j]/a-1/(a+x0[j])
            # update "a^{(k)}"
            a = a+alpha*np.sign(grad)
        # projection to [x_min,x_max]
        if a<xmin: xMLE[i]=xmin
        elif a>xmax: xMLE[i]=xmax
        else: xMLE[i]=a

    return xMLE
```

```
xMLE=coordinateMLE(x0,G,Y,min(x),max(x))
print(x0)
print(xMLE)
print(x0-xMLE)
```

```
[1.22440733 0.90509149 1.06241023 0.80809095]
[1.20499993 0.88509149 1.06241023 0.82809095]
[ 0.0194074  0.02       0.        -0.02      ]
```

We employ a simplified version of gradient ascent where we take only the sign of the gradient and set $\alpha = 0.01$. This way, we can do an efficient update. Otherwise, the range of the gradient scales with n, yielding an unstable update. We also apply the projection of the MLE solution so that the estimate is within $[x_{\min}, x_{\max}]$. Notice in the above that the coordinate-wise MLE is slightly different from the initial estimate.

Next, we employ $\delta^{(t)}$ to decide whether to take the MLE solution:

$$x_i^{(t+1)} = \begin{cases} x_i^{\text{MLE}}, & |x_i^{\text{MLE}} - x_i^{(t)}| > \delta^{(t)}; \\ x^{(t)}, & |x_i^{\text{MLE}} - x_i^{(t)}| \leq \delta^{(t)} \end{cases} \tag{3.58}$$

where

$$\delta^{(t)} = c\left\{\sqrt{\frac{\ln n}{npL}} + \frac{1}{2^t}\left(\sqrt{\frac{\ln n}{pL}} - \sqrt{\frac{\ln n}{npL}}\right)\right\}$$

We set the hyperparameters $c = 0.1$ and $T_{\text{iter}} = 7$. In this case, the range of $\delta^{(t)}$ is:

```
Titer=7
t=np.arange(Titer)
c=0.1
delta= c*(np.sqrt(np.log(n)/(n*p*L)) \
  + 1/(2**t)*(np.sqrt(np.log(n)/(p*L)) \
  - np.sqrt(np.log(n)/(n*p*L))))
print(delta)
```

```
[0.01414214 0.0106066  0.00883883 0.00795495 0.00751301
0.00729204 0.00718155]
```

Under this setting, we iterate the coordinate-wise MLE T_{iter} times.

```
def iter_coordinateMLE(x0,G,Y,p,L,Titer,xmin,xmax):
    t=np.arange(Titer)
    n=len(G[0])
    c=0.01
    delta= c*(np.sqrt(np.log(n)/(n*p*L)) \
      + 1/(2**t)*(np.sqrt(np.log(n)/(p*L)) \
      -np.sqrt(np.log(n)/(n*p*L))))
```

```
    xt=x0
    for it in range(Titer):
        xt1=np.zeros_like(xt)
        xMLE=coordinateMLE(xt,G,Y,xmin,xmax)
        for i in range(n):
            if np.abs(xMLE[i]-xt[i])>delta[it]:
              xt1[i]=xMLE[i]
            else: xt1[i]=xt[i]
        xt=xt1

    return xt

x_est=iter_coordinateMLE(x0,G,Y,p,L,7,min(x),max(x))

print(x0)
print(x_est)
print(x0-x_est)
```

```
[1.22440733 0.90509149 1.06241023 0.80809095]
[1.20499993 0.88509149 1.04241023 0.80809095]
[0.0194074  0.02       0.02       0.        ]
```

We see that the MLE solution is further away from the initial estimate.

Now we compare the performance of the PageRank variant and the advanced algorithm for a setting where $n = 1000$ and p spans the claimed limit.

```
import numpy as np
from scipy.stats import bernoulli
from numpy.random import binomial
import matplotlib.pyplot as plt

n=1000
K=5 # top-K ranking
L=20
Titer=7
p_range=np.linspace(0.02,0.12,30)
#ln(n)/n ~0.007
#nln(n)/(DeltaK**2)/(n*(n-1)*L/2)~0.0089

gap=0.1
# generate the ground truth score vector
x,DeltaK=scoreVector(gap,n,K)

MSE1 = np.zeros_like(p_range) # for pagerank
MSE2 = np.zeros_like(p_range) # for advanced algorithm
ITER=20
```

```
for idx,p in enumerate(p_range):
    for k in range(ITER):
        # Generate a comparison graph
        G,dmax=genGraph(n,p)
        # Transition probability matrix
        P,Y=transitionMatrix(x,n,L,G,dmax)
        # Compute the stationary distribution
        x0=power_method(P)
        # score scaling
        x0=x0/sum(x0)*n

        xest= \
        iter_coordinateMLE(x0,G,Y,p,L,Titer,min(x),max(x))
        # Compute MSE for pagerank and advanced algorithm
        MSE1[idx] += \
        sum(np.square(x-x0))/sum(np.square(x))/ITER
        MSE2[idx] += \
        sum(np.square(x-xest))/sum(np.square(x))/ITER

plimit=n*np.log(n)/(DeltaK**2)/(n*(n-1)*L/2)
p_norm=p_range/plimit

plt.figure(figsize=(5,5), dpi=100)
plt.plot(p_norm,MSE1,label='PageRank variant')
plt.plot(p_norm,MSE2,label='Advanced algorithm')
plt.yscale('log')
plt.title('Normalized MSE')
plt.legend()
plt.grid(linestyle=':', linewidth=0.5)
plt.show()
```

Fig. 3.32 shows the MSE performances of the PageRank variant and the advanced algorithm. We see an improvement with local refinement.

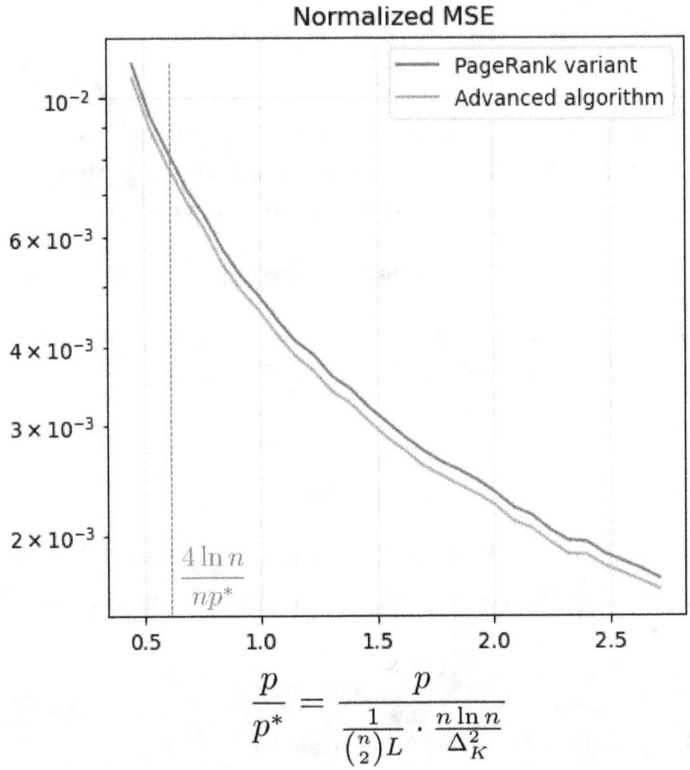

Figure 3.32. The MSE performances of the PageRank variant and the advanced algorithm as a function of observation probability p: $n = 1000$, $L = 20$, $K = 5$ and $\Delta_K = 0.125$.

Look ahead In the previous sections, we discussed the application of top-K ranking and studied an efficient algorithm that achieves the minimum sample complexity with some constant factor gap. Additionally, we implemented the algorithm using Python. As noted in the beginning of Part III, information-theoretic concepts like KL divergence and mutual information play a significant role in machine learning. In the next section, we will explore one such application.

Problem Set 9

Prob 9.1 (Ranking from pairwise comparisons) Suppose there exists a ground-truth ranking, say $\mathcal{R}$, of n items: e.g., $\mathcal{R} = \{1 \succ 2 \succ \cdots \succ n\}$. We wish to identify the ranking from pairwise comparisons, e.g., item i is preferred over item j. Let x_{ij} be a pairwise comparison between items i and j:

$$x_{ij} = \mathbf{1}\{\text{item } i \succ \text{item } j\}.$$

Suppose pairwise comparisons are observed uniformly at random. Each comparison for items (i,j) is observed with probability $p \in [0,1]$ independently over all (i,j)'s:

$$y_{ij} = \begin{cases} x_{ij}, & \text{w.p. } p; \\ \mathsf{e}, & \text{w.p. } 1-p. \end{cases}$$

Given y_{ij}'s, we wish to decode $\mathcal{R}$. Let $\hat{\mathcal{R}}$ be a decoded ranking. Let P_e be the probability of error: $P_e := \mathbb{P}(\hat{\mathcal{R}} \neq \mathcal{R})$. Consider an optimization problem. Given p,

$$P_e^*(p) := \min_{\text{algorithm},n} P_e.$$

For $\epsilon \in (0,1]$, what is $P_e^*(1-\epsilon)$? Also explain why.

Prob 9.2 (Ranking from pairwise comparisons) Suppose there exists a ground-truth ranking, say $\mathcal{R}$, of n items. We wish to identify the ranking of the n items from pairwise comparisons, e.g., item i is preferred over j. Suppose pairwise comparisons are given uniformly at random: items (i,j) are compared with probability $p \in [0,1]$ independently over all (i,j). Let $\hat{\mathcal{R}}$ be a decoded ranking. Let P_e be the probability of error: $P_e := \mathbb{P}(\hat{\mathcal{R}} \neq \mathcal{R})$. What is the number of pairwise comparisons on average required to make the probability of error arbitrarily close to 0 as $n \to \infty$?

Prob 9.3 (Top-K ranking) Consider the BTL model (Bradley and Terry, 1952) in Section 3.9. Let $\boldsymbol{x} := [x_1, x_2, \ldots, x_n]$ be the ground-truth score vector of n items where $x_i \in \mathbb{R}^+$. Assume that each pair of any two items is observed uniformly at random w.p. p and the observed pair has L independent copies:

$$y_{ij}^{(\ell)} = \begin{cases} \mathsf{Bern}\left(\frac{x_i}{x_i+x_j}\right), & \text{w.p. } p\,; \\ \mathsf{e}, & \text{w.p. } 1-p \end{cases}$$

where $y_{ij}^{(\ell)}$'s are i.i.d for all pairs (i,j) and $\ell \in \{1,2,\ldots,L\}$. Given $y_{ij}^{(\ell)}$'s, we wish to decode the set $f(\boldsymbol{x})$ of top-K ranked items (top-K partitioning). Let $\widehat{f(\boldsymbol{x})}$ be a decoded top-K set. Let P_e be the probability of error: $P_e := \mathbb{P}(\widehat{f(\boldsymbol{x})} \neq f(\boldsymbol{x}))$. In Section 3.9, we claimed that the minimum sample complexity required for reliable top-K ranking is:

$$\Theta\left(\frac{n \ln n}{\Delta_K^2}\right)$$

where $\Delta_K := \frac{x_K - x_{K+1}}{x_K}$.

(*a*) Explain what the notation $\Theta(\cdot)$ means.
(*b*) In Section 3.9, we claimed that the minimum sample complexity is a promising result, when comparing it to the total ordering case. Explain why.
(*c*) Show that if $p < \frac{\ln n}{n}$, P_e cannot be made arbitrarily 0 no matter what we do and whatsoever.
(*d*) Given that the (i,j) pair is observed, compute:

$$\lim_{L \to \infty} \frac{1}{L} \sum_{\ell=1}^{L} y_{ij}^{(\ell)}.$$

(*e*) Describe the PageRank variant (that we learned in Section 3.10).
(*f*) In Section 3.10, we introduced an additional stage, called *local refinement*. Explain why this stage is employed.

Prob 9.4 (Concavity) Consider a function:

$$f(x) = \frac{b}{x} - \frac{1}{x+c} \tag{3.59}$$

where $0 < x \leq 1$, $0 \leq b \leq 1$ and $0 < c \leq 1$. Prove that $f(x)$ is concave in x.

Prob 9.5 (A bounding technique) Consider a system with an input $X \in \{1, 2, \ldots, M\}$ and an output $Y \in \mathcal{Y}$. Here $\mathcal{Y}$ denotes a discrete alphabet. Let $\mathbb{P}_i$ be the probability distribution w.r.t. Y conditioned on $X = i$. Suppose X is uniformly distributed.

(*a*) Show that

$$I(X;Y) \leq \frac{1}{M^2} \sum_{i=1}^{M} \sum_{j=1}^{M} \mathsf{KL}(\mathbb{P}_i \| \mathbb{P}_j)$$

where $\mathsf{KL}(\cdot\|\cdot)$ indicates the KL divergence defined w.r.t. log base 2.
(*b*) Find a condition under which the equality holds in the above.

Prob 9.6 (True or False?)

(*a*) Let $\mathbf{A} \in \mathbb{R}^{n \times n}$ be a matrix with m positive eigenvalues λ_i's and eigenvectors $\mathbf{v}_i$'s:

$$\mathbf{A} := \lambda_1 \mathbf{v}_1 \mathbf{v}_1^T + \lambda_2 \mathbf{v}_2 \mathbf{v}_2^T + \cdots + \lambda_m \mathbf{v}_m \mathbf{v}_m^T$$

where $\lambda_1 > \lambda_2 \geq \lambda_3 \geq \cdots \geq \lambda_m$ and $\mathbf{v}_i$'s are orthonormal: $\mathbf{v}_i^T \mathbf{v}_j = \mathbf{1}\{i = j\}$. Let $\mathbf{v} \in \mathbb{R}^n$ be some non-zero vector. Then,

$$\frac{\mathbf{A}^k \mathbf{v}}{\sqrt{\|\mathbf{A}^k \mathbf{v}\|^2}} \longrightarrow \mathbf{v}_1$$

as $k \to \infty$.

3.12 Supervised Learning: Connection with Information Theory

Three key notions Part I focused on three essential notions in information theory: entropy, mutual information, and Kullback-Leibler divergence. In Parts I and II, we examined these notions in the context of Shannon's source and channel coding theorems. Entropy provides a concise way to determine the highest possible compression rate for an information source, while mutual information is a suitable metric for defining channel capacity.

Role of the notions in data science Throughout the remainder of Part III, we will delve into three different applications of machine learning and deep learning where the key notions we have discussed are crucial. Specifically, we will examine: (i) how entropy plays a significant role in one of the most prevalent methods of machine learning, known as supervised learning; (ii) the importance of KL divergence in another popular technique, unsupervised learning; and (iii) how mutual information guides the design of machine learning algorithms that promote fairness for disadvantaged groups in comparison to advantaged ones.

Outline In the upcoming sections, our focus will be on the role of entropy in supervised learning. We will begin by exploring what supervised learning is and then formulate an optimization problem that corresponds to it. We will then demonstrate that entropy plays a central role in designing an objective function that leads to an optimal architecture for the optimization problem. Finally, we will learn how to solve this optimization problem using a powerful algorithm known as *gradient descent*, which is a slight variation of gradient ascent that we covered in Section 3.11. To aid in the implementation, we will use TensorFlow, which is one of the most intuitive and widely-used deep learning frameworks. If you are unfamiliar with TensorFlow, please refer to a tutorial in Appendix B.

Machine learning Machine learning is about an algorithm that a computer system can execute by following a set of instructions. More formally, it refers to the study of algorithms used to train a computer system, enabling it to perform a specific task. A pictorial illustration of this concept can be found in Fig. 3.33. The goal is to develop a computer system (referred to as a machine) that can take in an input, denoted as x, and produce an output, denoted as y. This system or function is designed to carry out a task of interest. For instance, if a task is legitimate-emails

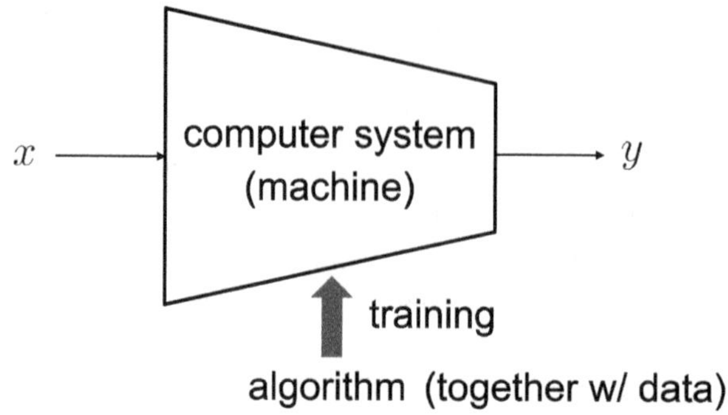

Figure 3.33. Machine learning is the study of algorithms which provide a set of instructions to a computer system so that it can perform a specific task of interest. Let the input x indicate information employed to perform a task. Let the output y denote a task result.

filtering against spams, x could be multi-dimensional quantities[4]: (i) frequency of a keyword like dollar signs $$$; and (ii) frequency of another keyword, say winner. And y could be an email entity, e.g., $y = +1$ indicates a legitimate email while $y = -1$ denotes a spam. In machine learning, such y is called a *label*. Or if an interested task is cat-vs-dog classification, x could be image-pixel values and y is a binary value indicating whether the fed image is a cat (say $y = 1$) or a dog ($y = 0$).

The essence of machine learning lies in designing algorithms that can train a computer system to perform a desired task effectively. This involves using *data* as a crucial component in the algorithm design process.

A remark on the naming From a machine's perspective, a *machine learns* the task from data. Hence, it is called machine learning, ML for short. This naming was originated in 1959 by Arthur Lee Samuel (Samuel, 1967). See Fig. 3.34.

Arthur Samuel is one of the pioneers in Artificial Intelligence (AI), which encompasses machine learning as a sub-field. AI involves the study of creating intelligence in machines, which differs from the natural intelligence observed in intelligent beings such as humans and animals.

One of Samuel's notable accomplishments in the early days of AI was the development of a computer player for the board game checkers (see the right figure in Fig. 3.34). He introduced several algorithms and concepts during the process of creating this computer program. These algorithms ultimately served as the foundation for AlphaGo (Silver *et al.*, 2016), a computer program developed for the

4. In machine learning, such quantities are called *features*. These refer to key components of data that well describe their characteristics.

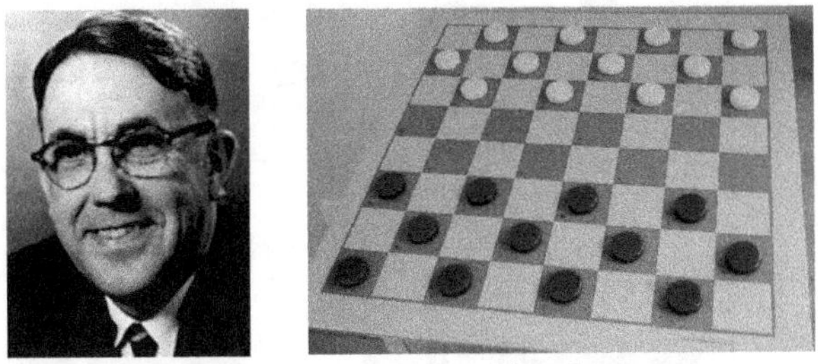

Figure 3.34. Arthur Lee Samuel is an American pioneer in artificial intelligence. One of his achievements in early days is to develop computer checkers which later formed the basis of AlphaGo.

board game Go. AlphaGo went on to defeat Lee Sedol, a professional 9-dan player, in 2016, winning 4 out of 5 games (News, 2016).

Mission of machine learning The ultimate objective of machine learning is to attain artificial intelligence. Thus, it can be regarded as one of the methodologies for achieving AI. As depicted in the block diagram in Fig. 3.33, the aim of ML is to develop an algorithm that enables the trained machine to exhibit behavior similar to intelligent beings.

Supervised learning There are some methodologies which help us to achieve the goal of ML. One specific yet popular method is:

Supervised Learning.

Supervised learning involves the process of learning a function $f(x)$ (which represents the machine's functionality) with the assistance of a supervisor, as illustrated in Fig. 3.35. The supervisor plays a crucial role in this process by providing input-output samples, which serve as the data used to train the machine. Typically, these input-output samples are represented as:

$$\{(x^{(i)}, y^{(i)})\}_{i=1}^{m} \tag{3.60}$$

where $(x^{(i)}, y^{(i)})$ indicates the ith input-output sample (or called a *training sample* or an *example*) and m denotes the number of samples. Using this notation (3.60), supervised learning is to:

$$\text{Estimate } f(\cdot) \text{ using the training samples } \{(x^{(i)}, y^{(i)})\}_{i=1}^{m}. \tag{3.61}$$

Optimization An effective approach to estimate $f(x)$ is through optimization. To fully grasp this concept, let's delve into the formal definition of optimization. Optimization refers to the selection of an optimization variable that minimizes or

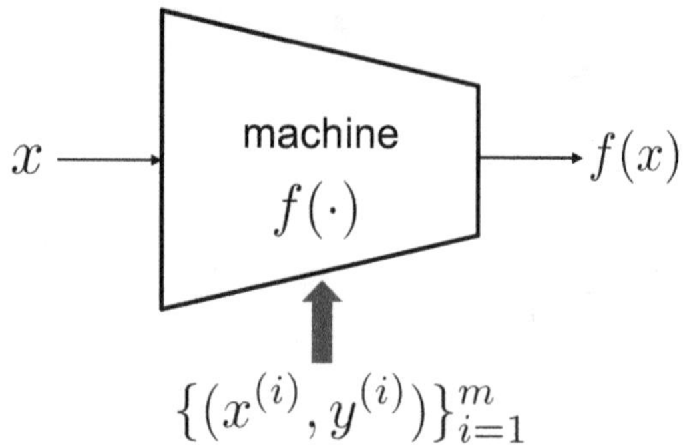

Figure 3.35. Supervised Learning: A methodology for designing a computer system $f(\cdot)$ with the help of a supervisor which offers input-output pair samples, called a train dataset $\{(x^{(i)}, y^{(i)})\}_{i=1}^{m}$.

maximizes a particular quantity of interest while taking any relevant constraints into account. Two significant factors come into play in this definition. Firstly, the optimization variable is a multi-dimensional quantity that can influence the quantity of interest and is subject to our design. Secondly, the quantity of interest that we aim to minimize or maximize is known as the objective function, and it is a one-dimensional scalar.

Objective function To figure this out, we need to know about the objective that supervised learning wishes to achieve. In view of the goal (3.61), what we want is:

$$y^{(i)} \approx f(x^{(i)}), \quad \forall i \in \{1, \ldots, m\}.$$

A natural question arises. How to quantify *closeness* (reflected in the "$\approx$" notation) between the two quantities: $y^{(i)}$ and $f(x^{(i)})$? One common way that has been used in the field is to employ a function, called a *loss* function, usually denoted by:

$$\ell(y^{(i)}, f(x^{(i)})). \tag{3.62}$$

One property that the loss function $\ell(\cdot, \cdot)$ should satisfy is that it should be small when the two arguments are close, while being zero when the two are identical. Using the loss function (3.62), one can formulate an optimization problem as:

$$\min_{f(\cdot)} \sum_{i=1}^{m} \ell(y^{(i)}, f(x^{(i)})). \tag{3.63}$$

How to introduce optimization variable? Unfortunately, there is no variable. Instead we have a different quantity that we can optimize over: *the function* $f(\cdot)$ that appears in (3.63). How to deal with such function optimization? There is one typical approach in the field. The approach is to specify a function class (e.g., linear or quadratic), represent the function with *parameters (or called weights)*, denoted by w, and then consider the weights as an optimization variable. Taking this approach, one can translate the problem (3.63) into:

$$\min_{w} \sum_{i=1}^{m} \ell(y^{(i)}, f_w(x^{(i)})) \tag{3.64}$$

where $f_w(x^{(i)})$ denotes the function $f(x^{(i)})$ parameterized by w.

The above optimization problem depends on how we define the two functions: (i) $f_w(x^{(i)})$ w.r.t. w; and (ii) the loss function $\ell(\cdot,\cdot)$. In machine learning, lots of works have been done for the choice of the functions.

Introduction of neural networks Around at the same time when the ML field was founded, one architecture was suggested for the first function $f_w(\cdot)$ in the context of simple binary classifiers in which y takes one among the two options. The architecture is called:

Perceptron,

and was invented in 1957 by one of the pioneers in AI, named Frank Rosenblatt (Rosenblatt, 1958). Frank Rosenblatt, a psychologist, was intrigued by the workings of the brains of intelligent beings. His research into this area led him to develop the Perceptron, which provided valuable insights into neural networks.

How brains work The architecture of Perceptron was inspired by the structure of the brain, which contains numerous electrically excitable cells, known as neurons; see Fig. 3.36. Each neuron is represented by a red circle in the figure, and there are three neurons shown. There are three key features of neurons that influenced the design of Perceptron. The first is that neurons possess electrical properties, and therefore have a voltage. The second feature is that neurons are interconnected with

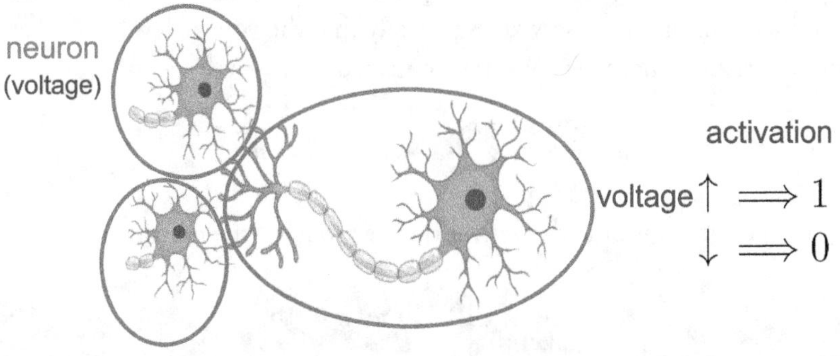

Figure 3.36. Neurons are electrically excitable cells and are connected through synapses.

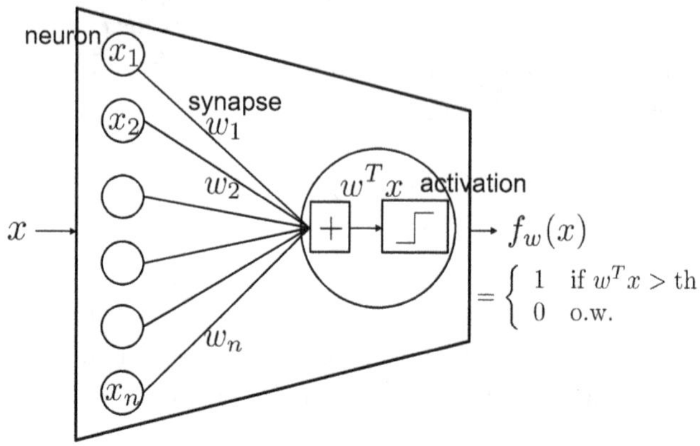

Figure 3.37. The architecture of Perceptron.

one another through channels called synapses, which facilitate the transmission of electrical voltage signals between neurons. Depending on the connectivity strength of a synapse, a voltage signal from one neuron to another can increase or decrease. Finally, a neuron produces an action, known as activation, by generating an all-or-nothing pulse depending on its voltage level. If the voltage level is above a specific threshold, it generates an impulse signal with a certain magnitude, say 1; otherwise, it produces nothing.

Perceptron Frank Rosenblatt proposed the Perceptron architecture based on the three properties mentioned earlier, as depicted in Fig. 3.37. Let x denote an n-dimensional real-valued signal, where x is represented as $[x_1, x_2, \ldots, x_n]^T$. Each component x_i is distributed to a neuron, and x_i is interpreted as the voltage level of the ith neuron. The voltage signal x_i is transmitted through a synapse to another neuron located on the right in the figure (indicated by a large circle). The voltage level can either increase or decrease, depending on the strength of the synapse's connectivity. To account for this, a weight w_i is multiplied to x_i, so that $w_i x_i$ is the delivered voltage signal at the terminal neuron. Rosenblatt introduced an adder that aggregates all voltage signals from multiple neurons to model the voltage signal at the terminal neuron. He observed empirically that the voltage level at the terminal neuron increases as more neurons are connected.

$$w_1 x_1 + w_2 x_2 + \cdots + w_n x_n = w^T x. \tag{3.65}$$

Lastly in an effort to mimic the activation, he modeled the output signal as

$$f_w(x) = \begin{cases} 1 & \text{if } w^T x > \text{th}, \\ 0 & \text{o.w.} \end{cases} \tag{3.66}$$

where "th" indicates a certain threshold level. It can also be simply denoted as

$$f_w(x) = \mathbf{1}\{w^T x > \text{th}\}. \tag{3.67}$$

Activation Taking the Perceptron architecture in Fig. 3.37, one can formulate the optimization problem (3.64) as:

$$\min_w \sum_{i=1}^{m} \ell(y^{(i)}, \mathbf{1}\{w^T x^{(i)} > \text{th}\}). \tag{3.68}$$

This is an initial optimization problem that was proposed, but an issue in solving this optimization was discovered. The problem lies in the objective function containing an indicator function, which makes it non-differentiable. This non-differentiability presents a difficulty in solving the problem. The reason for this is that the prominent algorithm that we learned in Section 3.11, gradient ascent (or descent), involves derivative operations, which cannot be applied when the function is non-differentiable.

To address this problem, one common approach that has been taken in the field is to approximate the activation function. There are several ways to achieve this approximation. From below, we will explore one popular method.

Logistic regression The popular approximation approach is to take a *smooth* transition from 0 to 1 for the abrupt indicator function:

$$f_w(x) = \frac{1}{1 + e^{-w^T x}}. \tag{3.69}$$

Notice that $f_w(x) \approx 0$ when $w^T x$ is very small; it then grows exponentially with an increase in $w^T x$; later grows logarithmically; and finally saturates as 1 when $w^T x$ is very large. See Fig. 3.38. The function (3.69) is a very popular one used in statistics,

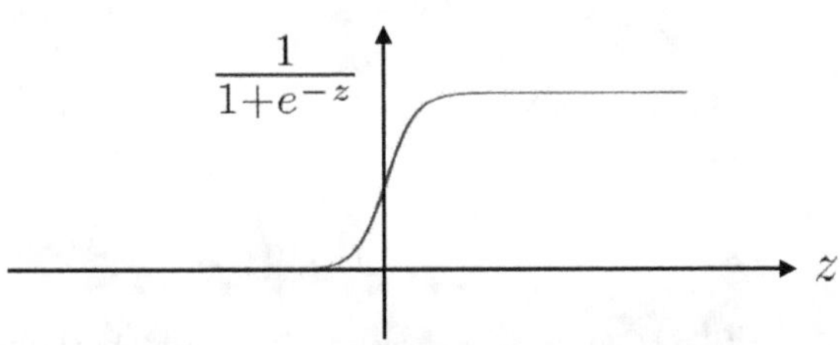

Figure 3.38. Logistic function: $\sigma(z) = \frac{1}{1+e^{-z}}$.

called the *logistic*[5] function (Garnier and Quetelet, 1838). There is another name: the *sigmoid*[6] function.

There are two good things about the logistic function. First it is differentiable. Second, it can serve as the probability for the output in the binary classifier, e.g., $\mathbb{P}(y = 1)$ where y denotes the ground-truth label in the binary classifier. So it is interpretable.

Look ahead Assuming the logistic activation function, what would be an appropriate loss function? In a certain sense, the design of an optimal loss function is closely related to the concept of entropy. In the following section, we will explore how this entropy-related concept is used to design the optimal loss function.

5. The word *logistic* comes from a Greek word which means a slow growth, like a logarithmic growth.

6. Sigmoid means resembling the lower-case Greek letter sigma, S-shaped.

3.13 Supervised Learning: Logistic Regression and Cross Entropy

Recap In the previous section, we formulated an optimization problem for supervised learning based on the Perceptron architecture:

$$\min_{w} \sum_{i=1}^{m} \ell(y^{(i)}, f_w(x^{(i)})). \tag{3.70}$$

As an activation function, we considered a logistic function:

$$f_w(x) = \frac{1}{1 + e^{-w^T x}}. \tag{3.71}$$

We then claimed that an entropy-related concept plays a role in the design of the *optimal* loss function.

Outline This section will demonstrate the validity of the claim in three steps. Firstly, we will explore the definition of the optimal loss function. Secondly, we will examine the role of entropy in the design of the optimal loss function. Finally, we will discuss how to solve the optimization problem.

Optimality in a sense of maximizing likelihood Logistic regression is a binary classifier that uses the logistic function (3.71). Fig. 3.39 provides a visual representation of logistic regression. It should be noted that the output $\hat{y}$ of logistic regression falls between 0 and 1.

$$0 \leq \hat{y} \leq 1.$$

Hence, one can interpret this as a probability quantity. The optimality of a classifier can be defined under the following assumption inspired by the probabilistic interpretation:

$$\text{Assumption} : \hat{y} = \mathbb{P}(y = 1|x). \tag{3.72}$$

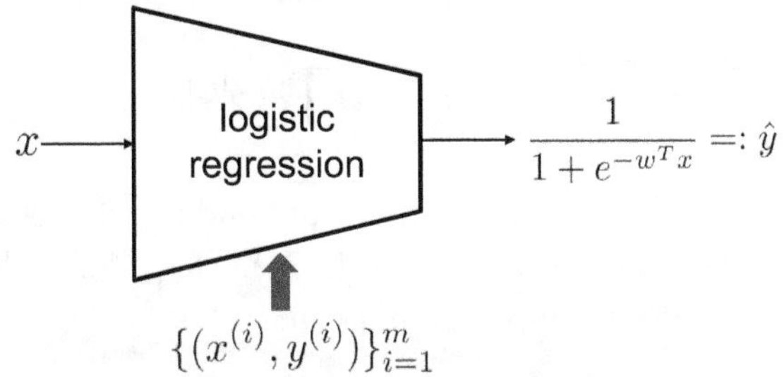

Figure 3.39. Logistic regression.

To understand what it means, consider the likelihood of the ground-truth classifier:

$$\mathbb{P}(\{y^{(i)}\}_{i=1}^{m} | \{x^{(i)}\}_{i=1}^{m}). \tag{3.73}$$

Notice that the classifier output $\hat{y}$ is a function of weights w. Hence, assuming (3.72), the likelihood (3.73) is also a function of w.

We are now ready to define the optimal w. The optimal weight, say w^*, is defined as the one that *maximizes the likelihood* (3.73):

$$w^* := \arg\max_{w} \mathbb{P}(\{y^{(i)}\}_{i=1}^{m} | \{x^{(i)}\}_{i=1}^{m}). \tag{3.74}$$

There are other ways to define the optimality. Here, we employ the maximum likelihood principle, the most popular choice. This is exactly where the definition of the *optimal loss function*, say $\ell^*(\cdot,\cdot)$ kicks in. We say that $\ell^*(\cdot,\cdot)$ is the one that satisfies:

$$\arg\min_{w} \sum_{i=1}^{m} \ell^*(y^{(i)}, \hat{y}^{(i)}) = \arg\max_{w} \mathbb{P}(\{y^{(i)}\}_{i=1}^{m} | \{x^{(i)}\}_{i=1}^{m}). \tag{3.75}$$

As mentioned earlier, an entropy-related concept appears in $\ell^*(\cdot,\cdot)$. From below, we will figure this out.

Finding the optimal loss function $\ell^*(\cdot,\cdot)$ Usually samples are obtained from different data $x^{(i)}$'s. Hence, it is reasonable to assume that such samples are independent with each other:

$$\{(x^{(i)}, y^{(i)})\}_{i=1}^{m} \text{ are independent over } i. \tag{3.76}$$

Under this assumption, we can rewrite the likelihood (3.73) as:

$$\begin{aligned}
\mathbb{P}(\{y^{(i)}\}_{i=1}^{m} | \{x^{(i)}\}_{i=1}^{m}) &\overset{(a)}{=} \frac{\mathbb{P}(\{(x^{(i)}, y^{(i)})\}_{i=1}^{m})}{\mathbb{P}(\{x^{(i)}\}_{i=1}^{m})} \\
&\overset{(b)}{=} \frac{\prod_{i=1}^{m} \mathbb{P}(x^{(i)}, y^{(i)})}{\prod_{i=1}^{m} \mathbb{P}(x^{(i)})} \\
&\overset{(c)}{=} \prod_{i=1}^{m} \mathbb{P}(y^{(i)} | x^{(i)})
\end{aligned} \tag{3.77}$$

where (a) and (c) are due to the definition of conditional probability; and (b) comes from the independence assumption (3.76). Here $\mathbb{P}(x^{(i)}, y^{(i)})$ denotes the probability distribution of the input-output pair:

$$\mathbb{P}(x^{(i)}, y^{(i)}) := \mathbb{P}(X = x^{(i)}, Y = y^{(i)}) \tag{3.78}$$

where X and Y indicate random variables of the input and the output, respectively.

Recall the assumption (3.72) made with regard to $\hat{y}$:

$$\hat{y} = \mathbb{P}(y = 1|x).$$

This implies that:

$$\begin{aligned} y = 1: &\quad \mathbb{P}(y|x) = \hat{y}; \\ y = 0: &\quad \mathbb{P}(y|x) = 1 - \hat{y}. \end{aligned}$$

Hence, one can represent $\mathbb{P}(y|x)$ as:

$$\mathbb{P}(y|x) = \hat{y}^{y}(1 - \hat{y})^{1-y}.$$

Using the notations of $(x^{(i)}, y^{(i)})$ and $\hat{y}^{(i)}$, we get:

$$\mathbb{P}(y^{(i)}|x^{(i)}) = (\hat{y}^{(i)})^{y^{(i)}}(1 - \hat{y}^{(i)})^{1-y^{(i)}}.$$

Plugging this into (3.77), we get:

$$\begin{aligned} &\mathbb{P}(\{y^{(i)}\}_{i=1}^{m}|\{x^{(i)}\}_{i=1}^{m}) \\ &= \prod_{i=1}^{m} \mathbb{P}(x^{(i)}) \prod_{i=1}^{m} (\hat{y}^{(i)})^{y^{(i)}}(1 - \hat{y}^{(i)})^{1-y^{(i)}}. \end{aligned} \tag{3.79}$$

This together with (3.74) yields:

$$\begin{aligned} w^* &:= \arg\max_{w} \prod_{i=1}^{m} (\hat{y}^{(i)})^{y^{(i)}}(1 - \hat{y}^{(i)})^{1-y^{(i)}} \\ &\overset{(a)}{=} \arg\max_{w} \sum_{i=1}^{m} y^{(i)} \log \hat{y}^{(i)} + (1 - y^{(i)}) \log(1 - \hat{y}^{(i)}) \\ &\overset{(b)}{=} \arg\min_{w} \sum_{i=1}^{m} -y^{(i)} \log \hat{y}^{(i)} - (1 - y^{(i)}) \log(1 - \hat{y}^{(i)}) \end{aligned} \tag{3.80}$$

where (a) comes from the fact that $\log(\cdot)$ is a non-decreasing function and $\prod_{i=1}^{m} (\hat{y}^{(i)})^{y^{(i)}}(1 - \hat{y}^{(i)})^{1-y^{(i)}}$ is positive; and (b) is due to changing the sign of the objective while replacing max with min.

The term inside the summation in the last equality in (3.80) respects the formula of another key notion in information theory: *cross entropy*. In the context of a loss function, it is named cross entropy loss:

$$\ell_{\mathsf{CE}}(y, \hat{y}) := -y \log \hat{y} - (1 - y) \log(1 - \hat{y}). \tag{3.81}$$

Hence, the optimal loss function that yields the maximum likelihood solution is cross entropy loss:

$$\ell^*(\cdot,\cdot) = \ell_{\mathsf{CE}}(\cdot,\cdot).$$

Remarks on cross entropy loss (3.81) Let us say a few words about why the loss function (3.81) is called cross entropy loss. This naming comes from the definition of cross entropy. The cross entropy is defined w.r.t. two random variables. For simplicity, consider two binary random variables, say $X \sim \mathsf{Bern}(p)$ and $Y \sim \mathsf{Bern}(q)$. For such two random variables, cross entropy is defined as:

$$H(p,q) := -p\log q - (1-p)\log(1-q). \tag{3.82}$$

Notice that the formula of (3.81) is exactly the same as the term inside summation in (3.80), except for having different notations. Hence, it is called *cross entropy loss*. You may wonder why $H(p,q)$ in (3.82) is called cross entropy. The rationale comes from the following fact (check this in Prob 10.4):

$$H(p,q) \geq H(p) := -p\log p - (1-p)\log(1-p) \tag{3.83}$$

where $H(p)$ denotes the entropy of $\mathsf{Bern}(p)$. In Prob 10.4, you will be asked to verify that the equality holds when $p = q$. One can interpret $H(p,q)$ as an *entropic-*measure of discrepancy a*cross* distributions. Hence, it is called *cross entropy*.

How to solve logistic regression? In view of (3.80), the optimization problem for logistic regression can be written as:

$$\min_{w} \sum_{i=1}^{m} -y^{(i)} \log \frac{1}{1+e^{-w^T x^{(i)}}} - (1-y^{(i)}) \log \frac{e^{-w^T x^{(i)}}}{1+e^{-w^T x^{(i)}}}. \tag{3.84}$$

Let $J(w)$ be the normalized version of the objective function:

$$J(w) := \frac{1}{m}\sum_{i=1}^{m} -y^{(i)}\log \hat{y}^{(i)} - (1-y^{(i)})\log(1-\hat{y}^{(i)}). \tag{3.85}$$

It turns out the above optimization belongs to *convex optimization*. In other words, $J(w)$ is convex in optimization variable w. For the rest of this section, we will prove the convexity, and then discuss how to solve the problem.

Proof of convexity First, one can readily show that convexity preserves under addition. Why? Think about the definition of convex functions. So it suffices to prove the following two:

$$\text{(i)} - \log \frac{1}{1 + e^{-w^T x}} \text{ is convex in } w;$$

$$\text{(ii)} - \log \frac{e^{-w^T x}}{1 + e^{-w^T x}} \text{ is convex in } w.$$

Since the second function in the above can be represented as the sum of a linear function and the first function:

$$-\log \frac{e^{-w^T x}}{1 + e^{-w^T x}} = w^T x - \log \frac{1}{1 + e^{-w^T x}},$$

it suffices to prove the convexity of the first function.

The first function can be rewritten as:

$$-\log \frac{1}{1 + e^{-w^T x}} = \log(1 + e^{-w^T x}). \tag{3.86}$$

In fact, proving the convexity of the function (3.86) is a bit involved if one relies on the definition of convex functions. There is another way to prove. It is based on the second derivative of a function, called the Hessian. How to compute the Hessian? What is the dimension of the Hessian? For a function $f : \mathbb{R}^d \to \mathbb{R}$, the gradient $\nabla f(x) \in \mathbb{R}^d$ and the Hessian $\nabla^2 f(x) \in \mathbb{R}^{d \times d}$. If you are not familiar, check it from the vector calculus course or from wikipedia.

A well-known fact says that if the Hessian of a function is positive semi-definite (PSD)[7], then the function is convex. Check this in Prob 10.5. If this proof is too much, you may want to remember this fact only. No need to prove, but the statement itself is useful. Here we will use this fact to prove the convexity of the function (3.86).

Taking a derivative of the RHS formula in (3.86) w.r.t. w, we get:

$$\nabla_w \log(1 + e^{-w^T x}) = \frac{1}{\ln 2} \frac{-x e^{-w^T x}}{1 + e^{-w^T x}}.$$

This is due to a chain rule of derivatives and the fact that $\frac{d}{dz} \ln z = \frac{1}{z}$, $\frac{d}{dz} e^z = e^z$ and $\frac{d}{dw} w^T x = x$. Taking another derivative of the above, we obtain a Hessian as

7. We say that a symmetric matrix, say $Q = Q^T \in \mathbf{R}^{d \times d}$, is positive semi-definite if $v^T Q v \geq 0, \forall v \in \mathbf{R}^d$, i.e., all the eigenvalues of Q are non-negative. It is simply denoted by $Q \succeq 0$.

follows:

$$
\begin{aligned}
\nabla_w^2 \log(1+e^{-w^T x}) &= \nabla_w \left(\frac{1}{\ln 2} \frac{-xe^{-w^T x}}{1+e^{-w^T x}} \right) \\
&\overset{(a)}{=} \frac{1}{\ln 2} \frac{xx^T e^{-w^T x}(1+e^{-w^T x}) - xx^T e^{-w^T x} e^{-w^T x}}{(1+e^{-w^T x})^2} \\
&= \frac{1}{\ln 2} \frac{xx^T e^{-w^T x}}{(1+e^{-w^T x})^2} \\
&\succeq 0
\end{aligned}
\tag{3.87}
$$

where (a) is due to the derivative rule of a quotient of two functions: $\frac{d}{dz}\frac{f(z)}{g(z)} = \frac{f'(z)g(z)-f(z)g'(z)}{g^2(z)}$. You may wonder why $\frac{d}{dw}(-xe^{-w^T x}) = xx^T e^{-w^T x}$. Why not xx, $x^T x^T$ or $x^T x$ in front of $e^{-w^T x}$? One rule-of-thumb is to simply try all the candidates and choose the one which does not have a syntax error (matrix dimension mismatch). For instance, xx (or $x^T x^T$) is just an invalid operation. $x^T x$ is not a right one because the Hessian must be an d-by-d matrix. The only candidate left without any syntax error is xx^T. We see that xx^T has the single eigenvalue of $\|x\|^2$. Why? Since the eigenvalue $\|x\|^2$ is non-negative, the Hessian is PSD, and therefore we prove the convexity.

Gradient descent How to solve the convex optimization problem (3.80)? Since there is no constraint in the optimization, w^* must be the *stationary* point, i.e., the one such that

$$\nabla J(w^*) = 0. \tag{3.88}$$

However, we face a challenge in determining the optimal point w^* because there is no closed-form solution and it cannot be analytically derived. Various algorithms have been developed to overcome this challenge and find the optimal point without the need for a closed-form solution. One such algorithm is *gradient ascent*, which we learned about in Section 3.11. In this case, since the function of interest is convex, we use *gradient descent* instead. The process of gradient descent is similar to that of gradient ascent, with the only difference being that the estimate is updated in the opposite direction of the gradient.

$$w^{(t+1)} \longleftarrow w^{(t)} - \alpha \nabla J(w^{(t)}) \tag{3.89}$$

where $w^{(t)}$ the tth estimate of w^* and $\alpha > 0$ indicates the learning rate. We take the opposite direction because the sign of the gradient of a convex function is flipped relative to that w.r.t. a concave function. See Fig. 3.40.

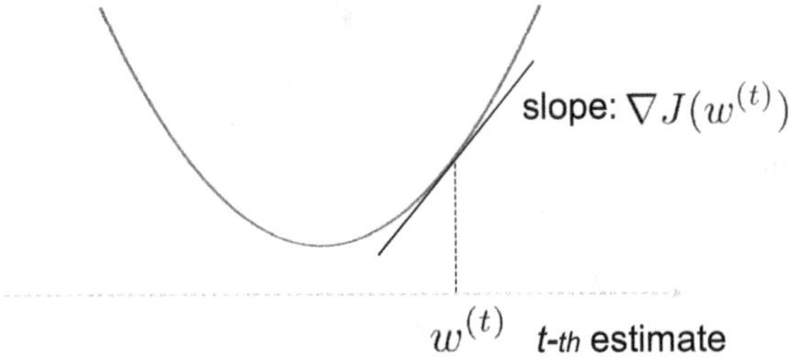

$$w^{(t+1)} \leftarrow w^{(t)} - \alpha \nabla J(w^{(t)})$$

Figure 3.40. How gradient descent works.

Look ahead We have established an optimization problem for supervised learning and observed the crucial role of cross entropy in designing the optimal loss function. Moreover, we acquired knowledge about solving the problem with the widely-used gradient descent algorithm. In the following section, we will put this algorithm into practice using TensorFlow to create a simple classifier.

3.14 Supervised Learning: TensorFlow Implementation

Recap In the previous sections, we have formulated an optimization problem for supervised learning:

$$\min_{w} \sum_{i=1}^{m} \ell_{\mathsf{CE}}(y^{(i)}, \hat{y}^{(i)}) \tag{3.90}$$

where $\hat{y} := \frac{1}{1+e^{-w^T x}}$ indicates the prediction output with logistic activation and ℓ_{CE} denotes cross entropy loss:

$$\ell_{\mathsf{CE}}(y, \hat{y}) = -y \log \hat{y} - (1-y) \log(1-\hat{y}). \tag{3.91}$$

We proved that cross entropy loss $\ell_{\mathsf{CE}}(\cdot, \cdot)$ is the optimal loss function in a sense of maximizing the likelihood. We also showed that the normalized version $J(w)$ of the above objection function is convex in w, and therefore, it can be solved via gradient descent:

$$J(w) := \frac{1}{m} \sum_{i=1}^{m} -y^{(i)} \log \hat{y}^{(i)} - (1-y^{(i)}) \log(1-\hat{y}^{(i)}). \tag{3.92}$$

Outline In this section, we will explore the implementation of the algorithm using a software tool for a simple classifier. This section is divided into three parts. Firstly, we will examine the setting of the simple classifier that we will focus on. In the second part, we will discuss four implementation details regarding the classifier. The first detail is the dataset used for training and testing. In machine learning, testing refers to evaluating the performance of a trained model. For this purpose, we will use an unseen dataset called the *test dataset*, which has never been used during training. The second detail is how to build a deep neural network model with the ReLU activation function. The Perceptron, introduced in Section 3.12, is the first neural network. A deep neural network is an extended version of the Perceptron, with at least one hidden layer placed between the input and output layers (Ivakhnenko, 1971). The ReLU is a popular activation function often used in hidden layers (Glorot *et al.*, 2011). It stands for Rectified Linear Unit, and its operation is given by $\mathsf{ReLU}(x) = \max(0, x)$. The third implementation detail concerns the softmax activation used at the output layer, which is a natural extension of the logistic activation for multiple classes. The fourth implementation detail pertains to the Adam optimizer, which is an advanced version of gradient descent and widely used in practice (Kingma and Ba, 2014).

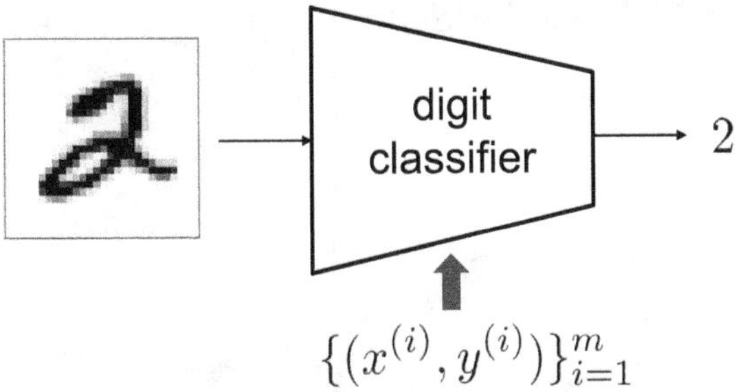

Figure 3.41. Handwritten digit classification.

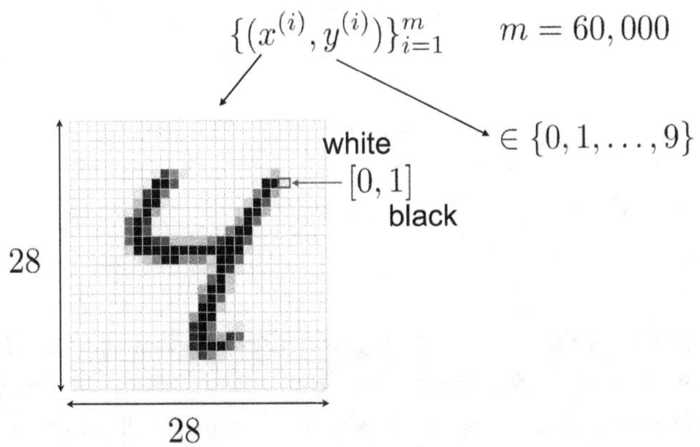

Figure 3.42. MNIST dataset: An input image is of 28-by-28 pixels, each indicating an intensity from 0 (white) to 1 (black); each label with size 1 takes one of the 10 classes from 0 to 9.

The final part of this section will focus on programming the classifier using TensorFlow, a popular deep learning framework. Specifically, we will be using Keras, a high-level programming language that is fully integrated with TensorFlow.

Handwritten digit classification We will be focusing on a simple classifier that aims to recognize handwritten digits from images, as shown in Fig. 3.41. For training our model, we will be using a widely popular dataset known as the MNIST (Modified National Institute of Standards and Technology) dataset (LeCun *et al.*, 1998), which contains $m = 60,000$ training images and $m_{\text{test}} = 10,000$ testing images. This dataset was created by remixing the examples from NIST's original dataset and was named after its creator, Yann LeCun. Each image, denoted as $x^{(i)}$, comprises of a 28×28 pixel matrix, with each pixel representing a grayscale level ranging from 0 (white) to 1 (black). Additionally, each image has a corresponding label, denoted as $y^{(i)}$, which falls under one of the ten classes, $y^{(i)} \in \{0, 1, \ldots, 9\}$, as shown in Fig. 3.42.

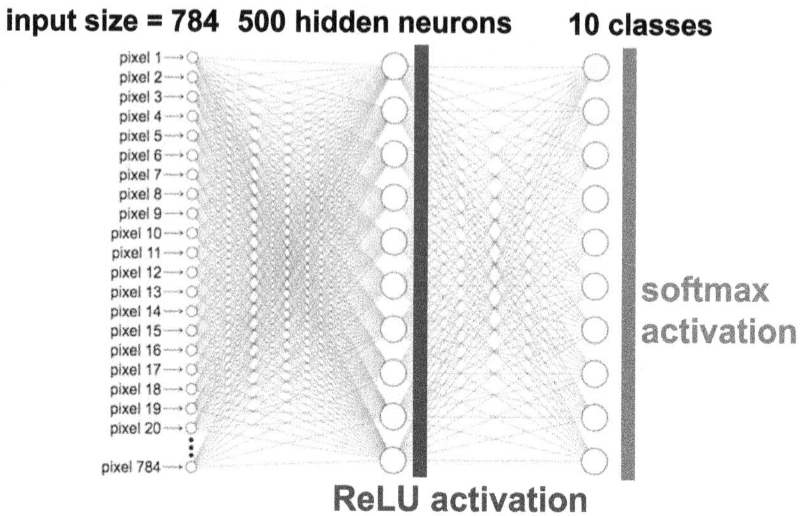

Figure 3.43. A two-layer fully-connected neural network where input size is 28 × 28 = 784, the number of hidden neurons is 500 and the number of classes is 10. We employ ReLU activation at the hidden layer, and softmax activation at the output layer; see Fig. 3.44 for details.

A deep neural network model We will use an advanced version of logistic regression as our model for the handwritten digit classifier. The first reason for using this advanced version is that logistic regression is a linear classifier and its performance may not be optimal in many applications since the prediction function is restricted to a linear function class. To overcome this limitation, researchers developed a Perceptron-like neural network with multiple layers, known as a deep neural network (DNN). Although the DNN was invented in the 1960s (Ivakhnenko, 1971), its performance benefits started to be greatly appreciated only in the past decade, due to a big event in 2012. Geoffrey Hinton and his PhD students achieved human-level recognition performance on ImageNet recognition competition using a DNN, which was never achieved before (Krizhevsky *et al.*, 2012). This event marked the start of the deep learning revolution. Since a linear classifier does not perform well for digit classification, we will use a simple version of DNN with only two layers – a hidden layer and an output layer, as shown in Fig. 3.43. By convention, the input layer is not counted as a layer, so it is referred to as a two-layer neural network instead of a three-layer one.

Each neuron in the hidden layer respects the same procedure as that in the Perceptron: a linear operation followed by activation. For activation, the logistic function or its shifted version, called the tanh function (spanning -1 and $+1$), were frequently employed in early days. However, a significant number of experts and practitioners have discovered that the Rectified Linear Unit (ReLU) is a more powerful function that enables faster training and delivers better or equivalent

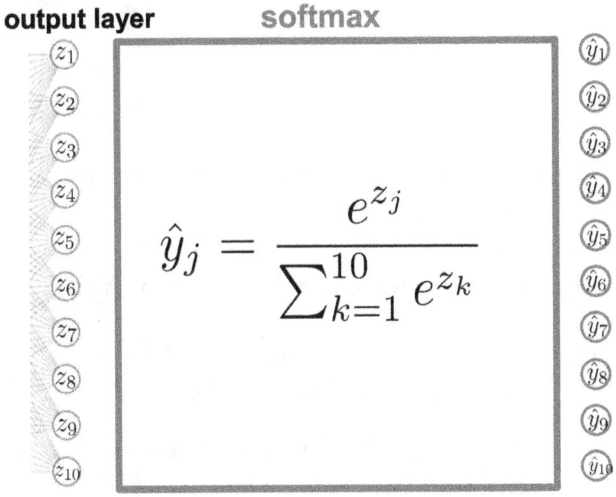

Figure 3.44. Softmax activation employed at the output layer. This is a natural extension of logistic activation intended for the two-class case.

performance (Glorot *et al.*, 2011). As stated previously, ReLU's function can be expressed as: $\text{ReLU}(x) = \max(0, x)$. In the deep learning community, a popular approach is to utilize ReLU activation in all hidden layers, and we will follow this standard as depicted in Fig. 3.43.

Softmax activation at the output layer We have another reason to use an advanced version of logistic regression, which pertains to the number of classes in our classifier. Since logistic regression is designed for binary classification, it cannot be directly used for our digit classifier, which has 10 classes. To address this, we need to use a generalized version of logistic function known as softmax. The operation of softmax is illustrated in Fig. 3.44.

Let z be the output of the last layer in a neural network prior to activation:

$$z := [z_1, z_2, \ldots, z_c]^T \in \mathbb{R}^c \tag{3.93}$$

where c denotes the number of classes. The softmax function is then defined as:

$$\hat{y}_j := [\text{softmax}(z)]_j = \frac{e^{z_j}}{\sum_{k=1}^{c} e^{z_k}} \quad j \in \{1, 2, \ldots, c\}. \tag{3.94}$$

Note that this is a natural extension of the logistic function: for $c = 2$,

$$\hat{y}_1 := [\text{softmax}(z)]_1 = \frac{e^{z_1}}{e^{z_1} + e^{z_2}}$$

$$= \frac{1}{1 + e^{-(z_1 - z_2)}}$$
$$= \sigma(z_1 - z_2) \tag{3.95}$$

where $\sigma(\cdot)$ is the logistic function. Viewing $z_1 - z_2$ as the binary classifier output $\hat{y}$, this coincides with the logistic function.

Here $\hat{y}_i$ can be interpreted as the probability that the ith example belongs to class i. Hence, like the binary classifier, one may want to assume:

$$\hat{y}_i = \mathbb{P}(y = [0, \ldots, \underbrace{1}_{i\text{th position}}, \ldots, 0]^T | x), \quad i \in \{1, \ldots, c\}. \tag{3.96}$$

Under this assumption, one can verify that the optimal loss function (in a sense of maximizing likelihood) is again cross entropy loss:

$$\ell^*(y, \hat{y}) = \ell_{\text{CE}}(y, \hat{y}) = \sum_{j=1}^{c} -y_j \log \hat{y}_j$$

where y indicates a label of one-hot vector type. For instance, in the case of label $= 2$ with $c = 10$, y takes:

$$y = \begin{bmatrix} 0 \\ 0 \\ 1 \\ 0 \\ 0 \\ 0 \\ 0 \\ 0 \\ 0 \\ 0 \end{bmatrix} \begin{matrix} 0 \\ 1 \\ 2 \\ 3 \\ 4 \\ 5 \\ 6 \\ 7 \\ 8 \\ 9 \end{matrix}$$

The proof is almost the same as that in the binary classifier. So we will omit the proof. Instead you will have a chance to prove it in Prob 10.3.

Due to the above rationales, softmax activation has been widely used for many classifiers. Hence, we will use the conventional activation in our digit classifier.

Adam optimizer (Kingma and Ba, 2014) We employ a specific algorithm. As mentioned earlier, we will use an advanced version of gradient descent, called the

Adam optimizer. To see how the optimizer operates, let us first recall the vanilla gradient descent:

$$w^{(t+1)} \leftarrow w^{(t)} - \alpha \nabla J(w^{(t)})$$

where $w^{(t)}$ indicates the estimated weight in the t-th iteration, and α denotes the learning rate. Notice that the weight update relies only on the *current* gradient, reflected in $\nabla J(w^{(t)})$. Hence, in case $\nabla J(w^{(t)})$ fluctuates too much over iterations, the weight update oscillates significantly, thereby bringing about unstable training. To address this, people often use a variant algorithm that exploits *past* gradients for the purpose of stabilization. That is, the Adam optimizer.

Here is how Adam works. The weight update takes the following formula instead:

$$w^{(t+1)} = w^{(t)} + \alpha \frac{m^{(t)}}{\sqrt{s^{(t)}} + \epsilon} \tag{3.97}$$

where $m^{(t)}$ indicates a weighted average of the current and past gradients:

$$m^{(t)} = \frac{1}{1-\beta_1^t}(\beta_1 m^{(t-1)} - (1-\beta_1)\nabla J(w^{(t)})). \tag{3.98}$$

Here $\beta_1 \in [0, 1]$ is a hyperparameter that captures the weight of past gradients, and hence it is called the *momentum* (Polyak, 1964). The notation m stands for momentum. The factor $\frac{1}{1-\beta_1^t}$ is applied in front, in an effort to stabilize training in initial iterations (small t). Check the detailed rationale behind this in Prob 10.7.

$s^{(t)}$ is a normalization factor that makes the effect of $\nabla J(w^{(t)})$ almost constant over t. In case $\nabla J(w^{(t)})$ is too big or too small, we may have significantly different scalings in magnitude. Similar to $m^{(t)}$, $s^{(t)}$ is defined as a weighted average of the current and past values (Hinton *et al.*, 2012):

$$s^{(t)} = \frac{1}{1-\beta_2^t}(\beta_2 s^{(t-1)} + (1-\beta_2)(\nabla J(w^{(t)}))^2) \tag{3.99}$$

where $\beta_2 \in [0, 1]$ denotes another hyperparameter that captures the weight of past values, and s stands for square.

Notice that the dimensions of $w^{(t)}$, $m^{(t)}$ and $s^{(t)}$ are the same. All the operations that appear in the above (including division in (3.97) and square in (3.99)) are component-wise. In (3.97), ϵ is a tiny value introduced to avoid division by 0 in practice (usually 10^{-8}).

TensorFlow: Loading MNIST data We will learn how to implement the simple digit classifier using TensorFlow programming. The first step is to load the MNIST dataset, which is a well-known dataset and is available in the following package:

tensorflow.keras.datasets. Even more, train and test datasets are already therein with a proper split ratio. So we do not need to worry about how to split them. The only script that we should write is:

```
from tensorflow.keras.datasets import mnist
(X_train, y_train), (X_test, y_test) = mnist.load_data()
X_train = X_train/255.
X_test = X_test/255.
```

We divide the input (X_train or X_test) by its maximum value 255 for the purpose of normalization. This procedure is done as a part of data preprocessing.

TensorFlow: A two-layer DNN In order to implement the simple DNN, illustrated in Fig. 3.43, we rely upon two major packages:

(i) tensorflow.keras.models;

(ii) tensorflow.keras.layers.

The models package contains several functionalities regarding a neural network. One major module is Sequential which is a neural network entity and hence can be described as a linear stack of layers. The layers package includes many elements of a neural network. Examples include fully-connected dense layers and activation functions. These two allow us to readily construct a model illustrated in Fig. 3.43.

```
from tensorflow.keras.models import Sequential
from tensorflow.keras.layers import Dense, Flatten

model = Sequential()
model.add(Flatten(input_shape=(28,28)))
model.add(Dense(500, activation='relu'))
model.add(Dense(10,  activation='softmax'))
```

Flatten is an entity that indicates a vector expanded from a higher dimensional one, like a 2D matrix. In this example, a digit image of size 28-by-28 is flattened into a vector of size $784(= 28 \times 28)$. add() is a method for attaching an interested layer to the last part in the sequential model. Dense refers to a fully-connected layer. The input size is automatically determined by the last part that it will be attached to. The only thing to specify is the number of output neurons. In this example, 500 refers to the number of hidden neurons. We can also set an activation function with another argument, like activation='relu'. The output layer comes with 10 neurons (coinciding with the number of classes) and softmax activation.

TensorFlow: Training a model For training, we need to first set up an algorithm (optimizer) to be employed. We use the Adam optimizer. As mentioned earlier, Adam has three key hyperparameters: (i) the learning rate α; (ii) β_1 (capturing the weight of past gradients); and (iii) β_2 (indicating the weight of the square of past gradients). The default choice reads: $(\alpha, \beta_1, \beta_2) = (0.001, 0.9, 0.999)$. These values would be set if nothing is specified.

Next, we specify a loss function. We employ the optimal loss function: cross entropy loss. A performance metric that we will look at during training and testing can also be specified. One common metric is accuracy. One can set all of these via another method compile.

```
model.compile(optimizer='adam',
              loss='sparse_categorical_crossentropy',
              metrics=['acc'])
```

The option optimizer='adam' sets the default choice of the learning rate and betas. For a manual choice, we define:

```
opt=tensorflow.keras.optimizers.Adam(
                    learning_rate=0.01,
                    beta_1 = 0.92,
                    beta_2 = 0.992)
```

We then replace the above option with optimizer=opt. As for the loss option in compile, we employ 'sparse_categorical_crossentropy', which indicates cross entropy loss beyond the binary case.

Now we can bring this to train the model on MNIST data. During training, we employ a part of the entire examples to compute a gradient of a loss function. The part is called a *batch*. Two more terminologies. One is the *step* which refers to a loss computation procedure spanning the examples only in a single batch. The other is the *epoch* which refers to the entire procedure associated with all the examples. In our experiment, we use the batch size of 64 and the number 20 of epochs.

```
model.fit(X_train,y_train,batch_size=64,epochs=20)
```

TensorFlow: Testing the trained model For testing, we need to make a prediction from the model ouput. To this end, we use the predict() function as follows:

```
model.predict(X_test).argmax(1)
```

Here argmax(1) returns the class w.r.t. the highest softmax output among the 10 classes. In order to evaluate the test accuracy, we use the evaluate() function:

```
model.evaluate(X_test, y_test)
```

Look ahead We have concluded the supervised learning part, but there are additional topics that may pique your interest. However, due to the desire to cover other subjects, we will end our discussion of supervised learning here. If you are interested in delving deeper into this topic, we recommend taking one of the many useful online deep learning courses, such as those offered by Coursera. Moving forward, we will shift our focus to unsupervised learning and explore one of the most popular machine learning frameworks, *Generative Adversarial Networks (GANs)*. In the upcoming section, we will delve into the intimate connections between KL divergence, mutual information, and GANs, providing detailed coverage of the subject matter.

Problem Set 10

Prob 10.1 (Basic concepts on machine learning)

(*a*) State the definition of an *algorithm.*
(*b*) State the definition of *machine learning.*
(*c*) State the definition of *artificial intelligence.*
(*d*) State the definition of *examples* (the terminology in machine learning).
(*e*) State the definition of *supervised learning.*

Prob 10.2 (KL divergence and cross entropy) The KL divergence is a valuable metric that measures the dissimilarity between two distributions. As demonstrated in Prob 1.13, it can also represent mutual information $I(X;Y)$ by considering two distributions as a joint distribution $\mathbb{P}(X,Y)$ and a product distribution $\mathbb{P}(X)\mathbb{P}(Y)$. Furthermore, as revealed in Prob 4.4, it plays a crucial role in characterizing the exponential decay rate of probability, which is essential in the statistical field known as Large Deviation Theory.

Additionally, in this problem, we will utilize it to represent cross entropy, a well-known concept in machine learning commonly used in classification problems, which is defined as follows:

$$H(p,q) := \sum_{x \in \mathcal{X}} p(x) \log \frac{1}{q(x)} = \mathbb{E}_p \left[\log \frac{1}{q(X)} \right]. \tag{3.100}$$

Show that

$$H(p,q) = H(p) + \mathsf{KL}(p \| q)$$

where $H(p)$ indicates the entropy of a random variable having the distribution of p.

Prob 10.3 (Softmax activation for multi-class classifiers) This problem explores a general classifier setting in which the number of classes is not limited to 2, say $c \in \mathbb{N}$. Let $z := [z_1, z_2, \ldots, z_c]^T \in \mathbb{R}^c$ be the output of a *multi-perceptron* prior to activation:

$$z_j = \mathbf{w}_j^T \mathbf{x} \tag{3.101}$$

where $\mathbf{x} \in \mathbb{R}^n$ indicates the input and $\mathbf{w}_j := [w_{j1}, \ldots, w_{jn}]^T \in \mathbb{R}^n$ denotes the weight vector associated with the jth neuron in the output.

In an attempt to make those real values z_j's being interpreted as probability quantities that lie in between 0 and 1, people usually employ the following activation function, called softmax:

$$\hat{y}_j := [\mathsf{softmax}(z)]_j = \frac{e^{z_j}}{\sum_{k=1}^{c} e^{z_k}} \quad j \in \{1, 2, \ldots, c\}. \tag{3.102}$$

This is a natural extension of the logistic function: for $c = 2$,

$$\begin{aligned} \hat{y}_1 := [\mathsf{softmax}(z)]_1 &= \frac{e^{z_1}}{e^{z_1} + e^{z_2}} \\ &= \frac{1}{1 + e^{-(z_1 - z_2)}} \\ &= \sigma(z_1 - z_2) \end{aligned} \tag{3.103}$$

where $\sigma(\cdot)$ is the logistic function. Viewing $z_1 - z_2$ as the binary classifier output $\hat{y}$, this coincides with logistic regression.

Let $y \in \{[1, 0, \ldots, 0]^T, [0, 1, 0, \ldots, 0]^T, \ldots, [0, \ldots, 0, 1]^T\}$ be a label of one-hot-encoded-vector type. Here $\hat{y}_i$ can be interpreted as the probability that the ith example is classified into class i. Hence, we assume that

$$\hat{y}_i = \mathbb{P}(y = [0, \ldots, \underbrace{1}_{i\text{th position}}, \ldots, 0]^T | x), \; i \in \{1, \ldots, c\}. \tag{3.104}$$

We also assume that examples $\{(x^{(i)}, y^{(i)})\}_{i=1}^m$ are independent over i.

(*a*) Derive the likelihood of training examples:

$$\mathbb{P}(y^{(1)}, \ldots, y^{(m)} | x^{(1)}, \ldots, x^{(m)}). \tag{3.105}$$

Express it in terms of $y^{(i)}$'s and $\hat{y}^{(i)}$'s.

(*b*) Derive the optimal loss function that maximizes the likelihood (3.105).

(*c*) What is the name of the optimal loss function derived in part (*b*)?

Prob 10.4 (Cross entropy) Let p and q be two distributions. In information theory, there is an important notion, called *cross entropy* (Cover and Joy, 2006):

$$H(p, q) := -\sum_{x \in \mathcal{X}} p(x) \log_2 q(x) = \mathbb{E}_p \left[\log_2 \frac{1}{q(X)} \right] \tag{3.106}$$

where $X \in \mathcal{X}$ is a discrete random variable. Show that

$$H(p, q) \geq H(p) \tag{3.107}$$

where $H(p)$ denotes the entropy of a random variable with p:

$$H(p) := -\sum_{x\in\mathcal{X}} p(x)\log_2 p(x) = \mathbb{E}_p\left[\log_2 \frac{1}{p(X)}\right]. \tag{3.108}$$

Also identify conditions under which the equality in (3.107) holds.
Hint: Think about Jensen's inequality in Prob 1.5.

Prob 10.5 (2nd-order condition of convexity) Suppose $f : \mathbb{R}^d \to \mathbb{R}$ is twice differentiable, i.e., its second derivative $\nabla^2 f$ (also called the Hessian) exists at each point in $\mathbf{dom}f$. A well-known fact w.r.t. convexity is: f is convex if and only if

$$\begin{aligned}&\mathbf{dom}f \text{ is convex;}\\ &\nabla^2 f(x) \text{ is positive semi-definite, i.e., } \nabla^2 f(x) \succeq 0 \quad \forall x \in \mathbf{dom}f\end{aligned} \tag{3.109}$$

where $\mathbf{dom}f$ denotes the domain of the function f. This problem explores the proof of this via the following subproblems.

(*a*) State the definition of a *positive semi-definite matrix*.
(*b*) Suppose $d = 1$. Show that if $f(x)$ is convex, then (3.109) holds.
(*c*) Suppose $d = 1$. Show that if (3.109) holds, $f(x)$ is convex.
(*d*) Prove the 2nd-order condition for arbitrary d.

Prob 10.6 (Gradient descent) Consider a function $J(w) = w^2 + 2w$ where $w \in \mathbb{R}$. Consider gradient descent with the learning rate $\alpha^{(t)} = \frac{1}{2^t}$ and $w^{(0)} = 2$.

(*a*) Describe how gradient descent works.
(*b*) Using Python, run gradient descent to plot $w^{(t)}$ as a function of t.

Prob 10.7 (Optimizers) Consider gradient descent:

$$w^{(t+1)} = w^{(t)} - \alpha\nabla J(w^{(t)})$$

where $w^{(t)}$ indicates the weights of an interested model at the t-th iteration; $J(w^{(t)})$ denotes the cost function evaluated at $w^{(t)}$; and α is the learning rate. Note that only the current gradient, reflected in $\nabla J(w^{(t)})$, affects the weight update.

(*a*) *(Momentum optimizer (Polyak, 1964))* In the literature, there is a prominent variant of gradient descent that takes into account *past* gradients as well. Using the past information, one can damp an oscillating effect in the weight update that may incur instability in training. To capture past gradients and therefore address the oscillation problem, another quantity, denoted by $m^{(t)}$, is introduced:

$$m^{(t)} = \beta m^{(t-1)} + (1-\beta)(-\nabla J(w^{(t)})) \tag{3.110}$$

where β denotes another hyperparameter that captures the weight of the past gradients, called the *momentum*. Here m stands for the momentum vector. The variant of the algorithm (called the *momentum optimizer*) takes the following update for $w^{(t+1)}$:

$$w^{(t+1)} = w^{(t)} + \alpha m^{(t)}. \tag{3.111}$$

Show that

$$w^{(t+1)} = w^{(t)} - \alpha(1-\beta)\sum_{k=0}^{t-1} \beta^k \nabla J(w^{(t-k)}) + \alpha\beta^t m^{(0)}.$$

(*b*) *(Bias correction)* Assuming that $\nabla J(w^{(t)})$ is the same for all t and $m^{(0)} = 0$, show that

$$w^{(t+1)} = w^{(t)} - \alpha(1-\beta^t)\nabla J(w^{(t)}).$$

Note: For a large value of t, $1-\beta^t \approx 1$, so it has almost the same scaling as that in the regular gradient descent. On the other hand, for a small value of t, $1-\beta^t$ can be small, being far from 1. For instance, when $\beta = 0.9$ and $t = 2$, $1-\beta^t = 0.19$. This motivates us to rescale the moment $m^{(t)}$ in (3.110) through division by $1-\beta^t$. Hence, in practice, we use:

$$\hat{m}^{(t)} = \frac{m^{(t)}}{1-\beta^t}; \tag{3.112}$$

$$w^{(t+1)} = w^{(t)} + \alpha\hat{m}^{(t)}. \tag{3.113}$$

This technique is called the *bias correction*.

(*c*) *(Adam optimizer (Kingma and Ba, 2014))* Notice in (3.110) that a very large or very small value of $\nabla J(w^{(t)})$ affects the weight update in quite a different scaling. In an effort to avoid such a different scaling problem, people in practice do*normalization* in the weight update (3.113) via a normalization factor, denoted by $\hat{s}^{(t)}$ (Hinton *et al.*, 2012):

$$w^{(t+1)} = w^{(t)} + \alpha\frac{\hat{m}^{(t)}}{\sqrt{\hat{s}^{(t)}} + \epsilon} \tag{3.114}$$

where the division is component-wise, and

$$\hat{m}^{(t)} = \frac{m^{(t)}}{1-\beta_1^t}, \tag{3.115}$$

$$m^{(t)} = \beta_1 m^{(t-1)} - (1-\beta_1)\nabla J(w^{(t)}), \tag{3.116}$$

$$\hat{s}^{(t)} = \frac{s^{(t)}}{1 - \beta_2^t}, \tag{3.117}$$

$$s^{(t)} = \beta_2 s^{(t-1)} + (1 - \beta_2)(\nabla J(w^{(t)}))^2. \tag{3.118}$$

Here $(\cdot)^2$ indicates a component-wise square; ϵ is a tiny value introduced to avoid division by 0 in practice (usually 10^{-8}); and s stands for square. This optimizer (3.114) is called the Adam optimizer. Explain the rationale behind the division by $1 - \beta_2^t$ in (3.118).

Prob 10.8 (TensorFlow implementation of a digit classifier) Consider a handwritten digit classifier in Section 3.14. In this problem, you are asked to build a classifier using a two-layer neural network with ReLU activation at the hidden layer and softmax activation at the output layer.

(*a*) *(MNIST dataset)* Use the following script (or otherwise), load the MNIST dataset:

```
from tensorflow.keras.datasets import mnist
(X_train,y_train),(X_test,y_test)=mnist.load_data()
X_train = X_train/255.
X_test = X_test/255.
```

What are m (the number of training examples) and m_{test}? What are the shapes of X_train and y_train?

(*b*) *(Data visualization)* Upon the code in part (*a*) being executed, report an output for the following:

```
import matplotlib.pyplot as plt
num_of_images = 60
for index in range(1,num_of_images+1):
    plt.subplot(6,10, index)
    plt.axis('off')
    plt.imshow(X_train[index], cmap = 'gray_r')
```

(*c*) *(Model)* Using a skeleton code provided in Section 3.14, write a script for a two-layer neural network model with 500 hidden units fed by MNIST data.

(*d*) *(Training)* Using a skeleton code in Section 3.14, write a script for training the model generated in part (*c*) with cross entropy loss. Use the Adam optimizer with:

learning rate = 0.001; $\quad (\beta_1, \beta_2) = (0.9, 0.999)$

and the number of epochs is 10. Also plot a training loss as a function of epochs.

(*e*) *(Testing)* Using a skeleton code in Section 3.14, write a script for testing the model (trained in part (*d*)). What is the test accuracy?

Prob 10.9 (True or False?)

(*a*) Consider an optimization problem for supervised learning:

$$\min_{w} \sum_{i=1}^{m} \ell(y^{(i)}, f_w(x^{(i)})) \tag{3.119}$$

where $\{(x^{(i)}, y^{(i)})\}_{i=1}^m$ indicate input-output example pairs, and

$$f_w(x) = \frac{1}{1 + e^{-w^T x}}. \tag{3.120}$$

The optimal loss function (in a sense of maximizing the likelihood) is:

$$\ell^*(y, \hat{y}) = -y \log \hat{y} - (1 - y) \log(1 - \hat{y}). \tag{3.121}$$

(*b*) For two arbitrary distributions, say p and q, consider cross entropy $H(p, q)$. Then,

$$H(p, q) \geq H(q) \tag{3.122}$$

where $H(q)$ is the entropy w.r.t. q.

(*c*) For two arbitrary distributions, say p and q, consider cross entropy:

$$H(p, q) := -\sum_{x \in \mathcal{X}} p(x) \log q(x) = \mathbb{E}_p \left[\log \frac{1}{q(X)} \right] \tag{3.123}$$

where $X \in \mathcal{X}$ is a discrete random variable. Then,

$$H(p, q) = H(p) := -\sum_{x \in \mathcal{X}} p(x) \log p(x). \tag{3.124}$$

only when $q = p$.

(*d*) Consider a binary classifier where we are given input-output example pairs $\{(x^{(i)}, y^{(i)})\}_{i=1}^m$. Let $0 \leq \hat{y}^{(i)} \leq 1$ be the classifier output for the ith example.

Let w be parameters of the classifier. Define:

$$w^*_{\mathsf{CE}} := \arg\min_{w} \frac{1}{m} \sum_{i=1}^{m} \ell_{\mathsf{CE}}(y^{(i)}, \hat{y}^{(i)})$$

$$w^*_{\mathsf{KL}} := \arg\min_{w} \frac{1}{m} \sum_{i=1}^{m} \mathsf{KL}(y^{(i)} \| \hat{y}^{(i)})$$

where $\ell_{\mathsf{CE}}(\cdot,\cdot)$ denotes cross entropy loss and $\mathsf{KL}(y^{(i)} \| \hat{y}^{(i)})$ indicates the KL divergence between two binary random variables with parameters $y^{(i)}$ and $\hat{y}^{(i)}$, respectively. Then,

$$w^*_{\mathsf{CE}} = w^*_{\mathsf{KL}}.$$

(*e*) Suppose we execute the following code:

```
import numpy as np
a = np.random.randn(4,3,3)
b = np.ones_like(a)
print(b[0].shape)
print(b.shape[0])
```

Then, the two prints yield the same results.

(*f*) Suppose that image is an MNIST image of numpy array type. Then, one can use the following commands to plot the image:

```
import matplotlib.pyplot as plt
plt.imshow(image.squeeze(), cmap='gray_r')
```

3.15 Unsupervised Learning: Generative Modeling

Recap Over the past three sections, we have covered basic contents related to supervised learning. The primary objective of supervised learning is to estimate a function $f(\cdot)$ of a computer system from input-output samples, as depicted in Fig. 3.45. To transform a function optimization problem, a natural form of supervised learning, into a parameterized optimization problem, we represented the function with weights (or parameters) based on a particular system architecture, namely, the Perceptron. By using the logistic function, we obtained logistic regression and proved that cross entropy is the optimal loss function for maximizing likelihood. We also examined the more expressive Deep Neural Network (DNN) architecture for $f(\cdot)$. As for a choice of activation functions, we used ReLU activation functions for all hidden neurons and logistic (or softmax) function for the output layer. To solve optimization, we investigated the widely used gradient descent algorithm and its more advanced version, the Adam optimizer, which utilizes past gradients for stable training. Lastly, we learned how to implement the algorithm through TensorFlow.

Unsupervised learning What comes next? In reality, there is a significant challenge that arises in supervised learning. In many practical scenarios, collecting labeled data is not a straightforward task. Typically, obtaining labeled data is a costly process that requires extensive human labor for annotations. Therefore, there is a growing need to explore solutions that do not rely on labeled data. Then, what can we do *only* with $\{x^{(i)}\}_{i=1}^m$?

This is where *unsupervised learning* becomes relevant. Unsupervised learning is a methodology for acquiring knowledge about data $\{x^{(i)}\}_{i=1}^m$ without relying on labeled data. One of the prominent targets for unsupervised learning is the

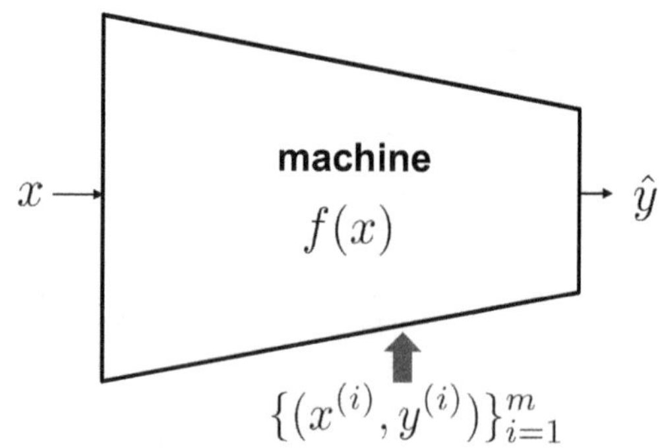

Figure 3.45. Supervised learning: Learning the function $f(\cdot)$ of an interested system from data $\{(x^{(i)}, y^{(i)})\}_{i=1}^m$.

Ian Goodfellow 2014

Figure 3.46. Ian Goodfellow, a young figure in the modern AI field. He is best known as the inventor of the Generative Adversarial Networks (GANs), which made a big wave in the AI history.

probability distribution, which is the most complex yet fundamental information. The probability distribution enables us to generate realistic signals according to our preferences. The generative modeling method is the unsupervised learning technique used to learn this fundamental entity and is widely recognized in the field. Therefore, we will concentrate on this approach.

Generative Adversarial Networks (GANs) Our focus will be on Generative Adversarial Networks (GANs) (Goodfellow *et al.*, 2014), a popular generative model in the literature. Ian Goodfellow, a research scientist in the AI field, invented GANs; see Fig. 3.46. GANs have proven to be instrumental in various applications, including image creation, human image synthesis, image inpainting, coloring, super-resolution image synthesis, speech synthesis, style transfer, and robot navigation. GANs are so effective that, as of October 3, 2019, the state of California passed a bill banning the use of GANs to create fake pornography without the consent of those depicted. From an information theory perspective, GANs are an intriguing framework since they are closely linked to KL divergence and mutual information.

Outline In the upcoming sections, we will delve deep into the connection between GANs, KL divergence, and mutual information. Our investigation will unfold in four parts. Firstly, we will explore the concept of generative modeling. Secondly, we will formulate an optimization problem, with a particular emphasis on the GAN framework. Then, we will establish a connection to the KL divergence and mutual information. Finally, we will learn how to solve the GAN optimization problem and implement it using TensorFlow. For this section, our focus will be on the first two parts.

Generative modeling Generative modeling refers to a method of producing synthetic data that follows a similar distribution as real data. The model parameters

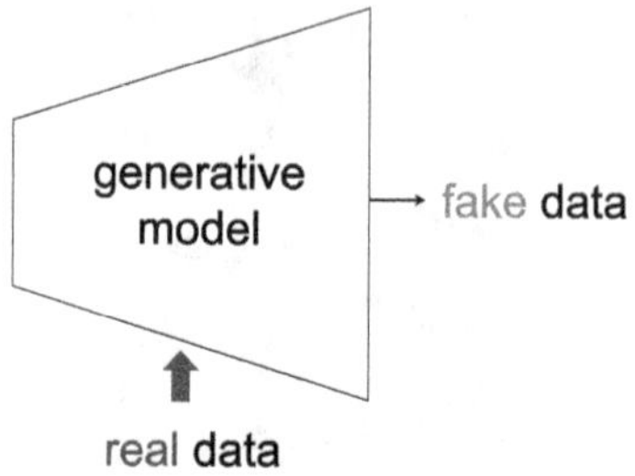

Figure 3.47. A generative model is the one that generates *fake* data which resembles *real* data. Here what *resembling* means in a mathematical language is that it has a *similar distribution*.

are trained using real data so that the model generates fake data that resembles the real data. Fig. 3.47 provides a visual illustration of this process. The input signal, which is not shown in the figure but should be fed into the model, can be either a randomly generated signal or a synthesized signal that serves as a seed for the fake data. The choice of input signal depends on the specific application. We will delve into this in further detail later on.

A remark on generative modeling The design of a generative model has been a classical age-old problem, and it is considered one of the most important problems in statistics. This is because the main objective of the statistics field is to determine the probability distribution of data, and the generative model serves as the underlying framework. Furthermore, the model can be utilized as a concrete function block, also known as the generator in the field, to generate realistic fake data. Density estimation is another common name for the problem in statistics, where the density pertains to the probability distribution.

Notations We relate generative modeling to optimization. We feed an input signal that one can arbitrary synthesize. A common way to generate the input is to use Gaussian or uniform distribution. For this *input*, we employ a conventional "x" notation, say $x \in \mathbb{R}^k$, where k is a dimension. To avoid the conflict in notation with real data $\{x^{(i)}\}_{i=1}^m$, we use a different notation, say $\{y^{(i)}\}_{i=1}^m$, for real data. Please don't be confused with labels. In fact, the convention in machine learning is to use a z notation for a fake input while maintaining $\{x^{(i)}\}_{i=1}^m$ for real data. This may be another way that you should take when writing papers. Let $\hat{y} \in \mathbb{R}^n$ be a fake output. Let $\{(x^{(i)}, \hat{y}^{(i)})\}_{i=1}^m$ be such fake input-output m pairs and let $\{y^{(i)}\}_{i=1}^m$ be real data examples. See Fig. 3.48.

Goal Let $G(\cdot)$ be a function of the generator. Then, the goal of the generative model can be stated as follows: Designing $G(\cdot)$ such that

$$\{\hat{y}^{(i)}\}_{i=1}^m \approx \{y^{(i)}\}_{i=1}^m \ \textit{in distribution}.$$

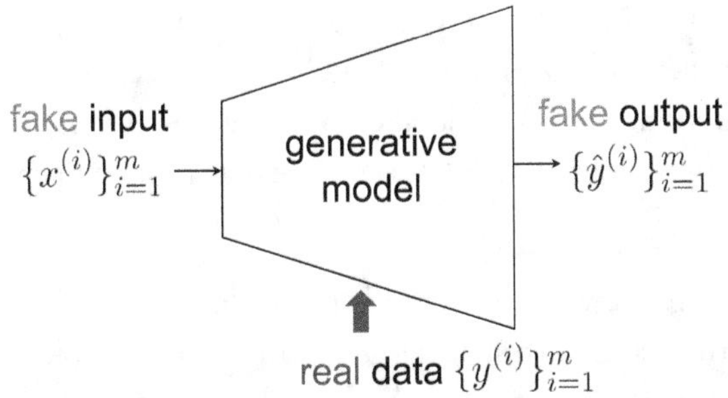

Figure 3.48. Problem formulation for generative modeling.

What does it mean by "in distribution"? To make it clear, we need to quantify closeness between two distributions. One natural yet prominent approach employed in statistics is to take the following two steps:

1. Compute empirical distributions or estimate distributions from $\{y^{(i)}\}_{i=1}^m$ and $\{(x^{(i)}, \hat{y}^{(i)})\}_{i=1}^m$. Let such distributions be:

$$\mathbb{Q}_Y, \mathbb{Q}_{\hat{Y}}$$

   for real and fake data, respectively.
2. Next employ a well-known *divergence measure* in statistics which can serve to quantify closeness of two distributions. Let $\mathsf{D}(\cdot,\cdot)$ be one such divergence measure. Then, the similarity between $\mathbb{Q}_Y$ and $\mathbb{Q}_{\hat{Y}}$ can be quantified as:

$$\mathsf{D}(\mathbb{Q}_Y, \mathbb{Q}_{\hat{Y}}).$$

Taking the above approach, one can state the goal as: Designing $G(\cdot)$ such that

$$\mathsf{D}(\mathbb{Q}_Y, \mathbb{Q}_{\hat{Y}}) \text{ is minimized.}$$

Optimization under the approach Under the approach, one can formulate an optimization problem as:

$$\min_{G(\cdot)} \mathsf{D}(\mathbb{Q}_Y, \mathbb{Q}_{\hat{Y}}). \tag{3.125}$$

As you may recognize, a couple of issues arise in solving the above problem (3.125). One issue is that it is a function optimization problem. As mentioned earlier, one common way to resolve this is to parameterize the function with a neural network:

$$\min_{G(\cdot)\in\mathcal{N}} \mathsf{D}(\mathbb{Q}_Y, \mathbb{Q}_{\hat{Y}}) \tag{3.126}$$

where $\mathcal{N}$ indicates a class of neural networks.

There are two more issues. First, the objective function $\mathsf{D}(\mathbb{Q}_Y, \mathbb{Q}_{\hat{Y}})$ is a complicated function of the knob $G(\cdot)$. Note that $\mathbb{Q}_{\hat{Y}}$ is a function of $G(\cdot)$, as $\hat{y} = G(x)$. The objective function is a twice folded composite function of $G(\cdot)$. The second is perhaps the most fundamental issue. It is not clear as to how to choose a divergence measure $\mathsf{D}(\cdot, \cdot)$.

Look ahead There are various methods to tackle the aforementioned concerns, and one of them leads to formulating an optimization problem for GANs. The following section will explore this method for deriving the optimization problem for GANs.

3.16 Generative Adversarial Networks (GANs) and KL Divergence

Recap In the previous section, we introduced unsupervised learning. The goal of unsupervised learning is to learn something about data, which we denoted by $\{y^{(i)}\}_{i=1}^m$, instead of $\{x^{(i)}\}_{i=1}^m$. There are several unsupervised learning methods available depending on the desired outcome. However, we have emphasized one particular approach, which is generative modeling, aimed at learning the probability distribution. We have also formulated an optimization problem for generative modeling:

$$\min_{G(\cdot)\in\mathcal{N}} \mathsf{D}(\mathbb{Q}_Y, \mathbb{Q}_{\hat{Y}}) \tag{3.127}$$

where $\mathbb{Q}_Y$ and $\mathbb{Q}_{\hat{Y}}$ indicate the empirical distributions for real and fake data, respectively; $G(\cdot)$ denotes the function of the generator; $\mathsf{D}(\cdot,\cdot)$ is a divergence measure; and $\mathcal{N}$ is a class of neural networks. Next, we brought up a few challenges in the optimization problem: (i) the optimization is a function optimization; (ii) the objective function involves complex dependencies on $G(\cdot)$; and (iii) the choice of $\mathsf{D}(\cdot,\cdot)$ is not straightforward.

We also pointed out that there are approaches to tackle these issues, one of which leads to an optimization problem for a highly effective generative model called Generative Adversarial Networks (GANs).

Outline This section will delve into the details on GANs. The section consists of three parts. Firstly, we will examine the path that leads to GANs. Secondly, we will derive an optimization problem for GANs. Finally, we will demonstrate that GANs have the close connection with the KL divergence and mutual information.

What is the way that leads to GANs? Remember one challenge that we are faced with in the optimization problem (3.127): $\mathsf{D}(\mathbb{Q}_Y, \mathbb{Q}_{\hat{Y}})$ is a complicated function of $G(\cdot)$. To address this, we take an *indirect* way to represent $\mathsf{D}(\mathbb{Q}_Y, \mathbb{Q}_{\hat{Y}})$. We first observe how $\mathsf{D}(\mathbb{Q}_Y, \mathbb{Q}_{\hat{Y}})$ should behave, and then based on the observation, we will come up with an indirect way to mimic the behaviour. It turns out the way leads us to explicitly compute $\mathsf{D}(\mathbb{Q}_Y, \mathbb{Q}_{\hat{Y}})$. Below are details.

How $\mathsf{D}(\mathbb{Q}_Y, \mathbb{Q}_{\hat{Y}})$ should behave? One observation that we can make is that if one can *easily discriminate* real data y from fake data $\hat{y}$, then the divergence must be *large*; otherwise, it should be small. This motivates us to:

Interpret $\mathsf{D}(\mathbb{Q}_Y, \mathbb{Q}_{\hat{Y}})$ as *the ability to discriminate.*

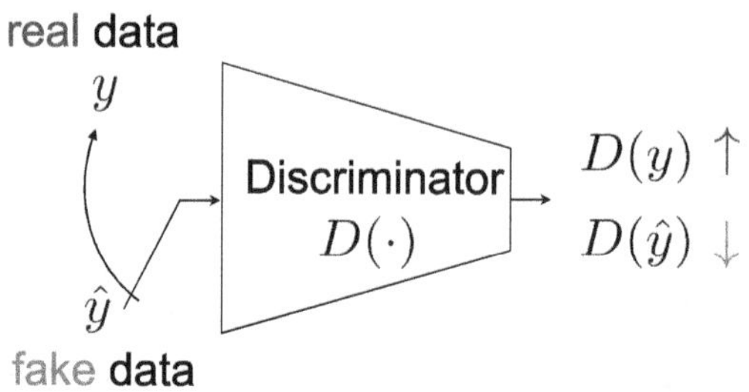

Figure 3.49. Discriminator wishes to output $D(\cdot)$ such that $D(y)$ is as large as possible while $D(\hat{y})$ is as small as possible.

We introduce an entity that can serve the discriminating function. This particular entity was introduced by Ian Goodfellow, the inventor of GAN, and he named it:

Discriminator.

Goodfellow considered a *binary-output* discriminator which takes as an input, either real data y or fake data $\hat{y}$. He then wanted to design $D(\cdot)$ such that $D(\cdot)$ well approximates the probability that the input $(\cdot)$ is real data:

$$D(\cdot) \approx \mathbb{P}((\cdot) = \text{real data}).$$

Noticing that

$$\mathbb{P}(y = \text{real}) = 1;$$
$$\mathbb{P}(\hat{y} = \text{real}) = 0,$$

he wanted to design $D(\cdot)$ such that:

$$D(y) \text{ is as large as possible, close to } 1;$$
$$D(\hat{y}) \text{ is as small as possible, close to } 0.$$

See Fig. 3.49.

How to quantity the ability to discriminate? Keeping the picture Fig. 3.49 in his mind, he wanted to quantify the ability to discriminate. To this end, he observed that if $D(\cdot)$ can easily discriminate, then we should have:

$$D(y) \uparrow; \quad 1 - D(\hat{y}) \uparrow .$$

Although one simplistic approach to capturing the ability is to add the above two terms, Goodfellow chose to use a logarithmic summation instead:

$$\log D(y) + \log(1 - D(\hat{y})). \tag{3.128}$$

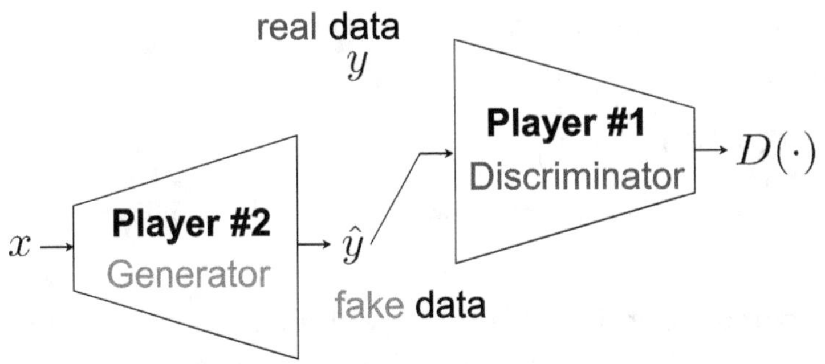

Figure 3.50. A two-player game for GAN: Discriminator $D(\cdot)$ wishes to maximize the quantified ability (3.129), while another player, generator $G(\cdot)$, wants to minimize (3.129).

During the NeurIPS 2016 conference, Goodfellow presented a tutorial on GANs and referenced a paper from AISTATS 2010 as the source of inspiration for the problem formulation (Gutmann and Hyvärinen, 2010). See Eq. (3) in the paper.

Making the particular choice, he quantified the ability to discriminate for m examples as:

$$\frac{1}{m}\sum_{i=1}^{m} \log D(y^{(i)}) + \log(1 - D(\hat{y}^{(i)})). \tag{3.129}$$

A two-player game Goodfellow then introduced a *two-player game* in which player 1, discriminator $D(\cdot)$, wishes to maximize the quantified ability (3.129), while player 2, generator $G(\cdot)$, wants to minimize (3.129). See Fig. 3.50 for illustration.

Optimization for GANs The two-player game motivated him to formulate the following min max optimization problem:

$$\min_{G(\cdot)} \max_{D(\cdot)} \frac{1}{m}\sum_{i=1}^{m} \log D(y^{(i)}) + \log(1 - D(\hat{y}^{(i)})). \tag{3.130}$$

You may be wondering why the order of "max min" was not used instead (i.e., first taking "min" and then "max"). While that is a possible approach, there is a specific reason why "min max" is preferred, which will become clear shortly. Note that the optimization is focused on two functions, $D(\cdot)$ and $G(\cdot)$, meaning that it is still a function optimization. Fortunately, the GAN paper was published in 2014, after the start of the deep learning revolution. This enabled Goodfellow to appreciate the power of neural networks: "Deep neural networks can represent any arbitrary function well." This inspired him to use neural networks to parameterize the two functions, resulting in the following optimization problem:

$$\min_{G(\cdot)\in\mathcal{N}} \max_{D(\cdot)\in\mathcal{N}} \frac{1}{m}\sum_{i=1}^{m} \log D(y^{(i)}) + \log(1 - D(\hat{y}^{(i)})) \tag{3.131}$$

where $\mathcal{N}$ denotes a class of neural networks. This is the optimization problem for GANs.

Related to original optimization? Remember what we mentioned earlier. The way leading to the GAN optimization is an indirect way of solving the original optimization problem:

$$\min_{G(\cdot)} \mathsf{D}(\mathbb{Q}_Y, \mathbb{Q}_{\hat{Y}}). \tag{3.132}$$

What is the relationship between the two problems, (3.131) and (3.132)? These problems are closely linked, and this is precisely where the selection of "min max" (rather than "max min") becomes important. The alternative approach cannot establish a connection. Research has demonstrated that, assuming deep neural networks can effectively represent any function, the GAN optimization problem (3.131) can be transformed into the original optimization form (3.132). Below we will demonstrate this.

Simplification & manipulation Let us start by simplifying the GAN optimization (3.131). Since we assume that $\mathcal{N}$ can represent any arbitrary function, the problem (3.131) becomes unconstrained:

$$\min_{G(\cdot)} \max_{D(\cdot)} \frac{1}{m}\sum_{i=1}^{m} \log D(y^{(i)}) + \log(1 - D(\hat{y}^{(i)})). \tag{3.133}$$

The objective is a function of $D(\cdot)$, and the two functions $D(\cdot)$'s appear but with different arguments: one is $y^{(i)}$; the other is $\hat{y}^{(i)}$. So in the current form (3.133), the inner (max) optimization is not quite tractable to solve. In an attempt to make it tractable, let us express it in a different manner using the following notations.

Define a random vector Y which takes one of the m real examples with probability $\frac{1}{m}$ (uniform distribution):

$$Y \in \{y^{(1)}, \ldots, y^{(m)}\} =: \mathcal{Y}; \quad \mathbb{Q}_Y(y^{(i)}) = \frac{1}{m}, \quad i \in \{1, 2, \ldots, m\}$$

where $\mathbb{Q}_Y$ indicates the probability distribution of Y. Similarly define $\hat{Y}$ for fake examples:

$$\hat{Y} \in \{\hat{y}^{(1)}, \ldots, \hat{y}^{(m)}\} =: \hat{\mathcal{Y}}; \quad \mathbb{Q}_{\hat{Y}}(\hat{y}^{(i)}) = \frac{1}{m}, \quad i \in \{1, 2, \ldots, m\}$$

where $\mathbb{Q}_{\hat{Y}}$ indicates the probability distribution of $\hat{Y}$. Using these notations, one can rewrite the problem (3.133) as:

$$\min_{G(\cdot)} \max_{D(\cdot)} \sum_{i=1}^{m} \mathbb{Q}_Y(y^{(i)}) \log D(y^{(i)}) + \mathbb{Q}_{\hat{Y}}(\hat{y}^{(i)}) \log(1 - D(\hat{y}^{(i)})). \tag{3.134}$$

Still we have different arguments in the two $D(\cdot)$ functions. To address this, we introduce another notation. Let $z \in \mathcal{Y} \cup \hat{\mathcal{Y}}$. Newly define $\mathbb{Q}_Y(\cdot)$ and $\mathbb{Q}_{\hat{Y}}(\cdot)$ such that:

$$\mathbb{Q}_Y(z) := 0 \text{ if } z \in \hat{\mathcal{Y}} \setminus \mathcal{Y}; \tag{3.135}$$

$$\mathbb{Q}_{\hat{Y}}(z) := 0 \text{ if } z \in \mathcal{Y} \setminus \hat{\mathcal{Y}}. \tag{3.136}$$

Using the z notation, one can rewrite the problem (3.134) as:

$$\min_{G(\cdot)} \max_{D(\cdot)} \sum_{z \in \mathcal{Y} \cup \hat{\mathcal{Y}}} \mathbb{Q}_Y(z) \log D(z) + \mathbb{Q}_{\hat{Y}}(z) \log(1 - D(z)). \tag{3.137}$$

We see that the same arguments appear in the two $D(\cdot)$ functions.

Solving the inner optimization We are ready to solve the inner optimization in (3.137). Key observations are: $\log D(z)$ is concave in $D(\cdot)$; $\log(1 - D(z))$ is concave in $D(\cdot)$; and therefore, the objective function is concave in $D(\cdot)$. This implies that the objective has the unique maximum in the function space $D(\cdot)$. Hence, one can find the maximum by searching for the stationary point. Taking a derivative and setting it to zero, we get:

$$\frac{1}{\ln 2} \sum_{z} \left[\frac{\mathbb{Q}_Y(z)}{D^*(z)} - \frac{\mathbb{Q}_{\hat{Y}}(z)}{1 - D^*(z)} \right] = 0.$$

This then yields:

$$D^*(z) = \frac{\mathbb{Q}_Y(z)}{\mathbb{Q}_Y(z) + \mathbb{Q}_{\hat{Y}}(z)} \qquad z \in \mathcal{Y} \cup \hat{\mathcal{Y}}. \tag{3.138}$$

Plugging this into (3.137), we obtain:

$$\min_{G(\cdot)} \sum_{z \in \mathcal{Y} \cup \hat{\mathcal{Y}}} \mathbb{Q}_Y(z) \log \frac{\mathbb{Q}_Y(z)}{\mathbb{Q}_Y(z) + \mathbb{Q}_{\hat{Y}}(z)} + \mathbb{Q}_{\hat{Y}}(z) \log \frac{\mathbb{Q}_{\hat{Y}}(z)}{\mathbb{Q}_Y(z) + \mathbb{Q}_{\hat{Y}}(z)}. \tag{3.139}$$

Connection to KL divergence We massage the objective function in (3.139) to express it as:

$$\min_{G(\cdot)} \underbrace{\sum_{z\in\mathcal{Y}\cup\hat{\mathcal{Y}}} \mathbb{Q}_Y(z)\log\frac{\mathbb{Q}_Y(z)}{\frac{\mathbb{Q}_Y(z)+\mathbb{Q}_{\hat{Y}}(z)}{2}} + \mathbb{Q}_{\hat{Y}}(z)\log\frac{\mathbb{Q}_{\hat{Y}}(z)}{\frac{\mathbb{Q}_Y(z)+\mathbb{Q}_{\hat{Y}}(z)}{2}}} -2. \tag{3.140}$$

The above underbraced term can be expressed with a well-known divergence measure: the KL divergence. Hence, we get:

$$\min_{G(\cdot)} \sum_{z\in\mathcal{Y}\cup\hat{\mathcal{Y}}} \mathbb{Q}_Y(z)\log\frac{\mathbb{Q}_Y(z)}{\frac{\mathbb{Q}_Y(z)+\mathbb{Q}_{\hat{Y}}(z)}{2}} + \mathbb{Q}_{\hat{Y}}(z)\log\frac{\mathbb{Q}_{\hat{Y}}(z)}{\frac{\mathbb{Q}_Y(z)+\mathbb{Q}_{\hat{Y}}(z)}{2}} - 2 \tag{3.141}$$

$$= \min_{G(\cdot)} \mathsf{KL}(\mathbb{Q}_Y \| (\mathbb{Q}_Y+\mathbb{Q}_{\hat{Y}})/2) + \mathsf{KL}(\mathbb{Q}_{\hat{Y}} \| (\mathbb{Q}_Y+\mathbb{Q}_{\hat{Y}})/2) - 2. \tag{3.142}$$

Slightly manipulating the above, we obtain an equivalent optimization:

$$G^*_{\mathsf{GAN}} = \arg\min_{G(\cdot)} \frac{1}{2}\{\mathsf{KL}(\mathbb{Q}_Y \| (\mathbb{Q}_Y+\mathbb{Q}_{\hat{Y}})/2) + \mathsf{KL}(\mathbb{Q}_{\hat{Y}} \| (\mathbb{Q}_Y+\mathbb{Q}_{\hat{Y}})/2)\}. \tag{3.143}$$

Note that the objective coincides with Jensen-Shannon divergence that we introduced in Prob 1.13.

Connection to mutual information From (3.143), we can also make a connection to mutual information. Recall what you were asked to prove in Prob 1.13. That is, for two random variables, say T and $\bar{Y}$,

$$I(T;\bar{Y}) = \sum_{t\in\mathcal{T}} \mathbb{P}_T(t)\mathsf{KL}(\mathbb{P}_{\bar{Y}|t} \| \mathbb{P}_{\bar{Y}}) \tag{3.144}$$

where $\mathbb{P}_T$ and $\mathbb{P}_{\bar{Y}}$ denote the probability distributions of T and $\bar{Y}$, respectively; and $\mathbb{P}_{\bar{Y}|t}$ indicates the conditional distribution of $\bar{Y}$ given $T = t$.

Suppose that $T \sim \mathsf{Bern}(\frac{1}{2})$ and we define $\bar{Y}$ as:

$$\bar{Y} = \begin{cases} Y, & T = 1; \\ \hat{Y}, & T = 0. \end{cases}$$

Then, we get:

$$\mathbb{P}_{\bar{Y}|1} = \mathbb{Q}_Y,\ \mathbb{P}_{\bar{Y}|0} = \mathbb{Q}_{\hat{Y}},\ \mathbb{P}_{\bar{Y}} = (\mathbb{Q}_Y+\mathbb{Q}_{\hat{Y}})/2$$

where the last is due to the total probability law (why?). This together with (3.143) and (3.144) gives:

$$G^*_{\mathsf{GAN}} = \arg\min_{G(\cdot)} I(T; \bar{Y}). \tag{3.145}$$

Look ahead We have developed an optimization problem for GANs and established an intriguing link to the KL divergence and mutual information. In the subsequent section, we will explore a method for solving the GAN optimization problem (3.131) and implement it utilizing TensorFlow.

3.17 GANs: TensorFlow Implementation

Recap In the prior section, we investigated Goodfellow's approach to formulate an optimization problem for GANs. He began by quantifying the ability to discriminate real against fake samples:

$$\frac{1}{m}\sum_{i=1}^{m} \log D(y^{(i)}) + \log(1 - D(\hat{y}^{(i)})) \tag{3.146}$$

where $y^{(i)}$ and $\hat{y}^{(i)} := G(x^{(i)})$ indicate real and fake samples, respectively; $D(\cdot)$ denotes the output of discriminator; and m is the number of examples. He then introduced two players: (i) player 1, discriminator, who wishes to maximize the ability; (ii) player 2, generator, who wants to minimize it. This led to the optimization problem for GANs:

$$\min_{G(\cdot)\in\mathcal{N}} \max_{D(\cdot)\in\mathcal{N}} \frac{1}{m}\sum_{i=1}^{m} \log D(y^{(i)}) + \log(1 - D(G(x^{(i)}))) \tag{3.147}$$

where $\mathcal{N}$ denotes a class of neural networks. Lastly we demonstrated that the problem (3.147) can be stated in terms of the KL divergence or mutual information, thus making a connection to information theory.

Two natural questions arise. First, how to solve the problem (3.147)? Second, how to do TensorFlow implementation?

Outline This section will address two inquiries. We will cover four stuffs in detail. Initially, we will explore a practical approach to resolving problem (3.147). Next, we will conduct a case study to exercise the approach. Specifically, we will focus on generating MNIST-style handwritten digit images. We will then delve into one important implementation detail: Batch Normalization (Ioffe and Szegedy, 2015), which is known to be quite useful for deep neural networks. Finally, we will acquire the knowledge of scripting a TensorFlow program for software implementation.

Parameterization Solving the problem (3.147) starts with parameterizing the two functions $G(\cdot)$ and $D(\cdot)$ with neural networks:

$$\min_{w} \max_{\theta} \underbrace{\frac{1}{m}\sum_{i=1}^{m} \log D_\theta(y^{(i)}) + \log(1 - D_\theta(G_w(x^{(i)})))}_{=:J(w,\theta)} \tag{3.148}$$

where w and θ indicate parameters for $G(\cdot)$ and $D(\cdot)$, respectively. Is the parameterized problem (3.148) the one that we are familiar with? In other words, is $J(w,\theta)$ is *convex* in w? Is $J(w,\theta)$ is *concave* in θ? Unfortunately, it is not the case. In general, the objective is highly non-convex in w and highly non-concave in θ.

Then, what can we do? In fact, there is nothing we can do more beyond what we know. We only know how to find a stationary point via a method like gradient descent. One practical way is to simply look for a stationary point, say (w^*, θ^*), such that

$$\nabla_w J(w^*, \theta^*) = 0, \ \nabla_\theta J(w^*, \theta^*) = 0,$$

while cross-fingering that such a point yields a near optimal performance. Luckily, it is often the case in practice, especially when employing neural networks for parameterization. Huge efforts have been made by many smart theorists in figuring out why that is the case, e.g., (Arora *et al.*, 2017). However, a clear theoretical understanding is still lacking despite their efforts.

Alternating gradient descent One practical method to attempt to find (yet not necessarily guarantee to find) such a stationary point in the min-max optimization (3.148) is: *alternating gradient descent.*

Here is how it works. At the tth iteration, update generator's weight:

$$w^{(t+1)} \leftarrow w^{(t)} - \alpha_1 \nabla_w J(w^{(t)}, \theta^{(t)})$$

where $w^{(t)}$ and $\theta^{(t)}$ denote the weights of generator and discriminator at the tth iteration, respectively; and α_1 is the learning rate for generator. Given $(w^{(t+1)}, \theta^{(t)})$, we next update discriminator's weight as per:

$$\theta^{(t+1)} \leftarrow \theta^{(t)} + \alpha_2 \nabla_\theta J(w^{(t+1)}, \theta^{(t)})$$

where α_2 is the learning rate for discriminator. In the discriminator update, we perform gradient *ascent.* Lastly we repeat the above two until converged.

In practice, we may wish to control the frequency of discriminator weight update relative to that of generator. To this end, we often employ $k:1$ alternating gradient descent:

1. Update generator's weight:

$$w^{(t+1)} \leftarrow w^{(t)} - \alpha_1 \nabla_w J(w^{(t)}, \theta^{(t \cdot k)}).$$

2. Update discriminator's weight k times while fixing $w^{(t+1)}$: for i = 1:k,

$$\theta^{(t \cdot k+i)} \leftarrow \theta^{(t \cdot k+i-1)} + \alpha_2 \nabla_\theta J(w^{(t+1)}, \theta^{(t \cdot k+i-1)}).$$

3. Repeat the above.

You may wonder why we update discriminator more frequently than generator. Usually more updates in the *inner* optimization yield better performances in practice. Further, we employ the Adam optimizer together with batches. We leave details in Prob 11.6.

A practical tip on generator Let us say a few words about generator optimization. Given discriminator's parameter θ, the generator wishes to minimize:

$$\min_{w} \frac{1}{m_{\mathcal{B}}} \sum_{i \in \mathcal{B}} \log D_\theta(y^{(i)}) + \log(1 - D_\theta(G_w(x^{(i)})))$$

where $\mathcal{B}$ indicates a batch and $m_{\mathcal{B}}$ is the batch size (the number of examples in the batch). Notice that $\log D_\theta(y^{(i)})$ in the above is irrelevant of generator's weight w. Hence, it suffices to minimize:

$$\underbrace{\min_{w} \frac{1}{m_{\mathcal{B}}} \sum_{i \in \mathcal{B}} \log(1 - D_\theta(G_w(x^{(i)})))}_{\textit{generator loss}}$$

where the underbraced term is called "generator loss". However, in practice, instead of minimizing the generator loss directly, people rely on the following *proxy*:

$$\min_{w} \frac{1}{m_{\mathcal{B}}} \sum_{i \in \mathcal{B}} -\log D_\theta(G_w(x^{(i)})). \tag{3.149}$$

You may wonder why. There is a technical rationale behind the use of the proxy. Check this in Prob 11.2.

Task We introduce one case study for implementation. The task is related to the simple digit classifier that we implemented in Section 3.14. The task is to generate MNIST style handwritten digit images, as illustrated in Fig. 3.51. We intend to train generator so that it outputs an MNIST style fake image when fed by a random input signal.

Model for generator As a generator model, we employ a 5-layer fully-connected neural network with four hidden layers, as depicted in Fig. 3.52. For activation at each hidden layer, we employ ReLU. Remember that an MNIST image consists of 28-by-28 pixels, each indicating a gray-scaled value that spans from 0 to 1. Hence, for the output layer, we use 784 (= 28 × 28) neurons and logistic activation to ensure the range of [0, 1].

The employed network has five layers, so it is deeper than the two layer network that we used earlier. In practice, for a somewhat deep neural network, each layer's signals can exhibit quite different scalings. Such dynamically-swinged scaling yields

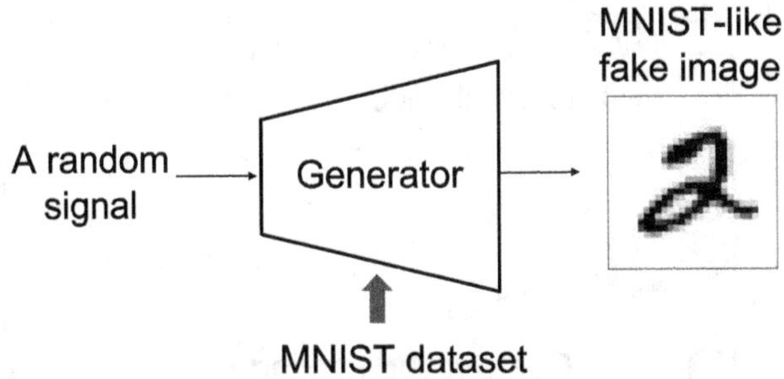

Figure 3.51. Generator for MNIST-style handwritten digit images.

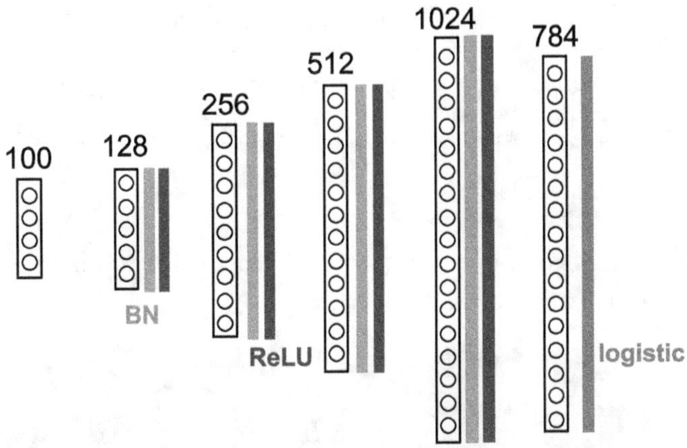

Figure 3.52. Generator: A 5-layer fully-connected neural network where the input size (the dimension of a fake input signal) is 100; the numbers of hidden neurons are 128, 256, 512, 1024; and the output size is 784 (=28 × 28). We employ ReLU activation at every hidden layer, and logistic activation at the output layer to ensure 0-to-1 output signals. We use Batch Normalization prior to ReLU at each hidden layer. See Fig. 3.53 for details.

a detrimental effect upon training: *unstable* training. Hence, people often apply an additional procedure (prior to ReLU), in order to control the scaling in our own manner. The procedure is called: Batch Normalization.

Batch Normalization (Ioffe and Szegedy, 2015) Here is how it works. See Fig. 3.53. For illustrative purpose, focus on one particular hidden layer. Let $z := [z_1, \ldots, z_n]^T$ be the output of the considered hidden layer prior to activation. Here n denotes the number of neurons in the hidden layer.

Batch Normalization (BN for short) consists of two steps. First we do zero-centering and normalization using the mean and variance w.r.t. examples in an associated batch $\mathcal{B}$:

$$\mu_{\mathcal{B}} = \frac{1}{m_{\mathcal{B}}} \sum_{i \in \mathcal{B}} z^{(i)}, \quad \sigma_{\mathcal{B}}^2 = \frac{1}{m_{\mathcal{B}}} \sum_{i \in \mathcal{B}} (z^{(i)} - \mu_{\mathcal{B}})^2 \tag{3.150}$$

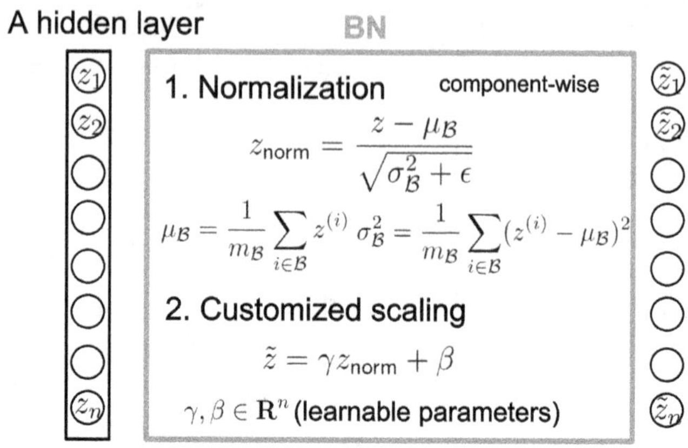

Figure 3.53. Batch Normalization (BN): First we do zero-centering and normalization with the mean $\mu_{\mathcal{B}}$ and the variance $\sigma_{\mathcal{B}}^2$ computed over the examples in an associated batch $\mathcal{B}$. Next we do a customized scaling by introducing two new parameters learnable during training: $\gamma \in \mathbb{R}^n$ and $\beta \in \mathbb{R}^n$.

where $(\cdot)^2$ indicates a component-wise square, hence $\sigma_{\mathcal{B}}^2 \in \mathbb{R}^n$. In other words, we generate the normalized output, say z_{norm}, as:

$$z_{\text{norm}} = \frac{z^{(i)} - \mu_{\mathcal{B}}}{\sqrt{\sigma_{\mathcal{B}}^2 + \epsilon}} \tag{3.151}$$

where division and multiplication are all component-wise. Here ϵ is a tiny value introduced to avoid division by 0 (typically 10^{-5}).

Second, we do a customized scaling as per:

$$\tilde{z}^{(i)} = \gamma \, z_{\text{norm}}^{(i)} + \beta \tag{3.152}$$

where γ, $\beta \in \mathbb{R}^n$ indicate two new scaling parameters which are learnable via training. Again, the operations in (3.152) are all component-wise.

BN lets the model learn the optimal scale and mean of the inputs for each hidden layer. This technique is quite instrumental in stabilizing and speeding up training especially for a very deep neural network. This has been verified experimentally by many practitioners.

Model for discriminator As a discriminator model, we use a 3-layer fully-connected network with two hidden layers; see Fig. 3.54. The input size must be the same as that of the flattened real (or fake) image. Again we employ ReLU at hidden layers and logistic activation at the output layer.

TensorFlow: How to use BN? Loading MNIST data is the same as before – so we omit it. Instead we discuss how to use BN. TensorFlow provides a built-in class

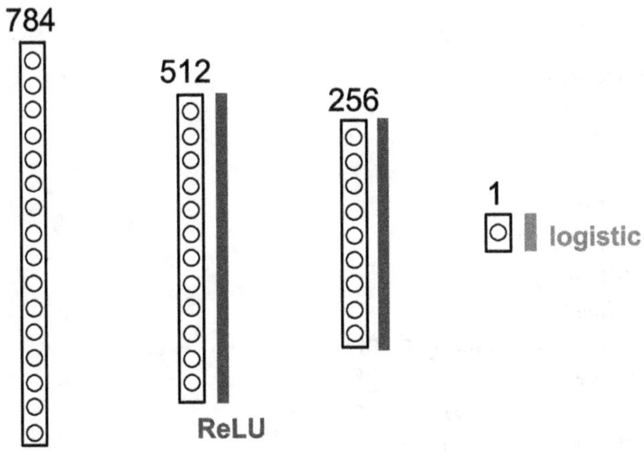

Figure 3.54. Discriminator: A 3-layer fully-connected neural network where the input size (the dimension of a flattened vector of a real (or fake) image) is 784 (=28 × 28); the numbers of hidden neurons are 512, 256; and the output size is 1. We employ ReLU activation at every hidden layer, and logistic activation at the output layer.

for BN:

BatchNormalization()

This is placed in tensorflow.keras.layers. Here is how to use the class in our setting:

```
from tensorflow.keras.models import Sequential
from tensorflow.keras.layers import Dense
from tensorflow.keras.layers import BatchNormalization
from tensorflow.keras.layers import ReLU

generator = Sequential()
generator.add(Dense(128,input_dim=latent_dim))
generator.add(BatchNormalization())
generator.add(ReLU())
generator.add(Dense(256))
generator.add(BatchNormalization())
# : : :
```

where latent_dim is the dimension of the fake input signal (which we set as 100).

TensorFlow: Models for generator & discriminator Using the deep neural networks for generator and discriminator illustrated in Figs. 3.52 and 3.54, we can implement a code as below.

```
from tensorflow.keras.models import Sequential
from tensorflow.keras.layers import Dense
from tensorflow.keras.layers import BatchNormalization
from tensorflow.keras.layers import ReLU
```

```
latent_dim =100
generator=Sequential()
generator.add(Dense(128,input_dim=latent_dim))
generator.add(BatchNormalization())
generator.add(ReLU())
generator.add(Dense(256))
generator.add(BatchNormalization())
generator.add(ReLU())
generator.add(Dense(512))
generator.add(BatchNormalization())
generator.add(ReLU())
generator.add(Dense(1024))
generator.add(BatchNormalization())
generator.add(ReLU())
generator.add(Dense(28*28,activation='sigmoid'))
```

```
from tensorflow.keras.models import Sequential
from tensorflow.keras.layers import Dense
from tensorflow.keras.layers import ReLU

discriminator=Sequential()
discriminator.add(Dense(512,input_shape=(784,)))
discriminator.add(ReLU())
discriminator.add(Dense(256))
discriminator.add(ReLU())
discriminator.add(Dense(1,activation= 'sigmoid'))
```

TensorFlow: Optimizers for generator & discriminator We use Adam optimizers with lr=0.0002 and (b1,b2)=(0.5,0.999). Since we have two models (generator and discriminator), we employ two optimizers accordingly:

```
from tensorflow.keras.optimizers import Adam
lr = 0.0002
b1 = 0.5
b2 = 0.999 # default choice
optimizer_G = Adam(learning_rate=lr, beta_1=b1)
optimizer_D = Adam(learning_rate=lr, beta_1=b1)
```

TensorFlow: Generator input As a generator input, we use a random signal with the Gaussian distribution. In particular, we use:

$$x \in \mathbb{R}^{\text{latent\_dim}} \sim \mathcal{N}(\mathbf{0}, \mathbf{I}_{\text{latent\_dim}}).$$

Here is how to generate the Gaussian random signal in TensorFlow:

```
from tensorflow.random import normal
x = normal([batch_size,latent_dim])
```

TensorFlow: Binary cross entropy loss Consider the batch version of the GAN optimization (3.148):

$$\min_{w} \max_{\theta} \frac{1}{m_{\mathcal{B}}} \sum_{i \in \mathcal{B}} \log D_\theta(y^{(i)}) + \log(1 - D_\theta(G_w(x^{(i)}))). \tag{3.153}$$

We introduce the ground-truth real-vs-fake indicator vector $[1, 0]^T$ (real $= 1$, fake $= 0$). Then, the term $\log D_\theta(y^{(i)})$ can be viewed as the minus binary cross entropy between the real/fake indicator vector and its prediction counterpart $[D_\theta(y^{(i)}), 1 - D_\theta(y^{(i)})]^T$:

$$\begin{aligned} \log D_\theta(y^{(i)}) &= 1 \cdot \log D_\theta(y^{(i)}) + 0 \cdot \log(1 - D_\theta(y^{(i)})) \\ &= -\ell_{\mathsf{BCE}}(1, D_\theta(y^{(i)})). \end{aligned} \tag{3.154}$$

On the other hand, another term $\log(1 - D_\theta(\hat{y}^{(i)}))$ can be interpreted as the minus binary cross entropy between the fake-vs-real indicator vector (fake $= 0$, real $= 1$) and its prediction counterpart:

$$\begin{aligned} \log(1 - D_\theta(\hat{y}^{(i)})) &= 0 \cdot \log D_\theta(\hat{y}^{(i)}) + 1 \cdot \log(1 - D_\theta(\hat{y}^{(i)})) \\ &= -\ell_{\mathsf{BCE}}(0, D_\theta(\hat{y}^{(i)})). \end{aligned} \tag{3.155}$$

We see that cross entropy plays a role in the computation of the objective function. TensorFlow offers a built-in class for cross entropy: BinaryCrossentropy(). This is placed in tensorflow.keras.losses. Here is how to use it in our setting:

```
from tensorflow.keras.losses import BinaryCrossentropy
CE_loss = BinaryCrossentropy(from_logits=False)
loss = CE_loss(real_fake_indicator, output)
```

where output denotes discriminator output, and real_fake_indicator is real/fake indicator vector (real $= 1$, fake $= 0$). Here output is the result *after* logistic activation; and real_fake_indicator is also a *vector* with the same dimension as output. The function BinaryCrossentropy() automatically detects the number of examples in an associated batch, thus yielding a normalized version (through division by $m_{\mathcal{B}}$).

TensorFlow: Generator loss Recall the proxy (3.149) for the generator loss that we will use:

$$\begin{aligned}&\min_w \frac{1}{m_{\mathcal{B}}} \sum_{i\in\mathcal{B}} -\log D_\theta(G_w(x^{(i)}))\\&\stackrel{(a)}{=} \min_w \frac{1}{m_{\mathcal{B}}} \sum_{i\in\mathcal{B}} \ell_{\mathsf{BCE}}(1, D_\theta(G_w(x^{(i)})))\end{aligned} \tag{3.156}$$

where (a) follows from (3.154). We can use the function CE_loss implemented above to write a code as below:

```
g_loss = CE_loss(valid, discriminator(gen_imgs))
```

where gen_imgs indicate fake images (corresponding to $G_w(x^{(i)})$'s) and valid denotes an all-1's vector with the same dimension as gen_imgs.

TensorFlow: Discriminator loss Recall the batch version of the optimization problem:

$$\max_\theta \frac{1}{m_{\mathcal{B}}} \sum_{i\in\mathcal{B}} \log D_\theta(y^{(i)}) + \log(1 - D_\theta(G_w(x^{(i)}))).$$

Taking the minus sign in the objective, we obtain the equivalent optimization:

$$\min_\theta \underbrace{\frac{1}{m_{\mathcal{B}}} \sum_{i\in\mathcal{B}} -\log D_\theta(y^{(i)}) - \log(1 - D_\theta(G_w(x^{(i)})))}_{\textit{discriminator loss}}$$

where the discriminator loss is defined as the minus version. Using (3.154) and (3.155), we can implement the discriminator loss as:

```
real_loss = CE_loss(valid, discriminator(real_imgs))
fake_loss = CE_loss(fake, discriminator(gen_imgs))
d_loss = real_loss + fake_loss
```

where real_imgs indicate real images (corresponding to $y^{(i)}$'s) and fake denotes an all-0's vector with the same dimension as gen_imgs.

TensorFlow: Training Using all of the above, one can implement a code for training. We leave details in Prob 11.6.

Look ahead Over the past sections, we have investigated two machine learning applications: one pertains to supervised learning, while the other relates to unsupervised learning. In the upcoming section, we will delve into the final application, which pertains to a societal issue in machine learning and is linked to mutual information: fair machine learning.

Problem Set 11

Prob 11.1 (Generative Adversarial Networks) Consider a GAN with generator $G(\cdot)$ and discriminator $D(\cdot)$. Let Y be a random variable that takes one of real samples $\{y^{(i)}\}_{i=1}^m$ with probability $\frac{1}{m}$, and $\mathbb{Q}_Y$ be such probability distribution. Similarly we define $\hat{Y}$ and $\mathbb{Q}_{\hat{Y}}$ for fake samples $\hat{y}^{(i)} := G(x^{(i)})$ where $x^{(i)}$ indicates an input to $G(\cdot)$, $i \in \{1, \ldots, m\}$.

Let $T \sim \mathsf{Bern}(\frac{1}{2})$ and

$$\bar{Y} = \begin{cases} Y, & T = 1; \\ \hat{Y}, & T = 0. \end{cases} \tag{3.157}$$

Let

$$G^*_{\mathsf{MI}} := \arg\min_{G(\cdot)} I(T; \bar{Y}). \tag{3.158}$$

(*a*) Show that G^*_{MI} is the same as

$$G^*_{\mathsf{JS}} := \arg\min_{G(\cdot)} \mathsf{JS}(\mathbb{Q}_Y \| \mathbb{Q}_{\hat{Y}}).$$

(*b*) Show that G^*_{JS} is the same as

$$G^*_{\mathsf{GAN}} := \arg\min_{G(\cdot)} \max_{D(\cdot)} \mathbb{E}_Y[\log D(Y)] + \mathbb{E}_{\hat{Y}}[\log(1 - D(\hat{Y}))].$$

(*c*) Suppose that the neural network class $\mathcal{N}$ can represent any arbitrary function. Show that the solution for $G(\cdot)$ to the following optimization

$$\min_{G(\cdot)\in\mathcal{N}} \max_{D(\cdot)\in\mathcal{N}} \frac{1}{m} \sum_{i=1}^m \log D(y^{(i)}) + \log(1 - D(\hat{y}^{(i)})) \tag{3.159}$$

converges to G^*_{GAN} as $m \to \infty$.

(*d*) In Section 3.16, we mentioned that (3.159) is the GAN optimization. Explain what the objective function in (3.159) means in the context of a two-player game in which one player (discriminator) wishes to discriminate real samples against fake ones while the other player (generator) wants to fool the discriminator.

Prob 11.2 (A proxy for the generator loss in GAN) Consider the optimization problem for GAN in Section 3.16:

$$\min_{G(\cdot)\in\mathcal{N}} \max_{D(\cdot)\in\mathcal{N}} \frac{1}{m} \sum_{i=1}^m \log D(y^{(i)}) + \log(1 - D(\hat{y}^{(i)})) \tag{3.160}$$

where $\mathcal{N}$ indicates a class of neural networks, and $y^{(i)}$ and $\hat{y}^{(i)} := G(x^{(i)})$ denote real and fake samples respectively. Here $x^{(i)}$ denotes an input to the generator, and m is the number of examples. Suppose that the inner optimization is solved to yield $D^*(\cdot)$. Then, the optimization problem becomes:

$$\min_{G(\cdot)\in\mathcal{N}} \frac{1}{m}\sum_{i=1}^{m} \log D^*(y^{(i)}) + \log(1 - D^*(\hat{y}^{(i)})). \tag{3.161}$$

(*a*) Show that the optimization problem (3.161) is equivalent to:

$$\min_{G(\cdot)\in\mathcal{N}} \frac{1}{m}\sum_{i=1}^{m} \log(1 - D^*(\hat{y}^{(i)})). \tag{3.162}$$

(*b*) Let w be the weights of the generator. Show that

$$\frac{d\log(1 - D^*(\hat{y}^{(i)}))}{dw} = \frac{1}{\ln 2}\frac{1}{D^*(\hat{y}^{(i)}) - 1}\frac{dD^*(\hat{y}^{(i)})}{d\hat{y}^{(i)}}\frac{d\hat{y}^{(i)}}{dw}, \tag{3.163}$$

$$\frac{d(-\log D^*(\hat{y}^{(i)}))}{dw} = \frac{1}{\ln 2}\frac{-1}{D^*(\hat{y}^{(i)})}\frac{dD^*(\hat{y}^{(i)})}{d\hat{y}^{(i)}}\frac{d\hat{y}^{(i)}}{dw}. \tag{3.164}$$

(*c*) Suppose that the discriminator works almost optimally, i.e., $D^*(\hat{y}^{(i)})$ is very close to 0. Which is larger in magnitude between (3.163) and (3.164)? Instead of solving (3.162), people prefer to solve the following for $G(\cdot)$:

$$\min_{G(\cdot)\in\mathcal{N}} \frac{1}{m}\sum_{i=1}^{m} -\log D^*(\hat{y}^{(i)}). \tag{3.165}$$

Explain the rationale behind this alternative.

Prob 11.3 (Batch normalization) Consider a deep neural network. Let $z^{(i)} := [z_1^{(i)}, \ldots, z_n^{(i)}]^T$ be the output of a hidden layer prior to activation for the ith example where $i \in \{1, 2, \ldots, m\}$ and m is the number of examples. Here n denotes the number of neurons in the hidden layer.

(*a*) Let

$$\mu = \frac{1}{m}\sum_{i=1}^{m} z^{(i)}, \quad \sigma^2 = \frac{1}{m}\sum_{i=1}^{m}(z^{(i)} - \mu)^2 \tag{3.166}$$

where $(\cdot)^2$ indicates a component-wise square, hence $\sigma^2 \in \mathbb{R}^n$. Consider

$$z_{\text{norm}}^{(i)} = \frac{z^{(i)} - \mu}{\sqrt{\sigma^2 + \epsilon}} \tag{3.167}$$

$$\tilde{z}^{(i)} = \gamma\, z_{\text{norm}}^{(i)} + \beta \tag{3.168}$$

where $\gamma, \beta \in \mathbb{R}^n$. Again the division and multiplication are all component-wise. Here ϵ is a tiny value introduced to avoid division by 0 (typically 10^{-5}). This is called a smoothing term. Assuming that ϵ is negligible and $z^{(i)}$'s are independent over i, what are the mean and variance of $\tilde{z}^{(i)}$?

(*b*) Many researchers employ $\tilde{z}^{(i)}$ instead of $z^{(i)}$ during training. These operations include zero-centering and normalization (hence it is named *batch normalization*), followed by rescaling and shifting with two new parameters (γ and β) which are learnable via training. In other words, these operations let the model learn the optimal scale and mean of the inputs for each layer. This technique plays a role in stabilizing and speeding up training especially for a very deep neural network. This has been verified experimentally by many practitioners.

In practice, this operation is done over the current mini-batch, so the whole procedure is summarized as follows: for the current mini-batch $\mathcal{B}$ with the size $m_{\mathcal{B}}$,

$$\begin{aligned} \mu_{\mathcal{B}} &= \frac{1}{m_{\mathcal{B}}}\sum_{i=1}^{m_{\mathcal{B}}} z^{(i)}, \quad \sigma_{\mathcal{B}}^2 = \frac{1}{m_{\mathcal{B}}}\sum_{i=1}^{m_{\mathcal{B}}} (z^{(i)} - \mu_{\mathcal{B}})^2, \\ z_{\text{norm}}^{(i)} &= \frac{z^{(i)} - \mu_{\mathcal{B}}}{\sqrt{\sigma_{\mathcal{B}}^2 + \epsilon}}, \quad \tilde{z}^{(i)} = \gamma\, z_{\text{norm}}^{(i)} + \beta. \end{aligned} \tag{3.169}$$

At *test* time, there is no mini-batch to compute the empirical mean and standard deviation. Then, what can we do? Suggest a way to handle this issue and explain the rationale behind your suggestion. You may want to consult with some well-known literature if you wish.

Prob 11.4 (Function optimization) Let $Y \sim \mathbb{P}_Y$ and $\hat{Y} \sim \mathbb{P}_{\hat{Y}}$. Consider:

$$\mathbb{E}_Y\left[\log \frac{\mathbb{P}_Y(Y)}{\mathbb{P}_Y(Y) + \mathbb{P}_{\hat{Y}}(Y)}\right] + \mathbb{E}_{\hat{Y}}\left[\log \frac{\mathbb{P}_{\hat{Y}}(\hat{Y})}{\mathbb{P}_Y(\hat{Y}) + \mathbb{P}_{\hat{Y}}(\hat{Y})}\right]. \tag{3.170}$$

(*a*) For $x < 1$, show that $\log(1 - x)$ is concave in x.

(*b*) Show that (3.170) is the same as:

$$\max_{D(\cdot)\in\mathbb{R}^+} \mathbb{E}_Y[\log D(Y)] + \mathbb{E}_{\hat{Y}}[\log(1 - D(\hat{Y}))]. \tag{3.171}$$

Prob 11.5 (A lower bound of mutual information) Let L and X be random variables. Define $\hat{Y} := G(X, L)$ for a function $G(\cdot, \cdot)$. Show that

$$I(L; \hat{Y}) \geq \mathbb{E}_{L,\hat{Y}}[\log Q(L|\hat{Y})] + H(L) \tag{3.172}$$

for some conditional distribution $Q(\cdot|\cdot)$.

Prob 11.6 (TensorFlow implementation of GAN) Consider Goodfellow's GAN that we learned in Section 3.16. In this problem, you are asked to build a simple GAN that generates MNIST style handwritten digit images. We employ a 5-layer neural network for generator with ReLU at all the hidden layers and logistic activation at the output layer.

(*a*) *(MNIST dataset loading)* Use the following script (or otherwise), load the MNIST dataset:

```
from tensorflow.keras.datasets import mnist
(X_train,y_train),(X_test, y_test)=mnist.load_data()
X_train = X_train/255.
X_test = X_test/255.
```

Explain the role of the following script:

```
import numpy as np
def get_batches(data, batch_size):
    batches = []
    for i in range(int(data.shape[0]//batch_size)):
        batch=data[i*batch_size:(i+1)*batch_size]
        batches.append(batch)
    return np.asarray(batches)
```

(*b*) *(Data visualization)* Assume that the code in part (*a*) is executed. Using a skeleton code provided in Prob 10.8(*b*), write a script that plots 60 images in the first batch of X_train in one figure. Also plot the figure.

(*c*) *(Generator)* Draw a block diagram for generator implemented by the following:

```
from tensorflow.keras.models import Sequential
from tensorflow.keras.layers import Dense
from tensorflow.keras.layers import BatchNormalization
from tensorflow.keras.layers import ReLU

latent_dim =100
```

```
generator=Sequential()
generator.add(Dense(128,input_dim=latent_dim))
generator.add(BatchNormalization())
generator.add(ReLU())
generator.add(Dense(256))
generator.add(BatchNormalization())
generator.add(ReLU())
generator.add(Dense(512))
generator.add(BatchNormalization())
generator.add(ReLU())
generator.add(Dense(1024))
generator.add(BatchNormalization())
generator.add(ReLU())
generator.add(Dense(28*28,activation='sigmoid'))
```

(*d*) *(Generator check)* Upon the above codes being executed, report an output for the following:

```
from tensorflow.random import normal
import matplotlib.pyplot as plt

batch_size = 64
x = normal([batch_size,latent_dim])
gen_imgs = generator.predict(x)
gen_imgs = gen_imgs.reshape(-1,28,28)

num_of_images = 60
for index in range(1,num_of_images+1):
    plt.subplot(6,10, index)
    plt.axis('off')
    plt.imshow(gen_imgs[index], cmap = 'gray_r')
```

(*e*) *(Discriminator)* Draw a block diagram for discriminator implemented by the following:

```
from tensorflow.keras.models import Sequential
from tensorflow.keras.layers import Dense
from tensorflow.keras.layers import ReLU

discriminator=Sequential()
discriminator.add(Dense(512,input_shape=(784,)))
discriminator.add(ReLU())
discriminator.add(Dense(256))
discriminator.add(ReLU())
discriminator.add(Dense(1,activation='sigmoid'))
```

(*f*) (*Training*) Suppose we construct the generator and discriminator as follows:

```
from tensorflow.keras.layers import Input
from tensorflow.keras.models import Model
from tensorflow.keras.optimizers import Adam

adam = Adam(learning_rate=0.0002, beta_1=0.5)

# discriminator compile
discriminator.compile(loss='binary_crossentropy',
                      optimizer=adam)
# freeze disc's weights while training generator
discriminator.trainable = False

# define GAN with fake input and disc. output
gan_input = Input(shape=(latent_dim,))
x = generator(inputs=gan_input)
output = discriminator(x)
gan = Model(gan_input, output)
gan.compile(loss='binary_crossentropy',
            optimizer=adam)
```

where generator() and discriminator() are the classes designed in parts (*c*) and (*e*), respectively.

Explain how generator and discriminator are trained in the following code:

```
import numpy as np
from tensorflow.random import normal

EPOCHS = 50
k=2 # k:1 alternating gradient descent
d_losses = []
g_losses = []

for epoch in range(1,EPOCHS + 1):
    # train per each batch
    np.random.shuffle(X_train)
    for i, real_imgs in enumerate(get_batches(X_train,
                                              batch_size)):
        #######################
        # train discriminator
        #######################
        # fake input generation
        gen_input = normal([batch_size,latent_dim])
        # fake images
        gen_imgs = generator.predict(gen_input)
```

```
real_imgs = real_imgs.reshape(-1,28*28)
# input for discriminator
d_input = np.concatenate([real_imgs,gen_imgs])
# label for discriminator
# (first half: real (1); second half: fake (0))
d_label = np.zeros(2*batch_size)
d_label[:batch_size] = 1
# train Discriminator
d_loss = discriminator.train_on_batch(d_input, d_label)

######################
# train generator
######################
if i%k: # 1:k alternating gradient descent
    # fake input generation
    g_input = normal([batch_size,latent_dim])
    # label for fake image
    # Generator wants fake images to be treated
    # as real ones
    g_label = np.ones(batch_size)
    # train generator
    g_loss = gan.train_on_batch(g_iput, g_label)

    d_losses.append(d_loss)
    g_losses.append(g_loss)
```

(*g*) *(Training check)* For epoch $= 10, 30, 50, 70, 90$: plot a figure that shows 25 fake images from generator trained in part (*f*) or by other methods of yours. Also plot the generator loss and discriminator loss as a function of epochs. Include Python scripts as well.

Prob 11.7 (Minimax theorem) Let $f(x, y)$ be a continuous real-valued function defined on $\mathcal{X} \times \mathcal{Y}$ such that

$$\text{(i) } f(x, y) \text{ is } \textit{convex} \text{ in } x \in \mathcal{X}\ \forall y \in \mathcal{Y}; \textit{and}$$

$$\text{(ii) } f(x, y) \text{ in } \textit{concave} \text{ in } y \in \mathcal{Y}\ \forall x \in \mathcal{X}$$

where $\mathcal{X}$ and $\mathcal{Y}$ are convex and compact sets.
Note: You do not need to solve the *optional* problems below.

(*a*) Show that

$$\min_{x\in\mathcal{X}} \max_{y\in\mathcal{Y}} f(x, y) \geq \max_{y\in\mathcal{Y}} \min_{x\in\mathcal{X}} f(x, y). \tag{3.173}$$

Does (3.173) hold also for any arbitrary function $f(\cdot, \cdot)$?

(*b*) Suppose

$$\alpha \le \min_{x\in\mathcal{X}} \max_{y\in\mathcal{Y}} f(x,y) \Longrightarrow \alpha \le \max_{y\in\mathcal{Y}} \min_{x\in\mathcal{X}} f(x,y). \tag{3.174}$$

Then, argue that (3.174) implies:

$$\min_{x\in\mathcal{X}} \max_{y\in\mathcal{Y}} f(x,y) \le \max_{y\in\mathcal{Y}} \min_{x\in\mathcal{X}} f(x,y). \tag{3.175}$$

(*c*) Suppose that $\alpha \le \min_{x\in\mathcal{X}} \max_{y\in\mathcal{Y}} f(x,y)$. Then, show that there are finite $y_1, \ldots, y_n \in \mathcal{Y}$ such that

$$\alpha \le \min_{x\in\mathcal{X}} \max_{y\in\{y_1,\ldots,y_n\}} f(x,y). \tag{3.176}$$

(*d*) *(Optional)* Suppose that $\alpha \le \min_{x\in\mathcal{X}} \max_{y\in\{y_1,y_2\}} f(x,y)$ for any $y_1, y_2 \in \mathcal{Y}$. Then, show that there exists $y_0 \in \mathcal{Y}$ such that

$$\alpha \le \min_{x\in\mathcal{X}} f(x,y_0). \tag{3.177}$$

(*e*) *(Optional)* Suppose that $\alpha \le \min_{x\in\mathcal{X}} \max_{y\in\{y_1,\ldots,y_n\}} f(x,y)$ for any finite $y_1, \ldots, y_n \in \mathcal{Y}$. Then, show that there exists $y_0 \in \mathcal{Y}$ such that

$$\alpha \le \min_{x\in\mathcal{X}} f(x,y_0). \tag{3.178}$$

Hint: Use the proof-by-induction and part (*d*).

Note: (3.178) implies that $\alpha \le \max_{y\in\mathcal{Y}} \min_{x\in\mathcal{X}} f(x,y)$. This together with the results in parts (*b*) and (*c*) proves (3.175). Combining this with (3.173) proves the *minimax theorem*:

$$\min_{x\in\mathcal{X}} \max_{y\in\mathcal{Y}} f(x,y) = \max_{y\in\mathcal{Y}} \min_{x\in\mathcal{X}} f(x,y). \tag{3.179}$$

Prob 11.8 (Training instability) Consider a function:

$$f(x,y) = (2 + \cos x)(2 + \cos y) \tag{3.180}$$

where $x, y \in \mathbb{R}$.

(*a*) Solve the following optimization (i.e., find the optimal solution as well as the points that achieve it):

$$\min_{x} \max_{y} f(x,y). \tag{3.181}$$

(*b*) Solve the reverse version of the optimization:

$$\max_{y} \min_{x} f(x, y). \tag{3.182}$$

(*c*) Suppose that we perform 1 : 1 alternating gradient descent for $f(x, y)$ with an initial point $(x^{(0)}, y^{(0)}) = (\pi + 0.1, -0.1)$. Plot $f(x^{(t)}, y^{(t)})$ as a function of t where $(x^{(t)}, y^{(t)})$ denotes the estimate at the tth iteration. What are the limiting values of $(x^{(t)}, y^{(t)})$? Also explain why.
Note: You may want to set the learning rates properly so that the convergence behaviour is clear.

(*d*) Redo part (*c*) with a different initial point $(x^{(0)}, y^{(0)}) = (0.1, \pi - 0.1)$.

Prob 11.9 (Alternating gradient descent) Consider a function:

$$f(x, y) = x^2 - y^2 \tag{3.183}$$

where $x, y \in \mathbb{R}$.

(*a*) Solve the following optimization:

$$\min_{x} \max_{y} f(x, y). \tag{3.184}$$

(*b*) Suppose that we perform 1 : 1 alternating gradient descent for $f(x, y)$ with an initial point $(x^{(0)}, y^{(0)}) = (1, 1)$. Plot $f(x^{(t)}, y^{(t)})$ as a function of t where $(x^{(t)}, y^{(t)})$ denotes the estimate at the tth iteration. What are the limiting values of $(x^{(t)}, y^{(t)})$? Also explain why.
Note: You may want to set the learning rates properly so that the convergence behaviour is clear.

(*c*) Redo part (*c*) with a different initial point $(x^{(0)}, y^{(0)}) = (-1, -1)$.

Prob 11.10 (True or False?)

(*a*) Consider the following optimization:

$$\min_{x \in \mathbb{R}} \max_{y \in \mathbb{R}} x^2 - y^2.$$

With 1:1 alternating gradient descent with a proper choice of the learning rates, one can achieve the optimal solution to the above.

(*b*) Consider the following optimization:

$$\min_{x\in\mathbb{R}} \max_{y\in\mathbb{R}} (2+\cos x)(2+\cos y).$$

Suppose we perform 1:1 alternating gradient descent with a proper choice of the learning rates. Then, the converging points can be distinct depending on different initial points.

3.18 Fair Machine Learning and Mutual Information (1/2)

Recap Throughout the preceding sections, we have investigated two prominent methodologies for machine learning: (i) supervised learning; and (ii) unsupervised learning. We found that cross entropy plays a pivotal role in designing the optimal loss function for supervised learning. Additionally, we established a fascinating relationship between GANs (an unsupervised learning framework) and two information-theoretic notions: the KL divergence and mutual information.

Next application As the final application, we will delve into a recent topic in machine learning: Fair machine learning. There are three reasons for choosing this topic. Firstly, as machine learning becomes increasingly prevalent in various applications, such as medicine, finance, job hiring and criminal justice, it is essential to ensure fairness for all groups involved. This morally and legally motivated need has gained significant attention in the design of machine learning algorithms, particularly with regards to the fairness issue highlighted in the learning algorithm used in the US Supreme Court, which yielded unbalanced recidivism scores across different races (Larson *et al.*, 2016). Thus, this important societal topic will be discussed in this book. Secondly, we will explore the connection between information theory and fair machine learning, specifically the role of mutual information in formulating an optimization problem for these algorithms. Finally, we will examine how the associated optimization problem is closely related to the GAN optimization we learned in the past sections, creating a coherent sequence of applications from supervised learning, GANs to fair machine learning.

During upcoming lectures Over the next couple of sections, we will thoroughly explore fair machine learning. We will cover four parts. Firstly, we will

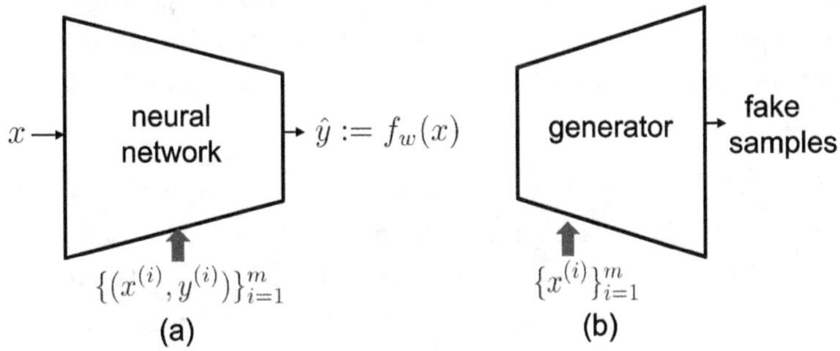

Figure 3.55. (a) Supervised learning: Learning the function $f(\cdot)$ of an interested system from input-output example pairs $\{(x^{(i)}, y^{(i)})\}_{i=1}^m$; (b) Generative modeling (an unsupervised learning methodology): Generating fake data that resemble real data, reflected in $\{x^{(i)}\}_{i=1}^m$.

Figure 3.56. Machine learning-based recidivism score predictor of the US Supreme Court: Black defendants were 77.3 percent more likely than white defendants to receive high recidivism scores.

define fair machine learning and its purpose. Following this, we will examine two widely used fairness concepts found in current literature. We will then formulate an optimization framework for fair machine learning algorithms that abide by the constraints of fairness inspired by these concepts. Furthermore, we will establish a connection between mutual information and the optimization, drawing parallels to GANs. Lastly, we will learn how to solve the optimization problem and implement it in TensorFlow. In this section, we will cover the first two.

Fair machine learning Fair machine learning refers to a specific area of machine learning that is concerned with fairness. It can be defined as a field of algorithms that train a machine to perform a given task in a fair manner. Fair machine learning can be divided into two main methodologies, just like traditional machine learning. The first is fair supervised learning, where the goal is to develop a fair classifier or predictor using a set of input-output sample pairs. The second is the unsupervised learning counterpart, which includes fair generative modeling. This aims to produce synthetic data that is both realistic and fair in terms of the statistics of the generated samples. In this book, we will focus on fair supervised learning.

Two major concepts on fairness To develop a fair classifier, it is necessary to grasp the meaning of fairness. The term "fairness" is rooted in law and has a lengthy and substantial history, with many concepts in the legal field. For our purposes, we will concentrate on two well-known concepts that have garnered significant attention in recent literature.

The first concept we will discuss is known as *disparate treatment (DT)*. This concept is centered around unequal treatment that results from sensitive attributes

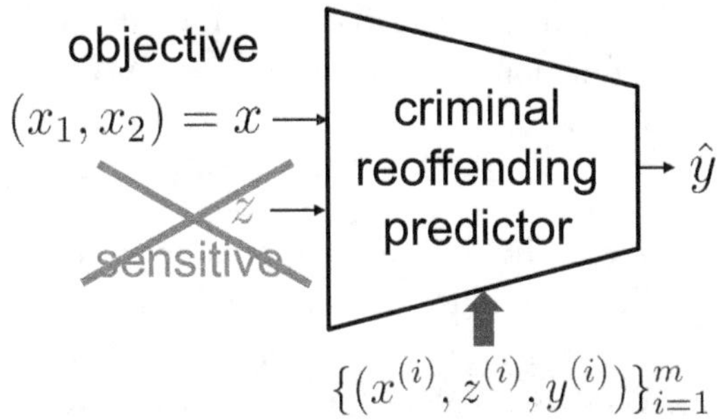

Figure 3.57. A criminal reoffending predictor.

such as race, sex, or religion. It is sometimes referred to as direct discrimination, as these attributes directly lead to discrimination.

The second concept we will focus on is *disparate impact (DI)*. This term is used to describe a situation where one group is adversely affected compared to another, even when neutral rules are in place. Neutral rules are those where sensitive attributes are not considered in classification, thereby preventing any instances of DT. Disparate impact is also known as indirect discrimination because the biased historical data leads to a disparate outcome indirectly.

Criminal reoffending predictor How can we design a fair classifier that meets both the disparate treatment and disparate impact fairness criteria? To make it easier, we will examine this in the context of a simple prediction scenario: forecasting criminal reoffending. The goal is to forecast if a person who has a criminal record is likely to reoffend within two years, and this has been used by the US Supreme Court in deciding parole.

A simple setting For illustrative purpose, we will examine a simplified version of the predictor and visualize it in Fig. 3.57. The predictor uses two types of data: (i) objective data; and (ii) sensitive data (sensitive attributes). For objective data denoted by x, we only consider two features, x_1 and x_2. The variable x_1 represents the number of prior criminal records, while x_2 represents the criminal type, such as misdemeanour or felony. For sensitive data, we use a different notation z. We consider a simple case in which z is binary, indicating only the race type of the individual, either white ($z = 0$) or black ($z = 1$). Let $\hat{y}$ be the classifier output which aims to represent the ground-truth conditional distribution $\mathbb{P}(y|x, z)$. Here y denotes the ground-truth label: $y = 1$ means reoffending within 2 years; $y = 0$ otherwise. This is a supervised learning setup, so we are given m example triplets: $\{(x^{(i)}, z^{(i)}, y^{(i)})\}_{i=1}^m$.

How to avoid disparate treatment? Firs of all, how to deal with disparate treatment? Recall the DT concept: An unequal treatment directly because of sensitive attributes. Hence, in order to avoid the DT, we should ensure that the prediction should not be a function of sensitive attributes. Mathematically, it means:

$$\mathbb{P}(y|x,z) = \mathbb{P}(y|x) \quad \forall z. \tag{3.185}$$

How to ensure the above? The solution is very simple: Not using the sensitive attribute z at all in prediction, as illustrated with a red-colored "x" mark in Fig. 3.57. Sensitive attributes are offered as part of training data although they are not used for an input. In other words, we employ $z^{(i)}$'s only in the design of an algorithm.

What about disparate impact? How about for the other fairness criterion regarding disparate impact? How to avoid the DI? Again recall the DI concept: An action that adversely affects one group against another even with formally neutral rules. Actually it is not that clear as to how to implement this.

To gain some insights, let us investigate the mathematical definition of DI. To this end, we introduce a few notations. Let Z be a random variable for a sensitive attribute. For instance, consider a binary case, say $Z \in \{0,1\}$. Let $\tilde{Y}$ be a binary hard-decision value of the predictor output $\hat{Y}$ at the middle threshold: $\tilde{Y} := \mathbf{1}\{\hat{Y} \geq 0.5\}$. Observe a ratio of likelihoods of positive example events $\tilde{Y} = 1$ for two cases: $Z = 0$ and $Z = 1$.

$$\frac{\mathbb{P}(\tilde{Y} = 1|Z = 0)}{\mathbb{P}(\tilde{Y} = 1|Z = 1)}. \tag{3.186}$$

One natural interpretation is that a classifier is more fair when the ratio is closer to 1; becomes unfair if the ratio is far away from 1. One quantification for the degree of fairness regarding the DI was proposed by (Zafar *et al.*, 2017):

$$\mathsf{DI} := \min\left(\frac{\mathbb{P}(\tilde{Y} = 1|Z = 0)}{\mathbb{P}(\tilde{Y} = 1|Z = 1)}, \frac{\mathbb{P}(\tilde{Y} = 1|Z = 1)}{\mathbb{P}(\tilde{Y} = 1|Z = 0)}\right). \tag{3.187}$$

Notice that $0 \leq \mathsf{DI} \leq 1$ and the larger DI, the more fair the situation is.

Two cases In view of the mathematical definition (3.187), reducing disparate impact means maximizing the quantity (3.187). How to design a classifier so as to maximize the DI? Depending on situations, the design methodology can be different. To see this, think about two extreme cases.

The first refers to a case in which training data already respects fairness:

$$\{(x^{(i)}, z^{(i)}, y^{(i)})\}_{i=1}^{m} \rightarrow \text{large } \mathsf{DI}.$$

In this case, a natural solution is to rely on a conventional classifier that aims to maximize prediction accuracy. Why? Because maximizing prediction accuracy would

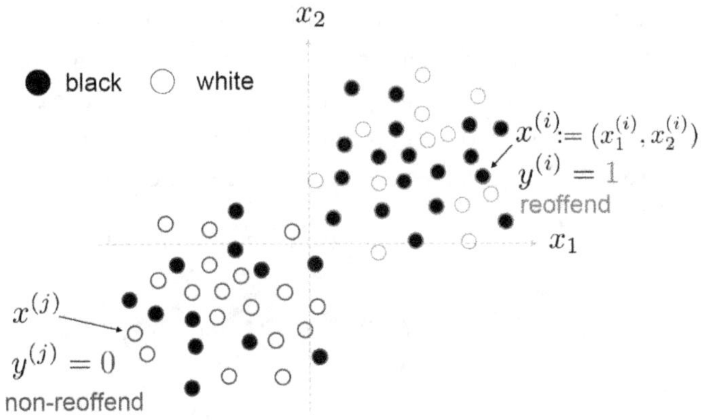

Figure 3.58. Visualization of a historically biased dataset: A hollowed (or black-colored-solid) circle indicates a data point of an individual with white (or black) race; the red (or blue) colored edge denotes $y = 1$ reoffending (or $y = 0$ non-reoffending) label.

well respect training data, which in turn yields a large DI. The second is a non-trivial case in which training data is far from being fair:

$$\{(x^{(i)}, z^{(i)}, y^{(i)})\}_{i=1}^{m} \rightarrow \text{small DI.}$$

In this case, the conventional classifier would yield a small DI. This is indeed a challenging scenario where we need to take some non-trivial action for ensuring fairness.

In reality, the second scenario is often observed due to the existence of biased historical records that form the basis of the training data. For example, the decisions made by the Supreme Court may be biased against certain races, and these decisions are likely to be included as part of the training data. Fig. 3.58 illustrates one such biased scenario, where a hollow or black-colored solid circle represents a data point for an individual of white or black race, respectively, and the red or blue colored edge denotes the event of the individual reoffending or not reoffending within two years. This is a biased situation, as there are more black-colored solid circles than hollow ones for positive examples where $y = 1$, indicating a bias in historical records favoring whites over blacks. Similarly, for negative examples where $y = 0$, there are more hollow circles than solid ones.

How to ensure a large DI? How can we guarantee a high level of DI in all scenarios, including the challenging one described above? To gain a better understanding, let us revisit an optimization problem we previously formulated in the development of a traditional classifier:

$$\min_{w} \frac{1}{m} \sum_{i=1}^{m} \ell_{\mathsf{CE}}(y^{(i)}, \hat{y}^{(i)}) \tag{3.188}$$

where $\ell_{\mathsf{CE}}(\cdot,\cdot)$ indicates binary cross entropy loss, and w denotes weights (parameters) of a classifier. One natural approach to encourage a large DI is to incorporate an DI-related constraint. Maximizing DI is equivalent to minimizing $1 - \mathsf{DI}$ (since $0 \leq \mathsf{DI} \leq 1$). We can resort to a well-known technique in optimization: *regularization*. That is to add the two objectives with different weights.

Regularized optimization Here is a regularized optimization:

$$\min_{w} \frac{1}{m} \sum_{i=1}^{m} \ell_{\mathsf{CE}}(y^{(i)}, \hat{y}^{(i)}) + \lambda \cdot (1 - \mathsf{DI}) \tag{3.189}$$

where λ denote a regularization factor that balances predication accuracy against the DI-associated objective (minimizing $1 - \mathsf{DI}$). However, an issue arises in solving the regularized optimization (3.189). Recalling the definition of DI

$$\mathsf{DI} := \min\left(\frac{\mathbb{P}(\tilde{Y}=1|Z=0)}{\mathbb{P}(\tilde{Y}=1|Z=1)}, \frac{\mathbb{P}(\tilde{Y}=1|Z=1)}{\mathbb{P}(\tilde{Y}=1|Z=0)}\right),$$

we see that DI is a complicated function of w. We have no idea as to how to express DI in terms of w.

Another way It is not feasible to express DI as a function of w, so we can consider an alternative approach inspired by information theory, specifically mutual information. If $\mathsf{DI} = 1$, then the sensitive attribute Z and the hard decision $\tilde{Y}$ are independent. Mutual information has a significant property that the mutual information between two random variables is zero when the two variables are independent, and it is a "sufficient and necessary condition." This motivates us to represent the constraint of $\mathsf{DI} = 1$ as follows:

$$I(Z; \tilde{Y}) = 0. \tag{3.190}$$

This captures the independence between Z and $\tilde{Y}$. Since the predictor output is $\hat{Y}$ (instead of $\tilde{Y}$), we consider another stronger condition that concerns $\hat{Y}$ directly:

$$I(Z; \hat{Y}) = 0. \tag{3.191}$$

The condition (3.191) is indeed stronger than (3.190), i.e., (3.191) implies (3.190). This is because

$$\begin{aligned} I(Z; \tilde{Y}) &\overset{(a)}{\leq} I(Z; \tilde{Y}, \hat{Y}) \\ &\overset{(b)}{=} I(Z; \hat{Y}) \end{aligned} \tag{3.192}$$

where (a) is due to the chain rule and non-negativity of mutual information; and (b) is because $\tilde{Y}$ is a function of $\hat{Y}$: $\tilde{Y} := \mathbf{1}\{\hat{Y} \geq 0.5\}$. Notice that (3.191) together with (3.192) gives (3.190).

Strongly regularized optimization In summary, the condition (3.191) indeed enforces the $\mathsf{DI} = 1$ constraint. This then motivates us to consider the following optimization:

$$\min_{w} \frac{1}{m} \sum_{i=1}^{m} \ell_{\mathsf{CE}}(y^{(i)}, \hat{y}^{(i)}) + \lambda \cdot I(Z; \hat{Y}). \tag{3.193}$$

How to express $I(Z; \hat{Y})$ in terms of classifier parameters w? Interestingly, there is a way to express it. The idea is intimately related to the GAN optimization that we learned.

Look ahead In the next section, we will review the GAN briefly and use it to formulate an optimization for a fair classifier.

3.19 Fair Machine Learning and Mutual Information (2/2)

Recap In the preceding section, we presented the last application of information theory: a fair classifier. We used a recidivism predictor as an instance of a fair classifier, which aims to forecast whether an individual with previous criminal records would reoffend within two years, as shown in Fig. 3.59. To prevent disparate treatment, a prominent fairness concept, we excluded the sensitive attribute from the input. To incorporate another fairness notion, disparate impact (DI for brevity), we added a regularized term to the conventional optimization that only considered prediction accuracy, resulting in the following expression:

$$\min_{w} \frac{1}{m} \sum_{i=1}^{m} \ell_{\mathsf{CE}}(y^{(i)}, \hat{y}^{(i)}) + \lambda \cdot I(Z; \hat{Y}) \tag{3.194}$$

where $\lambda \geq 0$ is a regularization factor that balances prediction accuracy (reflected in the binary cross entropy terms) against the fairness constraint, reflected in $I(Z; \hat{Y})$. Remember that $I(Z; \hat{Y}) = 0$ is a sufficient condition for $\mathsf{DI} = 1$. At the end, we claimed that one can express $I(Z; \hat{Y})$ in terms of an optimization parameter w, thereby enabling us to train the model parameterized by w. The idea for translation is to use the GAN trick that we learned in the past sections.

Outline In this section, we will support our claim. We will cover three parts in detail. Firstly, we will explain what it means by the "GAN trick". Secondly, we will utilize the "GAN trick" to construct an optimization problem in a simple scenario where the sensitive attribute is binary. Finally, we will generalize this approach to situations where the sensitive attribute can take on any value from an arbitrary alphabet.

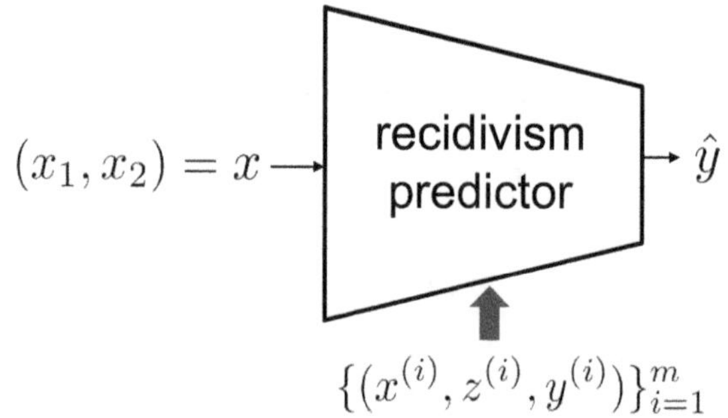

Figure 3.59. A simple recidivism predictor: Predicting a recidivism score $\hat{y}$ from $x = (x_1, x_2)$. Here x_1 indicates the number of prior criminal records; x_2 denotes a criminal type (misdemeanor or felony); and z is a race type among white ($z = 0$) and black ($z = 1$).

The GAN trick Recall the inner optimization in GANs:

$$\max_{D(\cdot)} \frac{1}{m} \sum_{i=1}^{m} \log D(y^{(i)}) + \log(1 - D(\hat{y}^{(i)}))$$

where $y^{(i)}$ and $\hat{y}^{(i)}$ indicate the ith real and fake samples, respectively; and $D(\cdot)$ is a discriminator output. Also recall the connection with mutual information:

$$I(T; \bar{Y}) = \max_{D(\cdot)} \frac{1}{m} \sum_{i=1}^{m} \log D(y^{(i)}) + \log(1 - D(\hat{y}^{(i)})) \tag{3.195}$$

where $T = \mathbf{1}\{\text{discriminator's input is real}\}$ and $\bar{Y}$ is a random variable that takes a real sample if $T = 1$; a fake sample if $T = 0$.

Here what we mean by the GAN trick is the other way around, taking the reverse order. We start with mutual information and then express it in terms of an optimization problem similarly to (3.195). Now let us apply this trick to our problem setting. For illustrative purpose, we start with a simple binary sensitive attribute setting.

Mutual information vs KL divergence In our optimization (3.194), Z is a sensitive attribute indicator, which plays the same role as T in (3.195). Similarly $\hat{Y}$ in (3.194) serves the same role as $\bar{Y}$ in (3.195). Hence, one can expect that $I(Z; \hat{Y})$ in (3.194) would be expressed similarly as in (3.195). A slight distinction lies in a detailed expression. To see the distinction, we start by manipulating $I(Z; \hat{Y})$ from scratch.

Starting with the relationship between mutual information and KL divergence, we get:

$$\begin{aligned}
I(Z; \hat{Y}) &= \mathsf{KL}(\mathbb{P}_{\hat{Y},Z} \| \mathbb{P}_{\hat{Y}} \mathbb{P}_Z) \\
&\overset{(a)}{=} \sum_{\hat{y} \in \hat{\mathcal{Y}}, z \in \mathcal{Z}} \mathbb{P}_{\hat{Y},Z}(\hat{y}, z) \log \frac{\mathbb{P}_{\hat{Y},Z}(\hat{y}, z)}{\mathbb{P}_{\hat{Y}}(\hat{y}) \mathbb{P}_Z(z)} \\
&= \sum_{\hat{y} \in \hat{\mathcal{Y}}, z \in \mathcal{Z}} \mathbb{P}_{\hat{Y},Z}(\hat{y}, z) \log \frac{\mathbb{P}_{\hat{Y},Z}(\hat{y}, z)}{\mathbb{P}_{\hat{Y}}(\hat{y})} \\
&\quad + \sum_{\hat{y} \in \hat{\mathcal{Y}}, z \in \mathcal{Z}} \mathbb{P}_{\hat{Y},Z}(\hat{y}, z) \log \frac{1}{\mathbb{P}_Z(z)} \\
&\overset{(b)}{=} \sum_{\hat{y} \in \hat{\mathcal{Y}}, z \in \mathcal{Z}} \mathbb{P}_{\hat{Y},Z}(\hat{y}, z) \log \frac{\mathbb{P}_{\hat{Y},Z}(\hat{y}, z)}{\mathbb{P}_{\hat{Y}}(\hat{y})}
\end{aligned}$$

$$
+\sum_{z\in\mathcal{Z}} \mathbb{P}_Z(z)\log\frac{1}{\mathbb{P}_Z(z)}
$$

$$
\stackrel{(c)}{=} \sum_{\hat{y}\in\hat{\mathcal{Y}},z\in\mathcal{Z}} \mathbb{P}_{\hat{Y},Z}(\hat{y},z)\log\frac{\mathbb{P}_{\hat{Y},Z}(\hat{y},z)}{\mathbb{P}_{\hat{Y}}(\hat{y})} + H(Z)
$$

where (a) is due to the definition of the KL divergence; (b) comes from the total probability law; and (c) is due to the definition of entropy.

Observation For the binary sensitive attribute case, we have:

$$
\begin{aligned}
I(Z;\hat{Y}) &= \sum_{\hat{y}\in\hat{\mathcal{Y}},z\in\mathcal{Z}} \mathbb{P}_{\hat{Y},Z}(\hat{y},z)\log\frac{\mathbb{P}_{\hat{Y},Z}(\hat{y},z)}{\mathbb{P}_{\hat{Y}}(\hat{y})} + H(Z)\\
&= \sum_{\hat{y}\in\hat{\mathcal{Y}}} \mathbb{P}_{\hat{Y},Z}(\hat{y},1)\log\underbrace{\frac{\mathbb{P}_{\hat{Y},Z}(\hat{y},1)}{\mathbb{P}_{\hat{Y}}(\hat{y})}}_{=:D^*(\hat{y})}\\
&\quad+\sum_{\hat{y}\in\hat{\mathcal{Y}}} \mathbb{P}_{\hat{Y},Z}(\hat{y},0)\log\underbrace{\frac{\mathbb{P}_{\hat{Y},Z}(\hat{y},0)}{\mathbb{P}_{\hat{Y}}(\hat{y})}}_{=1-D^*(\hat{y})} + H(Z).
\end{aligned} \tag{3.196}
$$

Notice that the first log-inside term, defined as $D^*(\hat{y})$, has the close relationship with the second log-inside term in the above: the sum of the two is 1. This reminds us of the objective function in the optimization (3.195). So one may conjecture that $I(Z;\hat{Y})$ can be expressed in terms of an optimization problem as follows:

Theorem 3.1. *The mutual information $I(Z;\hat{Y})$ can be represented as the following function optimization:*

$$
\begin{aligned}
I(Z;\hat{Y}) = \max_{D(\cdot)} \Bigg\{ &\sum_{\hat{y}\in\hat{\mathcal{Y}}} \mathbb{P}_{\hat{Y},Z}(\hat{y},1)\log D(\hat{y})\\
&+\sum_{\hat{y}\in\hat{\mathcal{Y}}} \mathbb{P}_{\hat{Y},Z}(\hat{y},0)\log\left(1-D(\hat{y})\right)\Bigg\} + H(Z).
\end{aligned} \tag{3.197}
$$

Proof. The optimization in (3.197) is convex in $D(\cdot)$, since the log function is concave and the concavity preserves under additivity. Hence, looking into the unique stationary point, we can prove the equivalence. Taking the derivative w.r.t. $D(\hat{y})$, we get:

$$\frac{1}{\ln 2}\left(\frac{\mathbb{P}_{\hat{Y},Z}(\hat{y},1)}{D_{\text{opt}}(\hat{y})}-\frac{\mathbb{P}_{\hat{Y},Z}(\hat{y},0)}{1-D_{\text{opt}}(\hat{y})}\right)=0 \quad \forall\hat{y}$$

where $D_{\text{opt}}(\hat{y})$ is the optimal solution to the optimization in (3.197). This gives:

$$D_{\text{opt}}(\hat{y})=\frac{\mathbb{P}_{\hat{Y},Z}(\hat{y},1)}{\mathbb{P}_{\hat{Y},Z}(\hat{y},1)+\mathbb{P}_{\hat{Y},Z}(\hat{y},0)}=\frac{\mathbb{P}_{\hat{Y},Z}(\hat{y},1)}{\mathbb{P}_{\hat{Y}}(\hat{y})}$$

where the second equality is due to the total probability law. Since $D_{\text{opt}}(\hat{y})$ is the same as $D^*(\hat{y})$ that we defined in (3.196), we complete the proof. □

How to express $I(Z;\hat{Y})$ in terms of w? The formula (3.197) contains two probability quantities $(\mathbb{P}_{\hat{Y},Z}(\hat{y},1),\mathbb{P}_{\hat{Y},Z}(\hat{y},0))$ which are not available. What we are given are: $\{(x^{(i)},z^{(i)},y^{(i)})\}_{i=1}^m$. We need to worry about what we can do with this information for computing the probability quantities. To this end, we rely upon the empirical distribution:

$$\mathbb{Q}_{\hat{Y},Z}(\hat{y}^{(i)},1)=\frac{1}{m};$$
$$\mathbb{Q}_{\hat{Y},Z}(\hat{y}^{(i)},0)=\frac{1}{m}.$$

In practice, the empirical distribution is likely to be uniform, since $\hat{y}^{(i)}$ is real-valued and hence the pair $(\hat{y}^{(i)},z)$ is unique with high probability. By applying these empirical distributions, we can approximate $I(Z;\hat{Y})$ as:

$$I(Z;\hat{Y})\approx\max_{D(\cdot)}\frac{1}{m}\left\{\sum_{i:z^{(i)}=1}\log D(\hat{y}^{(i)})+\sum_{i:z^{(i)}=0}\log\left(1-D(\hat{y}^{(i)})\right)\right\} \tag{3.198}$$
$$+H(Z).$$

Implementable optimization (Cho *et al.*, 2020) Recall the original optimization:

$$\min_{w}\frac{1}{m}\sum_{i=1}^{m}\ell_{\text{CE}}(y^{(i)},\hat{y}^{(i)})+\lambda\cdot I(Z;\hat{Y}).$$

Applying the approximation (3.198) into the above, we get:

$$\min_{w} \max_{\theta} \frac{1}{m} \left\{ \sum_{i=1}^{m} \ell_{\mathsf{CE}}(y^{(i)}, \hat{y}^{(i)}) + \lambda \left(\sum_{i:z^{(i)}=1} \log D_\theta(\hat{y}^{(i)}) + \sum_{i:z^{(i)}=0} \log\left(1 - D_\theta(\hat{y}^{(i)})\right) \right) \right\} \tag{3.199}$$

where $D(\cdot)$ is parameterized with θ. The sensitive-attribute entropy $H(Z)$ in (3.198) is removed, since it is irrelevant of the optimization parameters (θ, w). The objective function has an explicit relationship with the optimization parameters (θ, w). Hence, the parameters are trainable via a practical algorithm. In the next section, we will discuss details on the algorithm.

Extension to a non-binary sensitive attribute We previously focused on the binary sensitive attribute setting, but in practice, this may not always be the case. For example, there may be multiple race types, such as black, white, Asian, Hispanic, and multiple sensitive attributes, such as gender and religion. To account for these practical scenarios, we now consider a sensitive attribute with an arbitrary alphabet size. Multiple sensitive attributes can be represented as a single random variable with an arbitrary alphabet size. Therefore, we consider a setting where Z belongs to the set $\mathcal{Z}$ and the cardinality of $\mathcal{Z}$ is not limited to two.

By recalling the relationship between mutual information and the KL divergence, we can obtain:

$$\begin{aligned} I(Z; \hat{Y}) &= \mathsf{KL}\left(\mathbb{P}_{\hat{Y},Z} \| \mathbb{P}_{\hat{Y}} \mathbb{P}_Z\right) \\ &= \sum_{\hat{y} \in \hat{\mathcal{Y}}, z \in \mathcal{Z}} \mathbb{P}_{\hat{Y},Z}(\hat{y}, z) \log \frac{\mathbb{P}_{\hat{Y},Z}(\hat{y}, z)}{\mathbb{P}_{\hat{Y}}(\hat{y}) \mathbb{P}_Z(z)} \\ &= \sum_{\hat{y} \in \hat{\mathcal{Y}}, z \in \mathcal{Z}} \mathbb{P}_{\hat{Y},Z}(\hat{y}, z) \log \underbrace{\frac{\mathbb{P}_{\hat{Y},Z}(\hat{y}, z)}{\mathbb{P}_{\hat{Y}}(\hat{y})}}_{=:D^*(\hat{y}, z)} + H(Z). \end{aligned} \tag{3.200}$$

Defining the log-inside term in the above as $D^*(\hat{y}, z)$, we obtain:

$$\sum_{z \in \mathcal{Z}} D^*(\hat{y}, z) = 1 \quad \hat{y} \in \hat{\mathcal{Y}}.$$

This is due to the total probability law. Similar to Theorem 3.1, we can come up with the following equivalence.

Theorem 3.2. *The mutual information $I(Z;\hat{Y})$ can be represented as the following function optimization:*

$$I(Z;\hat{Y}) = \max_{D(\hat{y},z):\sum_{z\in\mathcal{Z}} D(\hat{y},z)=1} \sum_{\hat{y}\in\mathcal{Y},z\in\mathcal{Z}} \mathbb{P}_{\hat{Y},Z}(\hat{y},z)\log D(\hat{y},z) + H(Z). \tag{3.201}$$

Proof. It is a convex optimization problem, but we have multiple equality constraints. So we should take the Lagrange multiplier method which relies upon the KKT conditions (Karush, 1939; Kuhn and Tucker, 2014; Boyd and Vandenberghe, 2004). First define the Lagrange function:

$$\begin{aligned}\mathcal{L}(D(\hat{y},z),\nu(\hat{y})) = &\sum_{\hat{y}\in\hat{\mathcal{Y}},z\in\mathcal{Z}} \mathbb{P}_{\hat{Y},Z}(\hat{y},z)\log D(\hat{y},z)\\ &+\sum_{\hat{y}\in\hat{\mathcal{Y}}}\nu(\hat{y})\left(1-\sum_{z\in\mathcal{Z}} D(\hat{y},z)\right)\end{aligned}$$

where $\nu(\hat{y})$'s are Lagrange multipliers. There are the number $|\hat{\mathcal{Y}}|$ of Lagrange multipliers. We solve the problem via the KKT conditions:

$$\begin{aligned}\frac{d\mathcal{L}(D(\hat{y},z),\nu(\hat{y}))}{dD(\hat{y},z)} &= \frac{1}{\ln 2}\left(\frac{\mathbb{P}_{\hat{Y},Z}(\hat{y},z)}{D_{\text{opt}}(\hat{y},z)} - \nu^{\star}(\hat{y})\right) = 0 \quad \forall \hat{y}, z\\ \frac{d\mathcal{L}(D(\hat{y},z),\nu(\hat{y}))}{d\nu(\hat{y})} &= \frac{1}{\ln 2}\left(1-\sum_{z\in\mathcal{Z}} D_{\text{opt}}(\hat{y},z)\right) = 0 \quad \forall \hat{y}.\end{aligned}$$

Plugging the following:

$$D_{\text{opt}}(\hat{y},z) = \frac{\mathbb{P}_{\hat{Y},Z}(\hat{y},z)}{\mathbb{P}_{\hat{Y}}(\hat{y})}, \qquad \nu_{\text{opt}}(\hat{y}) = \mathbb{P}_{\hat{Y}}(\hat{y}),$$

we satisfy the KKT conditions. This implies that $D_{\text{opt}}(\hat{y},z)$ is indeed the optimal solution. Since $D_{\text{opt}}(\hat{y},z)$ is the same as $D^*(\hat{y},z)$ that we defined in (3.200), we complete the proof. □

Implementable optimization: General case (Cho *et al.*, 2020) Again for computation of $I(Z;\hat{Y})$, we rely on the empirical version of the true distribution $\mathbb{P}_{\hat{Y},Z}(\hat{y},z)$:

$$\mathbb{Q}_{\hat{Y},Z}(\hat{y}^{(i)},z^{(i)}) = \frac{1}{m} \qquad \forall i\in\{1,\ldots,m\}. \tag{3.202}$$

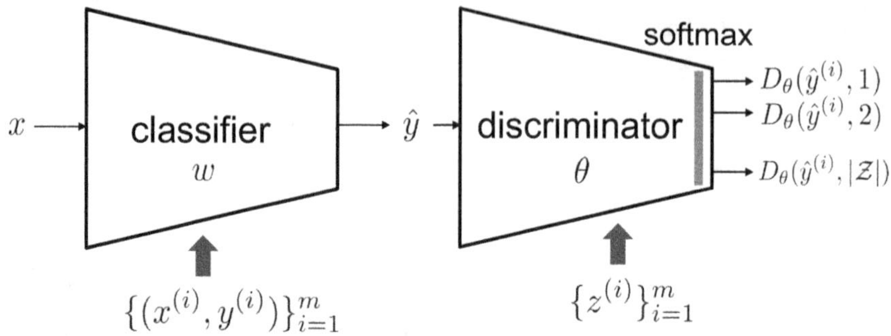

Figure 3.60. The architecture of the mutual information (MI)-based fair classifier. The prediction output $\hat{y}$ is fed into the discriminator wherein the goal is to figure out sensitive attribute z from $\hat{y}$. The discriminator output $D_\theta(\hat{y}^{(i)}, z^{(i)})$ can be interpreted as the probability that $\hat{y}$ belongs to the attribute z. Here the softmax function is applied to ensure the sum-up-to-one constraint.

So we get:

$$I(Z; \hat{Y}) \approx \max_{D(\hat{y},z):\sum_{z\in\mathcal{Z}} D(\hat{y},z)=1} \sum_{i=1}^{m} \frac{1}{m} \log D(\hat{y}^{(i)}, z^{(i)}) + H(Z). \tag{3.203}$$

By parameterizing $D(\cdot,\cdot)$ with θ and excluding $H(Z)$ (irrelevant of (θ, w)), we obtain the following optimization:

$$\min_{w} \max_{\theta:\sum_{z\in\mathcal{Z}} D_\theta(\hat{y},z)=1} \frac{1}{m} \left\{ \sum_{i=1}^{m} \ell_{\text{CE}}(y^{(i)}, \hat{y}^{(i)}) + \lambda \sum_{i=1}^{m} \log D_\theta(\hat{y}^{(i)}, z^{(i)}) \right\}. \tag{3.204}$$

The architecture of the fair classifier The architecture of the implementable optimization (3.204) is illustrated in Fig. 3.60. On top of a classifier, we introduce a new entity, called discriminator, which corresponds to the inner optimization. In discriminator, we wish to find θ^* that maximizes $\frac{1}{m}\sum_{i=1}^{m} \log D_\theta(\hat{y}^{(i)}, z^{(i)})$. On the other hand, the classifier wants to minimize the term. Hence, $D_\theta(\hat{y}^{(i)}, z^{(i)})$ can be viewed as the ability to figure out z from prediction $\hat{y}$. Notice that the classifier wishes to minimize the ability for the purpose of fairness, while the discriminator has the opposite goal. One natural interpretation that can be made on $D_\theta(\hat{y}^{(i)}, z^{(i)})$ is that it captures the probability that z is indeed the ground-truth sensitive attribute for $\hat{y}$. Here the softmax function is applied to ensure the sum-up-to-one constraint.

Analogy with GANs Since the classifier and the discriminator are competing, one can make an analogy with GANs, in which the generator and the discriminator also compete like a two-player game. While the fair classifier and the GAN bear

MI-based fair classifier	**GAN**
discriminator	discriminator
Goal: Figure out sensitive attribute from prediction	**Goal:** Distinguish real samples from fake ones.
classifier	generator
Maximize prediction accuracy	Generate realistic fake samples

Figure 3.61. MI-based fair classifier vs. GAN: Both bear similarity in structure (as illustrated in Fig. 3.60), yet distinctions in role.

strong similarity in their nature, these two are distinct in their roles. See Fig. 3.61 for the detailed distinctions.

Look ahead The optimization formulation of a fair classifier has been covered in this section. In the upcoming section, we will examine a method to solve the optimization problem (3.204) and how to implement it in TensorFlow.

3.20 Fair Machine Learning: TensorFlow Implementation

Recap Previously we formulated an optimization that respects two fairness constraints: disparate treatment (DT) and disparate impact (DI). Given m example triplets $\{(x^{(i)}, z^{(i)}, y^{(i)})\}_{i=1}^m$:

$$\min_w \frac{1}{m} \sum_{i=1}^{m} \ell_{\mathsf{CE}}(y^{(i)}, \hat{y}^{(i)}) + \lambda \cdot I(Z; \hat{Y})$$

where $\hat{y}^{(i)}$ indicates the classifier output, depending only on $x^{(i)}$ (not on the sensitive attribute $z^{(i)}$ due to the DT constraint); and λ is a regularization factor that balances prediction accuracy against the DI constraint, quantified as $I(Z; \hat{Y})$. Using the connection between mutual information and KL divergence, we could approximate $I(Z; \hat{Y})$ in the form of optimization:

$$I(Z; \hat{Y}) \approx H(Z) + \max_{\sum_z D(\hat{y},z)=1} \sum_{i=1}^{m} \frac{1}{m} \log D(\hat{y}^{(i)}, z^{(i)}). \tag{3.205}$$

We then parameterized $D(\cdot)$ with θ to obtain:

$$\min_w \max_{\theta: \sum_z D_\theta(\hat{y},z)=1} \frac{1}{m} \left\{ \sum_{i=1}^{m} \ell_{\mathsf{CE}}(y^{(i)}, \hat{y}^{(i)}) + \lambda \sum_{i=1}^{m} \log D_\theta(\hat{y}^{(i)}, z^{(i)}) \right\}. \tag{3.206}$$

Two questions that arise are: (i) how to solve the optimization (3.206)?; and (ii) how to implement it via TensorFlow?

Outline In this section, we will tackle the two questions. What we are going to do are four folded. Firstly, we will explore a practical algorithm to tackle optimization (3.206). Secondly, we will conduct a case study, focusing on recidivism prediction to exercise the algorithm. We will emphasize a specific implementation detail – synthesizing an unfair dataset. Thirdly, we will discuss how to implement this algorithm using TensorFlow. We will focus on a binary sensitive attribute setting for illustrative purposes.

Observation Let's begin by translating the optimization (3.206) into a version that is more friendly for programming:

$$\min_w \max_{\theta: \sum_z D_\theta(\hat{y},z)=1} \frac{1}{m} \left\{ \sum_{i=1}^{m} \ell_{\mathsf{CE}}(y^{(i)}, \hat{y}^{(i)}) + \lambda \sum_{i=1}^{m} \log D_\theta(\hat{y}^{(i)}, z^{(i)}) \right\}$$

$$
\begin{aligned}
&\overset{(a)}{=} \min_{w} \max_{\theta} \frac{1}{m} \left\{ \sum_{i=1}^{m} \ell_{\mathsf{CE}}(y^{(i)}, \hat{y}^{(i)}) \right. \\
&\qquad \left. +\lambda \left(\sum_{i:z^{(i)}=1} \log D_\theta(\hat{y}^{(i)}) + \sum_{i:z^{(i)}=0} \log(1 - D_\theta(\hat{y}^{(i)})) \right) \right\} \\
&\overset{(b)}{=} \min_{w} \max_{\theta} \frac{1}{m} \left\{ \sum_{i=1}^{m} \ell_{\mathsf{CE}}(y^{(i)}, \hat{y}^{(i)}) \right. \\
&\qquad \left. +\lambda \left(\sum_{i=1}^{m} z^{(i)} \log D_\theta(\hat{y}^{(i)}) + (1 - z^{(i)}) \log(1 - D_\theta(\hat{y}^{(i)})) \right) \right\} \\
&\overset{(c)}{=} \min_{w} \max_{\theta} \frac{1}{m} \left\{ \sum_{i=1}^{m} \ell_{\mathsf{CE}}(y^{(i)}, \hat{y}^{(i)}) - \lambda \sum_{i=1}^{m} \ell_{\mathsf{CE}}(z^{(i)}, D_\theta(\hat{y}^{(i)})) \right\} \\
&\overset{(d)}{=} \min_{w} \max_{\theta} \underbrace{\frac{1}{m} \left\{ \sum_{i=1}^{m} \ell_{\mathsf{CE}}(y^{(i)}, G_w(x^{(i)})) - \lambda \ell_{\mathsf{CE}}(z^{(i)}, D_\theta(G_w(x^{(i)}))) \right\}}_{=:J(w,\theta)}
\end{aligned}
$$

where (a) is because we consider a binary sensitive attribute setting and we denote $D_\theta(\hat{y}^{(i)}, 1)$ simply by $D_\theta(\hat{y}^{(i)})$; (b) is due to $z^{(i)} \in \{0, 1\}$; (c) follows from the definition of binary cross entropy loss $\ell_{\mathsf{CE}}(\cdot, \cdot)$; and (d) comes from $G_w(x^{(i)}) := \hat{y}^{(i)}$.

Notice that $J(w, \theta)$ contains two cross entropy loss terms, each being a non-trivial function of $G_w(\cdot)$ and/or $D_\theta(\cdot)$. Hence, in general, $J(w, \theta)$ is highly non-convex in w and non-concave in θ.

Alternating gradient descent Similar to the prior GAN setting in Section 3.17, what we can do is to apply the only technique that we are aware of: alternating gradient descent. And then hope for the best. We employ $k : 1$ alternating gradient descent:

1. Update classifier (generator)'s weight:

$$w^{(t+1)} \leftarrow w^{(t)} - \alpha_1 \nabla_w J(w^{(t)}, \theta^{(t \cdot k)}).$$

2. Update discriminator's weight k times while fixing $w^{(t+1)}$: for i=1:k,

$$\theta^{(t \cdot k+i)} \leftarrow \theta^{(t \cdot k+i-1)} + \alpha_2 \nabla_\theta J(w^{(t+1)}, \theta^{(t \cdot k+i-1)}).$$

3. Repeat the above.

Similar to the GAN setting, one can use the Adam optimizer possibly together with the batch version of the algorithm.

Optimization used in our experiments Here is the optimization that we will use in our experiments:

$$\min_{w} \max_{\theta} \frac{1}{m} \left\{ \sum_{i=1}^{m} (1-\lambda)\ell_{\mathsf{CE}}(y^{(i)}, G_w(x^{(i)})) - \lambda \ell_{\mathsf{CE}}(z^{(i)}, D_\theta(G_w(x^{(i)}))) \right\}. \tag{3.207}$$

In order to restrict the range of λ into $0 \le \lambda \le 1$, we apply the $(1-\lambda)$ factor to the loss term w.r.t. prediction accuracy.

Like the prior GAN setting, we define two loss terms. One is "classifier (or generator) loss":

$$\min_{w} \max_{\theta} \underbrace{\frac{1}{m} \left\{ \sum_{i=1}^{m} (1-\lambda)\ell_{\mathsf{CE}}(y^{(i)}, G_w(x^{(i)})) - \lambda \ell_{\mathsf{CE}}(z^{(i)}, D_\theta(G_w(x^{(i)}))) \right\}}_{\text{"classifier (generator) loss"}}.$$

Given w, discriminator wishes to maximize:

$$\max_{\theta} -\frac{\lambda}{m} \sum_{i=1}^{m} \ell_{\mathsf{CE}}(z^{(i)}, D_\theta(G_w(x^{(i)}))).$$

This is equivalent to minimizing the minus of the objective:

$$\min_{\theta} \underbrace{\frac{\lambda}{m} \sum_{i=1}^{m} \ell_{\mathsf{CE}}(z^{(i)}, D_\theta(G_w(x^{(i)})))}_{\text{"discriminator loss"}}. \tag{3.208}$$

This is how we define "discriminator loss".

Performance metrics We introduce a performance metric that captures the degree of fairness. To this end, we first define the hard-decision value of the prediction output w.r.t. a test example:

$$\tilde{Y}_{\mathsf{test}} := \mathbf{1}\{\hat{Y}_{\mathsf{test}} \ge 0.5\}.$$

The test accuracy is then defined as:

$$\frac{1}{m_{\mathsf{test}}} \sum_{i=1}^{m_{\mathsf{test}}} \mathbf{1}\{y_{\mathsf{test}}^{(i)} = \tilde{y}_{\mathsf{test}}^{(i)}\}$$

where m_{test} denotes the number of test examples. This is an empirical version of the ground truth $\mathbb{P}(Y_{\text{test}} = \tilde{Y}_{\text{test}})$.

How to define a fairness-related performance metric? Recall the mathematical definition of DI:

$$\mathsf{DI} := \min\left(\frac{\mathbb{P}(\tilde{Y}=1|Z=0)}{\mathbb{P}(\tilde{Y}=1|Z=1)}, \frac{\mathbb{P}(\tilde{Y}=1|Z=1)}{\mathbb{P}(\tilde{Y}=1|Z=0)}\right). \tag{3.209}$$

You may wonder how to compute two probabilities of interest: $\mathbb{P}(\tilde{Y} = 1|Z = 0)$ and $\mathbb{P}(\tilde{Y} = 1|Z = 1)$. Using their empirical versions together with the WLLN, we can estimate them. For instance,

$$\mathbb{P}(\tilde{Y}=1|Z=0) = \frac{\mathbb{P}(\tilde{Y}=1, Z=0)}{\mathbb{P}(Z=0)} \approx \frac{\sum_{i=1}^{m_{\text{test}}} \mathbf{1}\{\tilde{y}_{\text{test}}^{(i)} = 1, z_{\text{test}}^{(i)} = 0\}}{\sum_{i=1}^{m_{\text{test}}} \mathbf{1}\{z_{\text{test}}^{(i)} = 0\}}$$

where the first equality is due to the definition of conditional probability and the second approximation comes from the WLLN. The above approximation is getting more and more accurate as m_{test} gets larger. Similarly we can approximate the other interested probability $\mathbb{P}(\tilde{Y} = 1|Z = 1)$. This way, we can evaluate DI (3.209).

A case study Let us exercise what we have learned with a simple example. As a case study, we consider the same setting that we introduced earlier: recidivism prediction, wherein the task is to predict if an interested individual reoffends within two years, as illustrated in Fig. 3.62.

Synthesizing an unfair dataset In fair machine learning, we must be cautious about unfair datasets. To simplify matters, we will use a synthetic dataset instead of a real-world dataset. Although there is a real-world dataset for recidivism prediction called COMPAS (Angwin *et al.*, 2020), it contains many attributes, making it more complicated. Therefore, we will use a specific yet simple approach to synthesize a much simpler unfair dataset.

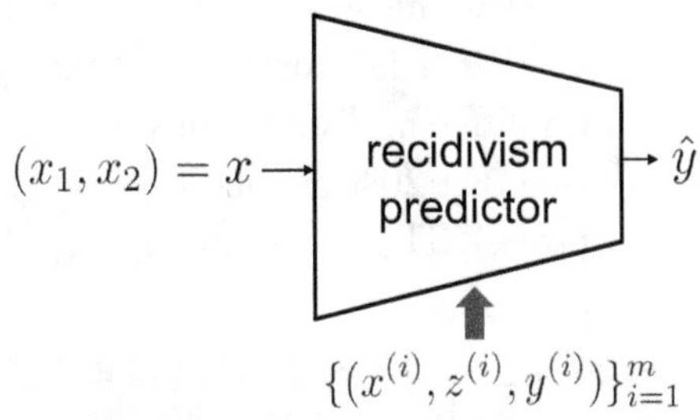

Figure 3.62. Predicting a recidivism score $\hat{y}$ from $x = (x_1, x_2)$. Here x_1 indicates the number of prior criminal records; x_2 denotes a criminal type: misdemeanor or felony; and z is a race type among white ($z = 0$) and black ($z = 1$).

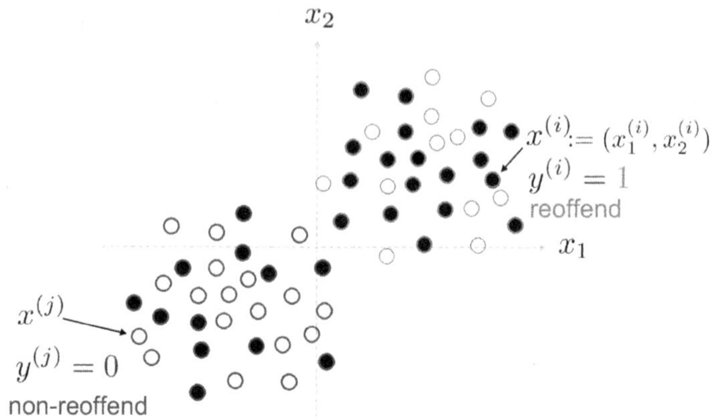

Figure 3.63. Visualization of a historically biased dataset: A hollowed (or black-colored-solid) circle indicates a data point of an individual with white (or black) race; the red (or blue) colored edge denotes $y = 1$ reoffending (or $y = 0$ non-reoffending) label.

Let us revisit the unfair data scenario visualization that we examined in Section 3.18, which will serve as the basis for our synthetic dataset (as explained in the sequel). The visualization is shown in Fig. 3.63, where a hollow (or black-colored solid) circle represents a data point corresponding to an individual of white (or black) race, and the red (or blue) colored edge (ring) denotes the event of the interested individual reoffending (or not reoffending) within two years. This scenario is inherently unfair: for $y = 1$, there are more black-colored solid circles than hollow circles, and conversely for $y = 0$, there are more hollow circles than solid circles.

To generate such an unfair dataset, we employ a simple method. See Fig. 3.64 for illustration of the method. We first generate m labels $y^{(i)}$'s so that they are i.i.d., each being according to $\text{Bern}(\frac{1}{2})$. For indices of positive examples ($y^{(i)} = 1$), we then generate i.i.d. $x^{(i)}$'s according to $\mathcal{N}((1,1), 0.5^2\mathbf{I})$; and i.i.d. $z^{(i)}$'s as per $\text{Bern}(0.8)$, meaning that 80% are blacks ($z = 1$) and 20% are whites ($z = 0$) among the positive individuals. Notice that the generation of $x^{(i)}$'s is not quite realistic. The first and second components in $x^{(i)}$ do not precisely capture the number of priors and a criminal type. You can view this generation as sort of a crude abstraction of the realistic data. On the other hand, for negative examples ($y^{(i)} = 0$), we generate i.i.d. $(x^{(i)}, z^{(i)})$'s with different distributions: $x^{(i)} \sim \mathcal{N}((-1,-1), 0.5^2\mathbf{I})$ and $z^{(i)} \sim \text{Bern}(0.2)$, meaning that 20% are blacks ($z = 1$) and 80% are whites ($z = 0$). This way, $z^{(i)} \sim \text{Bern}(\frac{1}{2})$. This is because

$$\begin{aligned}\mathbb{P}(Z=1) &\overset{(a)}{=} \mathbb{P}(Y=1)\mathbb{P}(Z=1|Y=1) + \mathbb{P}(Y=0)\mathbb{P}(Z=1|Y=0) \\ &\overset{(b)}{=} \frac{1}{2}\cdot 0.8 + \frac{1}{2}\cdot 0.2 = \frac{1}{2}\end{aligned}$$

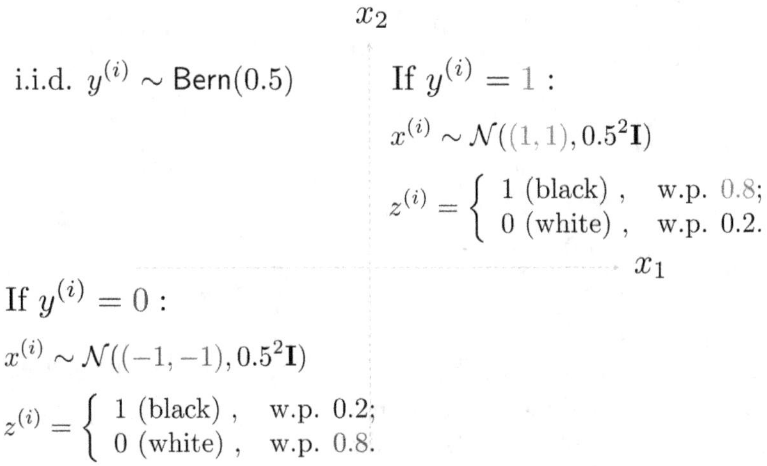

Figure 3.64. A simple way to synthesize an unfair dataset.

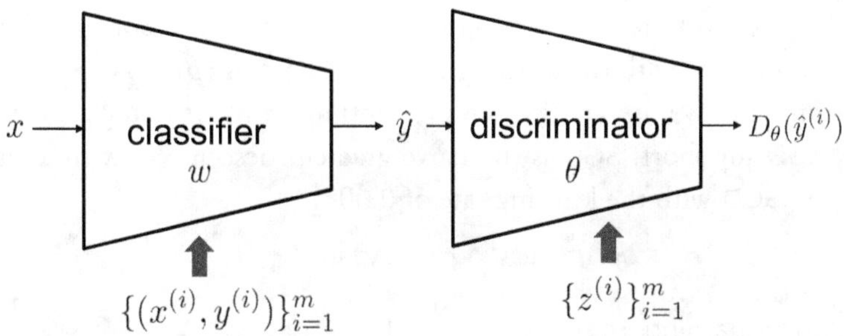

Figure 3.65. The architecture of the MI-based fair classifier.

where (*a*) follows from the total probability law and the definition of conditional probability; and (*b*) is due to the rule of the data generation method employed. Here Z and Y denote generic random variables for $z^{(i)}$ and $y^{(i)}$, respectively.

Model architecture Fig. 3.65 illustrates the architecture of the MI-based fair classifier. Since we focus on the binary sensitive attribute, the discriminator yields a single output $D_\theta(\hat{y})$. For models of the classifier and discriminator, we employ simple single-layer neural networks with logistic activation in the output layer; see Fig. 3.66.

TensorFlow: Synthesizing an unfair dataset First consider the synthesis of an unfair dataset. To generate i.i.d Bernoulli random variables for labels, we use:

```
import numpy as np
y_train = np.random.binomial(1,0.5,size=(train_size,))
```

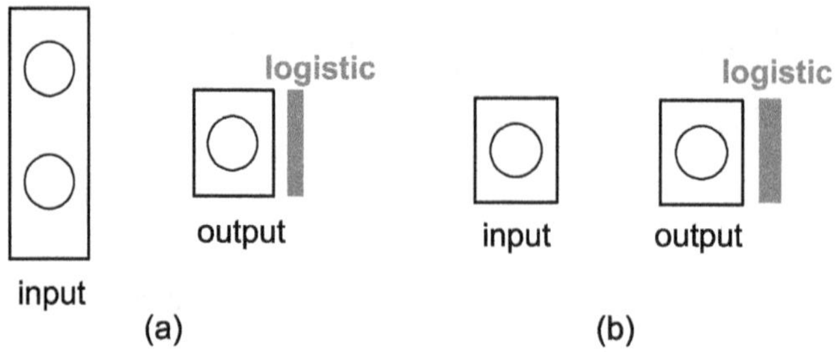

Figure 3.66. Models for (*a*) the classifier and (*b*) the discriminator.

where the first two arguments of (1,0.5) specify Bern(0.5); and the null space followed by train_size indicates a single dimension. Remember we generate i.i.d. Gaussian random variables for $x^{(i)}$'s. To this end, one can use:

```
x = np.random.normal(loc=(1,1),scale=0.5, size=(train_size,2))
```

TensorFlow: Optimizers for classifier & discriminator For classifier, we use the Adam optimizer with the learning rate of 0.005 and $(\beta_1, \beta_2) = (0.9, 0.999)$. For discriminator, we use another simpler optimizer, named Stochastic Gradient Descent, SGD for short. SGD is the naive gradient descent yet with a batch size of 1. We use SGD with the learning rate of 0.005.

```
from tensorflow.keras.optimizers import Adam
from tensorflow.keras.optimizers import SGD
adam=Adam(learning_rate=0.005,beta_1=0.9, beta_2=0.999)
sgd=SGD(learning_rate=0.005)
```

TensorFlow: Classifier (generator) loss Recall the optimization problem of interest:

$$\min_w \max_\theta \frac{1}{m}\left\{\sum_{i=1}^m (1-\lambda)\ell_{\text{CE}}(y^{(i)}, G_w(x^{(i)})) - \lambda \sum_{i=1}^m \ell_{\text{CE}}(z^{(i)}, D_\theta(G_w(x^{(i)})))\right\}.$$

To implement the classifier loss (the objective in the above), we use:

```
from tensorflow.keras.losses import BinaryCrossentropy
CE_loss = BinaryCrossentropy(from_logits=False)
p_loss = CE_loss(y_pred,y_train)
f_loss = CE_loss(discriminator(y_pred),z_train)
c_loss = (1-lamb)*p_loss - lamb*f_loss
```

where y_pred indicates the classifier output; y_train denotes a label; and z_train is a binary sensitive attribute.

TensorFlow: Discriminator loss Recall the discriminator loss that we defined in (3.208):

$$\min_{\theta} \frac{\lambda}{m} \sum_{i=1}^{m} \ell_{\mathsf{CE}}(z^{(i)}, D_{\theta}(G_w(x^{(i)}))).$$

To implement this, we use:

```
f_loss = CE_loss(discriminator(y_pred),z_train)
d_loss = lamb*f_loss
```

TensorFlow: Evaluation Recall the DI performance:

$$\mathsf{DI} := \min\left(\frac{\mathbb{P}(\tilde{Y}=1|Z=0)}{\mathbb{P}(\tilde{Y}=1|Z=1)}, \frac{\mathbb{P}(\tilde{Y}=1|Z=1)}{\mathbb{P}(\tilde{Y}=1|Z=0)}\right).$$

To evaluate the DI performance, we rely on the following approximation:

$$\mathbb{P}(\tilde{Y}=1|Z=0) \approx \frac{\sum_{i=1}^{m_{\text{test}}} \mathbf{1}\{\tilde{y}_{\text{test}}^{(i)}=1, z_{\text{test}}^{(i)}=0\}}{\sum_{i=1}^{m_{\text{test}}} \mathbf{1}\{z_{\text{test}}^{(i)}=0\}}.$$

Here is how to implement this in detail:

```
import numpy as np
y_tilde = (y_pred>0.5).int().squeeze()
z0_ind = (z_train == 0.0)
z1_ind = (z_train == 1.0)
z0_sum = int(np.sum(z0_ind))
z1_sum = int(np.sum(z1_ind))
P_y1_z0 = float(np.sum((y_tilde==1)[z0_ind]))/z0_sum
P_y1_z1 = float(np.sum((y_tilde==1)[z1_ind]))/z1_sum
```

Closing In Part I, we have explored three crucial notions in information theory, namely entropy, mutual information, and KL divergence. These notions play a significant role in representing the fundamental limit on the compression rate of an information source and proving the associated theorem, source coding theorem. Part II of the book has focused on investigating the fascinating phenomenon of the maximum transmission rate, *phase transition*, and highlighting the elegance of information theory, which deals with the laws governing the flow of information, much like physics deals with the laws governing the behavior of the physical universe. In addition, we have discovered the critical role of mutual information in defining the sharp threshold on the maximum transmission rate, known as channel capacity, as established in the channel coding theorem.

In Part III, we have showcased the modern applications of information theory in data science, emphasizing two main storylines. The first storyline focuses on the information theory of various systems that are of interest in data science, such as social networks, biological networks, and ranking systems. In this context, we have observed the occurrence of a phase transition in the amount of information required to perform various tasks, including community detection in social networks, Haplotype phasing in computational biology, and top-K ranking in search engines. To prove the achievability and converse of information-theoretic limits, we have utilized several powerful tools of information theory, including the union bound, MAP decoding, maximum likelihood decoding, Chernoff bound, Fano's inequality, and data processing inequality. The second storyline deals with the roles of information-theoretic notions in machine learning and deep learning. Specifically, we have explored the core role of cross entropy in designing a loss function for supervised learning, the fundamental role of KL divergence in the design of a powerful unsupervised learning framework known as GAN, and the recently discovered role of mutual information in the development of fair machine learning algorithms.

The topics discussed in this book encompass a range of classical and modern concepts in information theory. However, we acknowledge that there are still many other topics that we have not covered. Our approach has been to emphasize the development of logical and critical thinking skills, which we believe is more important than simply covering a wide range of topics. Information theory provides powerful principles and tools that have been successfully applied in various fields by many researchers. Although this book focuses on applications in data science, we believe that the principles discussed here have much broader applicability. We hope that you will find these principles and tools useful for your own purposes.

Problem Set 12

Prob 12.1 (Equalized Odds) In Section 3.18, we studied two fairness concepts: (i) disparate treatment; and (ii) disparate impact. In this problem, we explore another fairness notion that arises in the field: *Equalized Odds (EO for short)*. Let $Z \in \mathcal{Z}$ be a sensitive attribute. Let Y and $\hat{Y}$ be the ground-truth label and its prediction.

(*a*) For illustrative purpose, consider a simple setting where Z and Y are binary. Let $\tilde{Y} = \mathbf{1}\{\hat{Y} \geq 0.5\}$. The mathematical definition of the EO under this setting is:

$$\mathsf{EO} := \min_{y \in \{0,1\}} \min_{z \in \{0,1\}} \frac{\mathbb{P}(\tilde{Y} = 1 | Y = y, Z = 1 - z)}{\mathbb{P}(\tilde{Y} = 1 | Y = y, Z = z)}. \tag{3.210}$$

Show that $I(Z; \hat{Y}|Y) = 0$ implies $\mathsf{EO} = 1$.

(*b*) Suppose that Z and Y are not necessarily binary. The relationship between conditional mutual information and the KL divergence is:

$$I(Z; \hat{Y}|Y) = \mathsf{KL}(\mathbb{P}_{\hat{Y},Z|Y}, \mathbb{P}_{\hat{Y}|Y}\mathbb{P}_{Z|Y})$$

where $\mathbb{P}_{\hat{Y},Z|Y}, \mathbb{P}_{\hat{Y}|Y}$ and $\mathbb{P}_{Z|Y}$ indicate the conditional probability of $(\hat{Y}, Z)$, $\hat{Y}$, and Z, respectively, conditioned on Y. Using this definition, show that

$$I(Z; \hat{Y}|Y) = \sum_{y \in \mathcal{Y}, \hat{y} \in \hat{\mathcal{Y}}, z \in \mathcal{Z}} \mathbb{P}_{\hat{Y},Z,Y}(\hat{y}, z, y) \log \frac{\mathbb{P}_{\hat{Y},Z|y}(\hat{y}, z)}{\mathbb{P}_{\hat{Y}|y}(\hat{y})} + H(Z|Y) \tag{3.211}$$

where $\mathbb{P}_{\hat{Y},Z,Y}$ indicates the joint distribution of $(\hat{Y}, Z, Y)$; and $\mathbb{P}_{\hat{Y},Z|y}$ and $\mathbb{P}_{\hat{Y}|y}$ denote the conditional distributions of $(\hat{Y}, Z)$ and $\hat{Y}$, respectively, conditioned on $Y = y$.

(*c*) Show that

$$I(Z; \hat{Y}|Y) = H(Z|Y) + \max_{D(\hat{y},z,y): \sum_{z \in \mathcal{Z}} D(\hat{y},z,y) = 1} \sum_{\hat{y} \in \hat{\mathcal{Y}}, y \in \mathcal{Y}, z \in \mathcal{Z}} \mathbb{P}_{\hat{Y},Z,Y}(\hat{y}, z, y) \log D(\hat{y}, z, y). \tag{3.212}$$

(*d*) Explain the rationale behind the following approximation:

$$I(Z;\hat{Y}|Y) \approx H(Z|Y) + \max_{D(\hat{y},z,y):\sum_{z\in\mathcal{Z}} D(\hat{y},z,y)=1} \sum_{i=1}^{m} \frac{1}{m} \log D(\hat{y}^{(i)}, z^{(i)}, y^{(i)}). \tag{3.213}$$

(*e*) Formulate an optimization for a fair classifier that attempts to minimize both prediction accuracy and the approximated $I(Z;\hat{Y}|Y)$ (3.213). Use a notation λ for a regularization factor that balances prediction accuracy against the quantified fairness constraint. Also draw the classifier-&-discriminator architecture which represents the formulated optimization.

Prob 12.2 (A variant of the MI-based fair classifier) Let $Z \in \{0,1\}$ be a binary sensitive attribute. Let Y and $\hat{Y}$ be the ground-truth label and its prediction of a classifier. Let $\tilde{Y} = \mathbf{1}\{\hat{Y} \geq 0.5\}$.

(*a*) Show that $I(Z;\tilde{Y}) = 0$ is a necessary and sufficient condition for $\mathsf{DI} = 1$ where

$$\mathsf{DI} := \min\left(\frac{\mathbb{P}(\tilde{Y}=1|Z=0)}{\mathbb{P}(\tilde{Y}=1|Z=1)}, \frac{\mathbb{P}(\tilde{Y}=1|Z=1)}{\mathbb{P}(\tilde{Y}=1|Z=0)}\right).$$

(*b*) Approximate $I(Z;\tilde{Y})$ similarly to the formula claimed in part (*d*) of Prob 12.1. Also explain the rationale behind the approximation.

(*c*) Formulate an optimization for a fair classifier that attempts to minimize both prediction accuracy and the approximated $I(Z;\tilde{Y})$, derived in the prior part. Use a notation λ for a regularization factor that balances prediction accuracy against the fairness constraint. Also draw the classifier-&-discriminator architecture which respects the formulated optimization.

Prob 12.3 (TensorFlow implementation of the MI-based fair classifier) Consider the MI-based fair classifier in Sections 3.19 and 3.20. In this problem, you are asked to build a simple fair classifier that predicts recidivism scores of individuals with prior criminal records. See Fig. 3.67. We employ very simple single-layer neural networks for classifier (generator) and discriminator with logistic activation at the output layer.

(*a*) *(Unfair dataset synthesis)* Explain how an unfair dataset is generated in the following code:

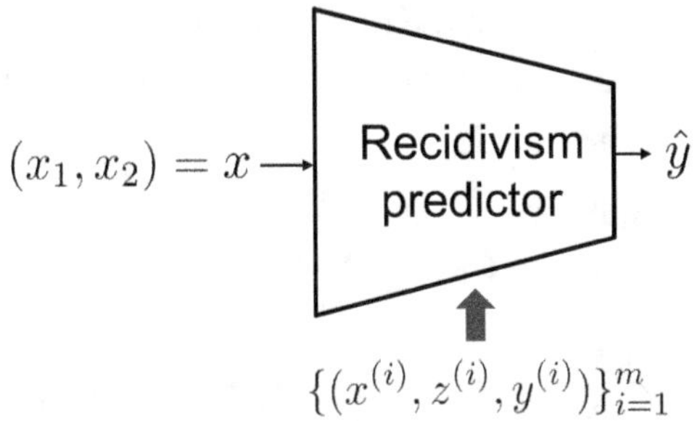

Figure 3.67. Predicting a recidivism score $\hat{y}$ from $x = (x_1, x_2)$. Here x_1 indicates the number of prior criminal records; x_2 denotes a criminal type: misdemeanor or felony; and z is a race type among white ($z = 0$) and black ($z = 1$).

```
import numpy as np
n_samples = 2000
p = 0.8
# numbers of positive and negative examples
n_Y1 = int(n_samples*0.5)
n_Y0 = n_samples - n_Y1
# generate positive samples
Y1 = np.ones(n_Y1)
X1 = np.random.normal(loc=[1,1],scale=0.5,
                      size=(n_Y1,2))
Z1 = np.random.binomial(1,p,size=(n_Y1,))
# generate negative samples
Y0 = np.zeros(n_Y0)
X0 = np.random.normal(loc=[-1,-1],scale=0.5,
                      size=(n_Y0,2))
Z0 = np.random.binomial(1,1-p,size=(n_Y0,))
# merge
Y = np.concatenate((Y1,Y0))
X = np.concatenate((X1,X0))
Z = np.concatenate((Z1,Z0))
Y = Y.astype(np.float32)
X = X.astype(np.float32)
Z = Z.astype(np.float32)
# shuffle and split into train & test data
shuffle = np.random.permutation(n_samples)
X_train = X[shuffle][:int(n_samples*0.8)]
Y_train = Y[shuffle][:int(n_samples*0.8)]
Z_train = Z[shuffle][:int(n_samples*0.8)]
X_test = X[shuffle][int(n_samples*0.8):]
Y_test = X[shuffle][int(n_samples*0.8):]
Z_test = X[shuffle][int(n_samples*0.8):]
```

(*b*) (*Data visualization*) Using the following code or otherwise, plot randomly sampled data points (say 200 random points) among the entire data points generated in part (*a*).

```
import matplotlib.pyplot as plt
# randomly select the number n_s of samples
n_s = 200
Xs = X_train[:n_s]
Ys = Y_train[:n_s]
Zs = Z_train[:n_s]
# choose part of X and Y assiciated with a certain Z
X_Z0 = Xs[Zs==0.0]
X_Z1 = Xs[Zs==1.0]
Y_Z0 = Ys[Zs==0.0]
Y_Z1 = Ys[Zs==1.0]
# plot
plt.figure(figsize=(14,10))
plt.scatter(
    X_Z0[Y_Z0==1.0][:,0], X_Z0[Y_Z0==1.0][:,1],
    color='red',marker='o',facecolors='none',
    s=120, linewidth=1.5, label='White reoffend')
plt.scatter(
    X_Z0[Y_Z0==0.0][:,0], X_Z0[Y_Z0==0.0][:,1],
    color='blue',marker='o',facecolors='none',
    s=120, linewidth=1.5, label='White non-reoffend')
plt.scatter(
    X_Z1[Y_Z1==1.0][:,0], X_Z1[Y_Z1==1.0][:,1],
    color='red',marker='o',facecolors='black',
    s=120, linewidth=1.5, label='Black reoffend')
plt.scatter(
    X_Z1[Y_Z1==0.0][:,0], X_Z1[Y_Z1==0.0][:,1],
    color='blue',marker='o',facecolors='black',
    s=120, linewidth=1.5, label='Black non-reoffend')
plt.legend(fontsize=16)
```

(*c*) (*Classifier & discriminator*) Draw block diagrams of the classifier and the discriminator implemented by the following code:

```
from tensorflow.keras.models import Sequential
from tensorflow.keras.layers import Dense

classifier=Sequential()
classifier.add(Dense(1,input_dim=2, activation='sigmoid'))
discriminator=Sequential()
discriminator.add(Dense(1,input_dim=1, activation='sigmoid'))
```

(*d*) *(Optimizers and loss functions)* Explain how the optimizers and loss functions of the discriminator and the classifier are implemented in the following code. Also draw a block diagram of the GAN model implemented as the name of gan.

```
from tensorflow.keras.layers import Input
from tensorflow.keras.models import Model
from tensorflow.keras.optimizers import Adam
from tensorflow.keras.optimizers import SGD
from tensorflow.keras.losses import BinaryCrossentropy
from tensorflow.keras.layers import Concatenate

# optimizers of classifier & discriminator
c_opt=Adam(learning_rate=0.005,beta_1=0.9,beta_2=0.999)
d_opt=SGD(learning_rate=0.005)

# define dicriminator loss
def d_loss(y_true,y_pred):
    CE_loss = BinaryCrossentropy(from_logits=False)
    lamb = 0.1
    return lamb*CE_loss(y_pred,y_true)
# discriminator compile
discriminator.compile(loss=d_loss, optimizer=d_opt)

# define classifier (generator) loss
def c_loss(y_true,y_pred):
    # y_true[:,0]: Y_train (label)
    # y_true[:,1]: Z_train (sensitive attribute)
    # y_pred[:,0]: classifier output G(x)
    # y_pred[:,1]: discriminator output fed by
    #            classifier output D(G(x))
    CE_loss = BinaryCrossentropy(from_logits=False)
    lamb = 0.1
    p_loss = CE_loss(y_pred[:,0],y_true[:,0])
    f_loss = CE_loss(y_pred[:,1],y_true[:,1])
    return (1-lamb)*p_loss - lamb*f_loss

# define the GAN model
# input: x
# output: [G(x), D(G(x))]
discriminator.trainable = False
gan_input = Input(shape=(2,))
Gx = classifier(inputs=gan_input)
DGx = discriminator(Gx)
output = Concatenate()([Gx,DGx])
gan = Model(gan_input, output)
```

```
# The GAN model compile
gan.compile(loss=c_loss, optimizer=c_opt)
```

(*e*) *(Training)* Explain how classifier and discriminator are trained in the following code:

```
import numpy as np

EPOCHS = 400
k=2 # k:1 alternating gradient descent
c_losses = []
d_losses = []

for epoch in range(1,EPOCHS+1):
    #######################
    # train discriminator
    #######################
    # input for discriminator
    d_input=classifier.predict(X_train)
    # label for discriminator
    d_label=Z_train
    # train discriminator
    d_loss=discriminator.train_on_batch(d_input,d_label)

    #######################
    # train classifier
    #######################
    if epoch % k == 0: # train once every k steps
        # label for classifier
        # 1st component: Y_train
        # 2nd component: Z_train (sensitive attribute)
        c_label = np.zeros((len(Y_train),2))
        c_label[:,0] = Y_train
        c_label[:,1] = Z_train
        # train classifier
        c_loss = gan.train_on_batch(X_train,c_label)

        c_losses.append(c_loss)
        d_losses.append(d_loss)
```

(*f*) *(Evaluation)* Suppose we train classifier and discriminator using the code in part (*e*) with EPOCHS=400. Plot the tradeoff performance between test accuracy and DI by sweeping λ from 0 to 1. Also include the Python script.

DOI: 10.1561/9781638281153.ch4

Appendix A

Python Basics

A.1 Jupyter Notebook

Outline To use Python, you will need to have another software platform, known as Jupyter notebook, installed on your system. In this section, we will cover some basic concepts related to Jupyter notebook. We will cover four parts in detail. First, we will explore the role of Jupyter notebook in light of Python. Next, we will provide guidance on how to install the software and launch a file for scripting a code. We will also examine some useful interfaces that simplify the process of scripting Python code. Finally, we will introduce several frequently-used shortcuts for writing and executing code.

What is Jupyter notebook? Jupyter notebook is a powerful tool that allows you to write and run Python code. One of its key advantages is that you can execute each line of code individually, rather than running the entire code all at once. This feature makes it easy to debug your code, particularly when dealing with lengthy programs.

```
a=1
b=2
a+b
```

```
3
```

Figure A.1. Three versions of Anaconda installers.

```
a=1
```

```
b=2
```

```
a+b
```

```
3
```

There are two common methods for using Jupyter notebook. The first involves running the code on a server or in the cloud, while the second involves using a local machine. In this section, we will focus on the latter approach.

Install & launch To use Jupyter notebook on a local machine, you need to install a software tool called Anaconda. The latest version can be downloaded from

https://www.anaconda.com/products/individual

There are three different versions available for different operating systems, as shown in Fig. A.1. During installation, you may encounter errors related to non-ASCII characters in the destination folder path or permission to access the path. To resolve these issues, ensure that the folder path does not include non-ASCII characters and run the installer under "run as administrator" mode.

To launch Jupyter notebook, you can use the Anaconda prompt for Windows or the terminal for Mac and Linux. Simply type "jupyter notebook" in the prompt and press Enter. The Jupyter notebook window will open automatically. If it does not appear, you can manually open it by copying and pasting the URL indicated by the arrow in Fig. A.2 into your web browser. Once properly launched, the Jupyter notebook window should look like Fig. A.3.

Generating a new notebook file is an easy process. Initially, navigate to the folder where you wish to save the notebook file. Then, select the New tab on the top right corner (highlighted in blue), and click on the Python 3 tab (indicated in red). Refer to Fig. A.4 to locate the tabs.

Figure A.2. How to launch Jupyter notebook in the Anaconda prompt.

Figure A.3. Web browser of a successfully launched Jupyter notebook.

Figure A.4. How to create a Jupyter notebook file on the web browser.

Interface Jupyter notebook contains two key components required to run a code. The first is a computational engine which executes the code. The engine is named Kernel and it can be controlled via several functions in the Kernel tab. See Fig. A.5 for details.

The second element is a component called a "cell," where you can write a script. The cell has two modes of operation: edit mode and command mode. In edit mode, you can type a code script for running a program or any text like a regular text editor.

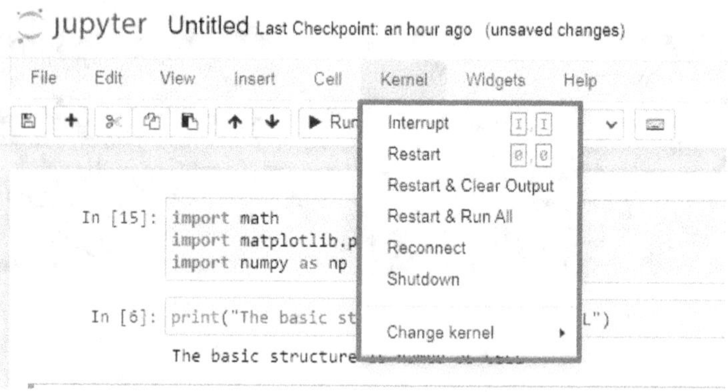

Figure A.5. Kernel is a computational engine which serves to run the code. There are several relevant functions under the Kernel tap.

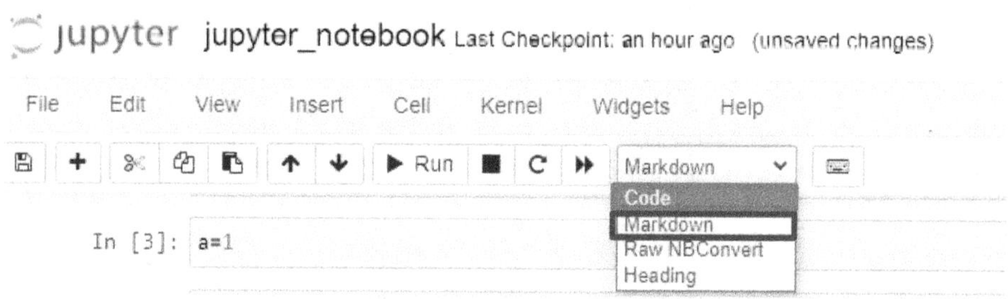

Figure A.6. How to choose the Code or Markdown option in the edit mode.

Code scripts are written under the Code tab, indicated by a red box in Fig. A.6, while text-editing is done under the Markdown tab, indicated by a blue box. In command mode, you can edit the notebook as a whole. This allows you to copy or delete cells, and move them around.

Shortcuts There are numerous shortcuts that are very useful for editing and navigating a Jupyter notebook. We will highlight three types of shortcuts that are commonly used. The first set is for changing between the edit and command modes. To switch from the edit to the command mode, we press the Esc key, while pressing Enter takes us back to the edit mode. The second set of shortcuts is for inserting or deleting a cell. Under the command mode, we can use the "a" shortcut to insert a new cell above the current cell, "b" to insert below, and "d+d" to delete the current cell. The final set of shortcuts is for executing a cell. We can use the arrow keys to move between cells, Shift + Enter to run the current cell and move to the next one, and Ctrl + Enter to stay in the current cell after execution.

A.2 Basic Syntaxes of Python

Outline This section will cover the basic Python syntax required to script for information-theoretic notions and algorithm implementation. Specifically, we will focus on three essential concepts: (i) class; (ii) package; and (iii) function. Additionally, we will introduce a range of Python packages that are relevant and useful for the topics covered in this book.

A.2.1 Data structure

There are two prominent data-structure components in Python: (i) list; and (ii) set.

(i) List List is a data type that is built-in in Python, which enables the storage of multiple elements in a single variable. The elements are listed in a specific order, and duplicates are allowed. Below are examples of how to use:

```
x = [1, 2, 3, 4]  # construct a simple list
print(x)
```

```
[1, 2, 3, 4]
```

```
x.append(5)  # add an item at the end
print(x)
```

```
[1, 2, 3, 4, 5]
```

```
x.pop()  # delete an item located in the last
print(x)
```

```
[1, 2, 3, 4]
```

```
# checking if a particular element exists in the list
if 3 in x:
    print(True)
if 5 in x:
    print(True)
else:
    print(False)
```

```
True
False
```

```
# A single-line construction of a list
y = [x for x in range(1,10)]
print(y)
```

```
[1, 2, 3, 4, 5, 6, 7, 8, 9]
```

```
# Retrieving all the elements through a "for" loop
for i in x:
    print(i)
```

1
2
3
4

(ii) Set Set is a built-in data type that has similarities with List, but with two key differences. Firstly, it is an unordered data type, and secondly, it does not allow for duplicate elements. Below are some examples of how to use.

```
x = set({1, 2, 3}) # construct a set
print(f"x: {x}, type of x: {type(x)}")
```

x: {1, 2, 3}, type of x: <class 'set'>

The f in front of strings in the print command tells Python to look at the values inside $\{\cdot\}$.

```
x.add(1) # add an existing item
print(x)
```

{1, 2, 3}

```
x.add(4) # add a new item
print(x)
```

{1, 2, 3, 4}

```
# checking if a particular element exists in the list
if 1 in x:
    print(True)
if 5 in x:
    print(True)
else:
    print(False)
```

True
False

```
# Retrieving all the elements through a "for" loop
for i in x:
    print(i)
```

1
2
3
4

A.2.2 Package

Let us present five packages that are essential for writing codes for the problems dealt in this book: (i) math; (ii) random; (iii) itertools; (iv) numpy; and (v) scipy.

(i) math The math module offers a range of useful mathematical expressions, such as exponential, logarithmic, square root, and power functions. Below are some examples to illustrate their usage.

```
import math
```

```
math.exp(1)  # exp(x)
```

```
2.718281828459045
```

```
print(math.log(1, 10))  # log(x, base)
print(math.log(math.exp(20))) # natural logarithm
print(math.log2(4))      # base-2 logarithm
print(math.log10(1000))   # base-10 lograithm
```

```
0.0
20.0
2.0
3.0
```

```
print(math.sqrt(16))  # square root
print(math.pow(2,4)) # x raised to y (same as x**y)
print(2**4)
```

```
4.0
16.0
16
```

```
print(math.cos(math.pi)) # cosine of x radians
print(math.dist([1,2],[3,4]))   # Euclidean distance
```

```
-1.0
2.8284271247461903
```

```
# The erf() function can be used to compute traditional
# statistical functions such as the CDF of
# the standard Gaussian distribution
def phi(x):
    # CDF of the standard Gaussian distribution
    return (1.0 + math.erf(x/math.sqrt(2.0)))/2.0

phi(1)
```

```
0.8413447460685428
```

(ii) random This module yields random number generation. See below for some examples.

```
import random
```

```
random.randrange(start=1, stop=10, step=1)
# a random number in range(start, stop, step)
random.randrange(10) # integer from 0 to 9 inclusive
```

```
5
```

```
# returns random integer n such that a<=n<=b
random.randint(1, 10)
```

```
7
```

(iii) itertools This package offers a concise method to explore all possible cases in various combinatorial situations.

```
from itertools import permutations, combinations
```

```
# generating all permutations of [1, 2, 3]
p = permutations([1, 2, 3])

for i in p:
    print (i)
```

```
(1, 2, 3)
(1, 3, 2)
(2, 1, 3)
(2, 3, 1)
(3, 1, 2)
(3, 2, 1)
```

```
# generating all length-2 combinations of [1, 2, 3]
c = combinations([1, 2, 3], 2)

for i in c:
    print (i)
```

```
(1, 2)
(1, 3)
(2, 3)
```

```
# generating all length-3 combinations of [1, 2, 3, 4, 5]
c = combinations([1, 2, 3, 4, 5], 3)
```

```
for i in c:
    print (i)
```

```
(1, 2, 3)
(1, 2, 4)
(1, 2, 5)
(1, 3, 4)
(1, 3, 5)
(1, 4, 5)
(2, 3, 4)
(2, 3, 5)
(2, 4, 5)
(3, 4, 5)
```

(iv) numpy Numpy is a widely used package for manipulating matrices and vectors. It provides numerous helpful functions, some of which are commonly utilized and listed below.

(a) numpy.array() numpy.array() is a specialized array data structure in numpy. This differs from Python data type array().

```
import numpy as np
```

```
np.array([1, 2, 3]) # construct an array
```

```
array([1, 2, 3])
```

```
np.array([[1, 2], [3, 4]]) # construct a 2D array
```

```
array([[1,  2],
       [3,  4]])
```

```
x = np.ones((2,2))
# construct an all-one matrix with size of 2-by-2
x = np.zeros((2,2))
# construct an all-zero matrix with size of 2-by-2
print(np.ones_like(x))
# all-one matrix with the same shape and type of input
print(np.zeros_like(x))
# all-zero matrix with the same shape and type of input
```

```
[[1. 1.]
 [1. 1.]]
[[0. 0.]
 [0. 0.]]
```

```
# range of x
x_grid=np.arange(0,1,0.0001)
# or one can use:
x_grid2=np.linspace(0,1,0.0001)
```

```
# concatenation of two numpy arrays
x1 = np.array([1,2])
x2 = np.array([3,4])
xc = np.concatenate((x1,x2)) # column-wise
xr = np.vstack((x1,x2)) # row-wise
print(xc)
print(xr)
```

```
[1 2 3 4]
[[1 2]
 [3 4]]
```

```
# sign function
x = np.array([1.2,-3,2,-4.2])
s = np.sign(x)
print(s)
```

```
[ 1. -1.  1. -1.]
```

(b) numpy.random() The purpose of this module is to generate random samples from different probability distributions. We provide some commonly used examples below, but for more information, you may want to refer to:

https://numpy.org/doc/1.16/reference/routines.random.html

```
# sampling a number from standard Gaussian distribution
np.random.normal(loc = 0, scale = 1)
# loc: mean, scale: standard deviation
np.random.randn() # plays the same role
```

```
-2.5459976698222495
```

```
# sampling multiple numbers as per the standard Gaussian
np.random.normal(0, 1, size = (2, 2))
# Here the size determines the output shape
np.random.randn(2,2) # plays the same role
```

```
array([[-1.8133258 , -1.01151295],
       [-0.37375747,  0.36005748]])
```

```
np.random.rand(2,2) # Uniform over [0,1]
```

```
array([[0.06535694, 0.2507505 ],
       [0.17559137, 0.60967901]])
```

```
# Uniform over [0.8,1]
np.random.uniform(0.8,1,(2,2))
```

```
array([[0.89902277, 0.85310313],
       [0.96578371, 0.85695091]])
```

```
# Binomial distribution
np.random.binomial(10000,0.5) # 10000 trials of Bern(0.5)
```

```
5042
```

(c) numpy.linalg Here are some of the useful linear-algebra related functions offered by this package.

```
from numpy import linalg
```

```
x = np.random.randn(2,2)
print(linalg.det(x))    # Determinant of a matrix x
print(linalg.inv(x))    # Inverse of a matrix x
print(linalg.norm(x))   # Matrix or vector norm
print(linalg.svd(x))    # Singular value decomposition
print(linalg.eig(x))    # Eigenvalue decomposition
```

```
0.7125655927348966
[[ 0.77007826 -0.38835738]
 [ 2.33455331  0.64504946]]
1.832010151997132
(array([[-0.2060815,  0.97853483],
       [ 0.97853483,  0.2060815 ]]),
array([1.78814528, 0.39849424]),
array([[-0.96330981,  0.2683919 ],
       [ 0.2683919 ,  0.96330981]]))
(array([0.50418566+0.67702467j, 0.50418566-0.67702467j]),
array([[0.02479485-0.37684352j, 0.02479485+0.37684352j],
       [0.92594502+0.j        , 0.92594502-0.j        ]]))
```

(d) numpy.fft One of the useful operations in communication and signal processing is Discrete Fourier Transform (DFT). Let $x[m]$'s be time-domain discrete signals where $m \in \{0, 1, \ldots, N-1\}$. Then, the corresponding frequency-domain signals read:

$$X[k] = \frac{1}{\sqrt{N}} \sum_{m=0}^{N-1} x[m] e^{-j\frac{2\pi}{N}mk} \qquad k \in \{0, 1, \ldots, N-1\}.$$

One can implement this using a built-in function fft in numpy.fft.

$$\mathrm{fft}(x) = \sum_{m=0}^{N-1} x[m]e^{-j\frac{2\pi}{N}mk}.$$

In order to align with the specified DFT, it is necessary to divide the fft by $\sqrt{N}$. Conversely, the inverse function ifft serves the opposite function and can be utilized in a similar manner.

$$\mathrm{ifft}(X) = \frac{1}{N}\sum_{k=0}^{N-1} X[k]e^{j\frac{2\pi}{N}mk}.$$

```
from numpy.fft import fft
from numpy.fft import ifft

x_time = np.random.randn(8)
X_freq = fft(x_time)/np.sqrt(8)
x_time_rec = np.sqrt(8)*ifft(X_freq)
print(x_time)
print(x_time_rec)
```

```
[ 0.19398987  0.92755053  1.14652418  1.05737049 -0.66500356
  0.43650243 1.04576987  -0.95167376]
[ 0.19398987+0.j  0.92755053+0.j  1.14652418+0.j  1.05737049+0.j
 -0.66500356+0.j  0.43650243+0.j  1.04576987+0.j  -0.95167376+0.j]
```

(e) resizing The resizing is used for transforming the dimension of one into another.

```
x = np.random.randn(4,4,1)
y = x.view(dtype=np.float_).reshape(-1,2)
# '-1' can be inferred from the context: Shape of (8,2)
print(y)
z = x.squeeze()
print(z.shape)
```

```
[[-0.85719316  2.99692221]
 [ 1.16327996 -0.11955541]
 [-0.76229609  0.79871494]
 [ 0.99757568  0.69329723]
 [-1.52198295 -0.74430996]
 [ 0.17174063  0.25343301]
 [ 0.07151011 -2.90945412]
 [ 1.1874155  -0.64209109]]
(4, 4)
```

(v) scipy This particular module offers an extensive collection of probability distributions and corresponding statistical metrics. Presented below are a few examples; for additional details, please refer to:

https://docs.scipy.org/doc/scipy/reference/stats.html

```
from scipy import stats
```

```
# A random variable with the standard Gaussian
X = stats.norm(loc = 0, scale = 1)
# loc:mean, scale:standard deviation
print(X.cdf(np.array([-1, 0, 1])))
# computes the CDF at each numpy array
print(X.rvs(size = 3))
# generating a sequence of random variables
```

```
[0.15865525 0.5        0.84134475]
[ 0.39460402 -0.8042592  -0.71404882]
```

```
# Another random variable with the uniform distribution
Y = stats.uniform(loc = 0, scale = 1)
# uniform distribution in [loc, loc + scale]
print(Y.cdf(np.array([-1, 0, 0.5, 1])))
print(Y.rvs(size = 3))
```

```
[0.  0.  0.5 1. ]
[0.72953474 0.67879248 0.47947748]
```

For binary random variables, we employ a built-in function, bernoulli in scipy.stats.

```
from scipy.stats import bernoulli
X = bernoulli(0.5)
X_samples = X.rvs(10)
print(X_samples)
```

```
[0 1 0 0 1 0 0 1 0 0]
```

It also contains built-in functions for entropy and the KL divergence.

```
from scipy.stats import entropy

pX1 = np.array([1/2, 1/2]) # numpy.array
pX2 = [1/2, 1/2]  # list
print(entropy(pX1, base=2))
print(entropy(pX2, base=2))
```

```
1.0
1.0
```

Here the input distribution can take either a numpy.array or a list.

```
from scipy.special import rel_entr

# Compute p(x,y)
pXY = np.array([1/4, 1/4, 1/3, 1/6])
# Compute p(x)p(y)
pXpY = np.array([pX[0]*pY[0],pX[0]*pY[1],
                 pX[1]*pY[0],pX[1]*pY[1]])
kl_builtin = rel_entr(pXY,pXpY)
print(sum(kl_builtin))
# To convert into log base 2
print(sum(kl_builtin)/np.log(2))
```

```
0.014362591564146779
0.020720839623908218
```

To calculate the KL divergence, scipy.special provides the function rel_entr. rel_entr uses natural logarithms instead of log base 2 and produces a list of values in the form of $p(x)\ln\frac{p(x)}{q(x)}$. Therefore, proper conversion is required.

In communication problems, it is common to compute the Q-function which is defined as:

$$Q(a) := \int_a^\infty \frac{1}{\sqrt{2\pi}} e^{-\frac{z^2}{2}}\, dz.$$

The analysis of communication error probability is aided by this process. The numerical computation of the required integration can be performed through the implementation of the erfc command provided by scipy.special.

$$\text{erfc}(x) := \int_x^\infty \frac{2}{\sqrt{\pi}} e^{-t^2}\, dt.$$

The relation between $Q(a)$ and $\text{erfc}(x)$ is:

$$\begin{aligned} Q(a) &:= \int_a^\infty \frac{1}{\sqrt{2\pi}} e^{-\frac{z^2}{2}}\, dz \\ &\overset{(a)}{=} \int_{\frac{a}{\sqrt{2}}}^\infty \frac{1}{\sqrt{\pi}} e^{-t^2}\, dt \\ &= \frac{1}{2} \cdot \text{erfc}\left(\frac{a}{\sqrt{2}}\right) \end{aligned}$$

where (a) comes from the change of variable $t := \frac{z}{\sqrt{2}}$ $(dz = \sqrt{2}dt)$.

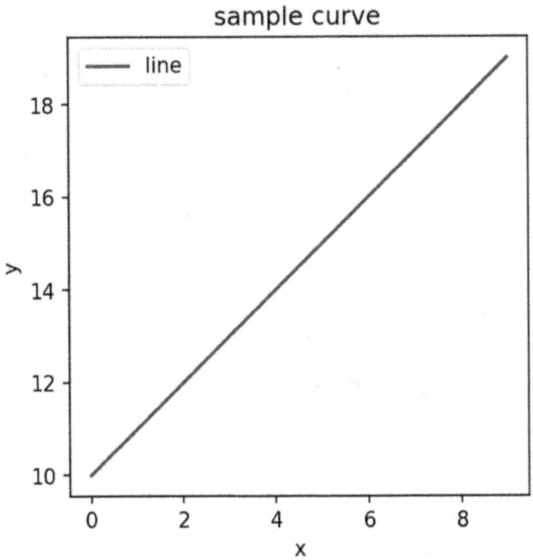

Figure A.7. Plotting a simple function via matplotlib.pyplot.

```
from scipy.special import erfc

a = 10
Qfunc = 1/2*erfc(a/np.sqrt(a/2))
print(Qfunc)
```

1.2698142947354283e-10

A.2.3 Visualization

matplotlib.pyplot is the most commonly used function for graph plotting. The following is a guide on how to utilize it.

```
import matplotlib.pyplot as plt
```

```
x_value = [x for x in range(10)]
y_value = [y for y in range(10, 20)]

plt.figure(figsize=(4,4),dpi=150) # figure size and resolution
plt.plot(x_value, y_value, color='blue', label='line')
plt.xlabel('x')            # labeling x-axis
plt.ylabel('y')            # labeling y-axis
plt.title('sample curve')
plt.legend()
plt.show() # No need to use show() in jupyter notebook.
```

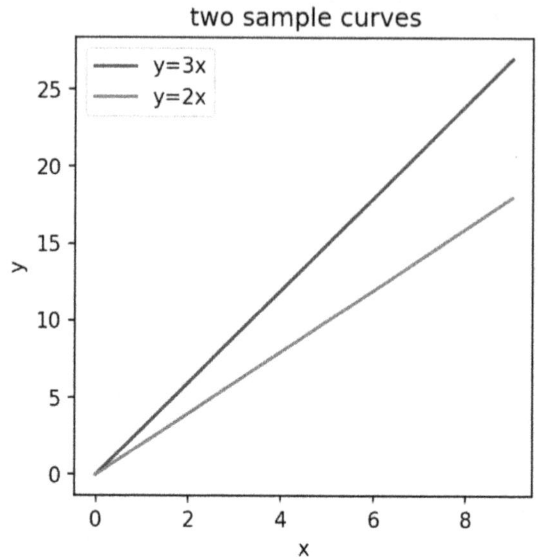

Figure A.8. Multiple functions and legend.

It is also possible to plot multiple curves in a single graph.

```
# we can plot multiple graphs at once

x = [x for x in range(10)]
y_1 = [3*y for y in range(10)]
y_2 = [2*y for y in range(10)]

plt.figure(figsize=(4,4), dpi=150)
plt.plot(x, y_1, color='blue', label='y=3x')  # plot_1
plt.plot(x, y_2, color='red', label='y=2x')  # plot_2
plt.xlabel('x')  # labeling x-axis
plt.ylabel('y')  # labeling y-axis
plt.title('two sample curves')
plt.legend()
plt.show()
```

To draw the probability distribution of a random variable, we often employ a stats visualization package, named seaborn.

```
import seaborn as sns
import matplotlib.pyplot as plt
from scipy.stats import bernoulli

X = bernoulli(0.5)
X_samples = X.rvs(1000)
```

```
plt.figure(figsize=(4,4), dpi=150)
sns.histplot(X_samples)
plt.xlabel('Values of a random variable')
plt.ylabel('Histogram')
plt.show()
```

In communication problems, it is common to plot the probability of error, which is often exceedingly small, such as 10^{-5}. To better differentiate between small probability values, a logarithmic scale of error probability is utilized. This can be achieved by employing the function plt.yscale('log').

```
import numpy as np
from scipy.special import erfc
import matplotlib.pyplot as plt

SNRdB = np.arange(0,21,1)
SNR = 10**(SNRdB/10)

# Q-function
Qfunc = 1/2*erfc(np.sqrt(SNR/2))

plt.figure(figsize=(4,4), dpi=150)
plt.plot(SNRdB, Qfunc, label='Q(sqrt(SNR))')
plt.yscale('log')
plt.xlabel('SNR (dB)')
plt.grid(linestyle=':', linewidth=0.5)
plt.title('Q function')
plt.legend()
plt.show()
```

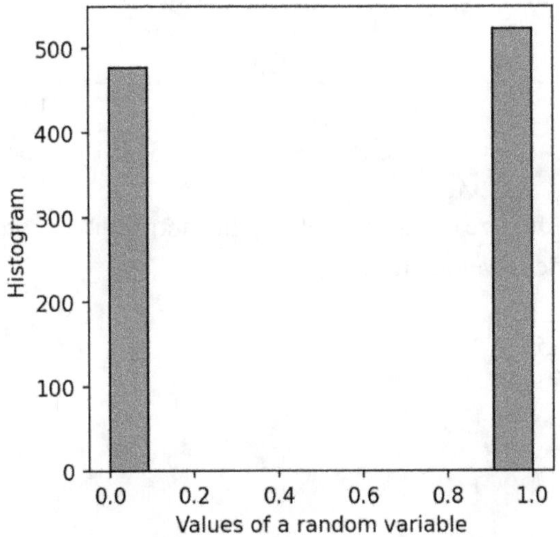

Figure A.9. Plotting a histogram of independent realizations of a random variable.

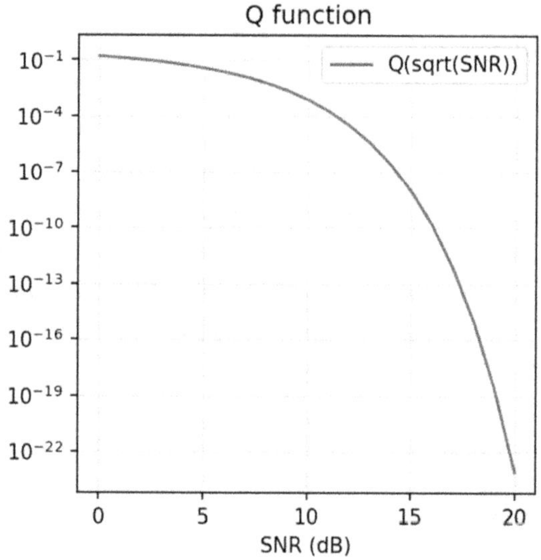

Figure A.10. Logarithmic scale of the Q-function as a function of SNR.

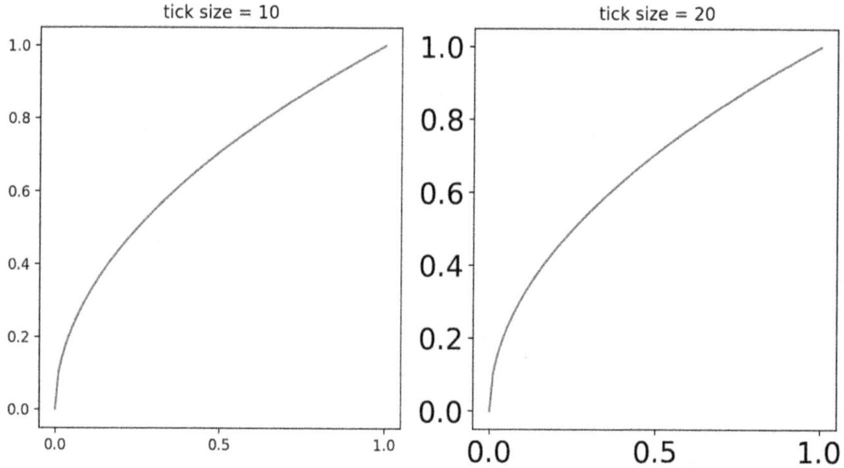

Figure A.11. How to adjust the font size of axis tick values.

To modify the font size of axis tick values, the matplotlib.rc command is utilized. The following code serves as an example:

```
# To adjust the font size of axis tick values
import matplotlib
import matplotlib.pyplot as plt
import numpy as np

x= np.linspace(0,1,100)
y=np.sqrt(x)
```

```
plt.figure(figsize=(10,5),dpi=200)
matplotlib.rc('xtick',labelsize=10)
matplotlib.rc('ytick',labelsize=10)
plt.subplot(1,2,1)
plt.plot(x,y)
plt.title('tick size = 10')
matplotlib.rc('xtick',labelsize=20)
matplotlib.rc('ytick',labelsize=20)
plt.subplot(1,2,2)
plt.plot(x,y)
plt.title('tick size = 20')
plt.show()
```

DOI: 10.1561/9781638281153.ch5

Appendix B

TensorFlow and Keras Basics

Outline Part III covered several applications in data science, such as machine learning and deep learning. Deep learning, a learning approach that utilizes a deep neural network (DNN) as a basic model for predictions, can be implemented using various software tools known as machine learning frameworks or application programming interfaces (APIs). TensorFlow, Keras, Pytorch, DL4J, Caffe, and mxnet are some examples of such frameworks. Each framework has its advantages and disadvantages, depending on the requirements of the deep learning model design, such as usability, training speed, functionality, and scalability in distributed training. This book prioritizes usability and therefore focuses on the high-level API with fast user experimentation, which is Keras.

The Keras API facilitates moving from idea to implementation with minimal steps, making it an ideal choice for this book. This appendix presents four basic contents related to Keras. Since Keras is fully integrated with TensorFlow, it comes packaged with the TensorFlow installation. In the first part, we will learn how to install TensorFlow. To implement deep learning, three key procedures are required: (i) data preparation and processing; (ii) neural network model building; and (iii) model training and testing. The second part will cover an easy way to handle data using Keras, followed by building a neural network model using popular packages such as keras.models and keras.layers. Finally, we will explore how to train and test a model accordingly. To illustrate these procedures easily, we will demonstrate them using a simple example.

Installation Installing Keras requires the installation of TensorFlow. Fortunately, the installation process is straightforward:

```
pip install tensorflow
```

Keras is fully supported by TensorFlow 2 packages. To ensure a proper installation, a pip version higher than 19.0 (or higher than 20.3 for macOS) is required. You may need to upgrade pip by running the command: "pip install -upgrade pip". To confirm a successful installation, try importing keras using the following command:

```
from tensorflow import keras
```

If there are no errors, then you are ready to start using Keras. However, if you do encounter any errors, you may want to refer to the installation guidelines found at:

https://www.tensorflow.org/install

A simple task We will be focusing on a simple task of classifying handwritten digits, where the objective is to identify a digit from an image of handwritten digits. An example of such an image is shown in Fig. B.1. The figure demonstrates a case where an image of the digit 2 is accurately identified.

Preparing and processing data The digit classification task is commonly associated with the MNIST (Modified National Institute of Standards and Technology) dataset, which contains 60, 000 training images and 10, 000 testing images. Each image, denoted by $x^{(i)}$, is a 28 × 28 pixel image with gray-scale levels ranging from 0 (white) to 1 (black). Additionally, each image has a label, denoted by $y^{(i)}$, that corresponds to one of the 10 classes, $y^{(i)} \in \{0, 1, \ldots, 9\}$. Please refer to Fig. B.2 for an illustration of the dataset.

Keras offers the advantage of having popular datasets, such as MNIST, readily available in a sub-package called keras.datasets. This sub-package includes both

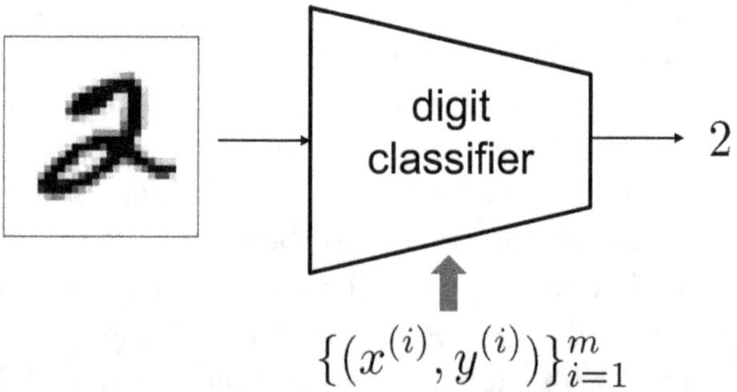

Figure B.1. Handwritten digit classification.

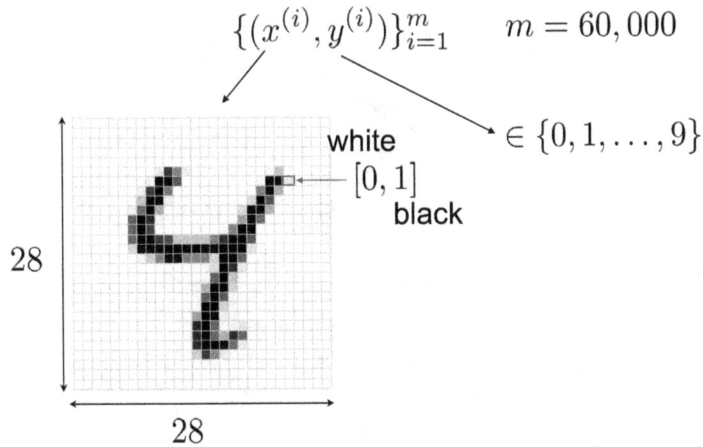

Figure B.2. MNIST dataset: An input image is of 28-by-28 pixels, each indicating an intensity from 0 (white) to 1 (black); each label with size 1 takes one of the 10 classes from 0 to 9.

the train and test datasets, which are already properly split. Hence, there is no need to worry about the splitting process. The only requirement is to write a script as follows:

```
from tensorflow.keras.datasets import mnist
(X_train, y_train), (X_test, y_test) = mnist.load_data()
X_train = X_train/255.
X_test = X_test/255.
```

```
Downloading data from https://storage.googleapis.com/
tensorflow/tf-keras-datasets/mnist.npz
11493376/11490434 [==========================] - 1s 0us/step
11501568/11490434 [==========================] - 1s 0us/step
```

Normalization is an essential data preprocessing step. Here to this end, we divide the input data (X_train or X_test) by its maximum value of 255. If the dataset we want to use is not available in the keras.datasets sub-package, we need to be familiar with other data preprocessing techniques. The pandas library offers one such technique that is useful in handling .csv files. However, this book does not cover the usage of pandas in detail. If you want to learn more about pandas, you can refer to:

https://pandas.pydata.org/

We use matplotlib.pyplot for data visualization. The following code shows how to plot a sample image:

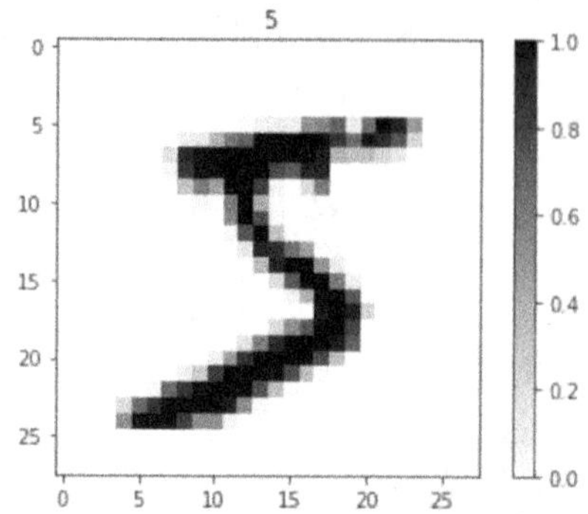

Figure B.3. A sample image in MNIST dataset.

```
import matplotlib.pyplot as plt

plt.imshow(X_train[0], cmap = 'gray_r')
plt.colorbar()
plt.title('{}'.format(y_train[0], fontsize=30))
```

The output of the code for plotting the sample image is shown in Fig. B.3. The 'gray_r' option is used to enable the white background and a black letter, while 'gray' is used for the flipped one, which is a white letter with a black background. The colorbar() function displays the color bar on the right, as seen in Fig. B.3. It is also possible to plot multiple images in a single figure. For instance, the following code shows how to display 60 images.

```
num_of_images = 60
for index in range(1,num_of_images+1):
    plt.subplot(6,10, index)
    plt.axis('off')
    plt.imshow(X_train[index], cmap = 'gray_r')
```

See Fig. B.4 for the output.

Building a neural network model We will use a two-layer neural network that was studied in Section 3.14. Specifically, we introduce a hidden layer with 500 neurons as shown in Fig. B.5. The ReLU activation function is used at the hidden layer and softmax is used at the output layer.

Keras includes two major packages:

(i) tensorflow.keras.models;

(ii) tensorflow.keras.layers.

The models package contains several functionalities regarding a neural network. One major module is Sequential which is a neural network entity and hence can

Figure B.4. Plotting many image samples in a single figure.

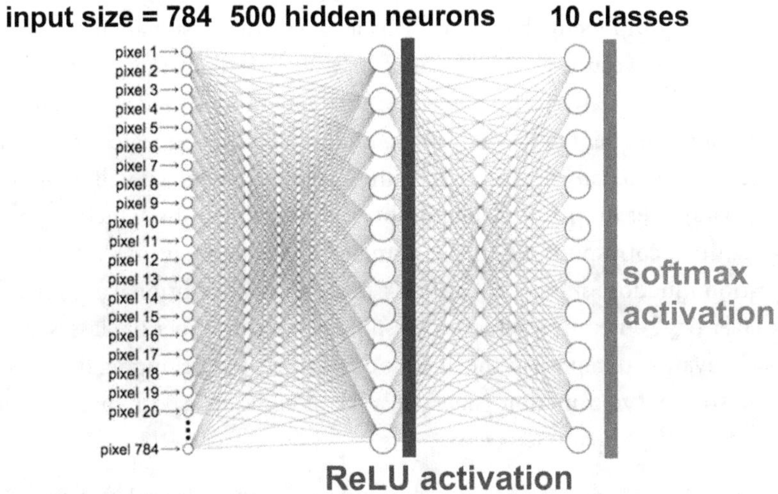

Figure B.5. A two-layer fully-connected neural network where input size is $28 \times 28 = 784$, the number of hidden neurons is 500 and the number of classes is 10. We employ ReLU activation at the hidden layer, and softmax activation at the output layer.

be described as a linear stack of layers. The layers package in Keras includes various elements required for constructing a neural network, such as fully-connected dense layers and activation functions. These components enable us to easily build a model as depicted in Fig. B.5.

```
from tensorflow.keras.models import Sequential
from tensorflow.keras.layers import Dense, Flatten

model = Sequential()
model.add(Flatten(input_shape=(28,28)))
model.add(Dense(500, activation='relu'))
model.add(Dense(10, activation='softmax'))
model.summary()
```

```
Model: "sequential\_1"
---------------------------------------------------------------
 Layer (type)                Output Shape              Param #
===============================================================
 flatten (Flatten)           (None, 784)               0

 dense (Dense)               (None, 500)               392500

 dense\_1 (Dense)             (None, 10)                5010
===============================================================
Total params: 397,510
Trainable params: 397,510
Non-trainable params: 0
```

Flatten is a component that transforms a higher dimensional entity, such as a 2D matrix, into a vector. In this instance, a 28-by-28 digit image is transformed into a vector of size $784 (= 28 \times 28)$. The add() method is used to append a layer to the end of the sequential model. Dense denotes a fully-connected layer, and the input size is automatically determined by the last layer to which it will be appended. The only parameter to specify is the number of output neurons, which is set to 500 in this example, corresponding to the number of hidden neurons. We can also set an activation function, such as activation='relu', with an additional argument. The output layer has 10 neurons, which corresponds to the number of classes, and uses softmax activation to represent the likelihood of an output belonging to a particular class. The summary() function generates a list of all layers, specifying the size and number of associated parameters.

Training a model First, we have to choose an optimizer algorithm. One popular algorithm is gradient descent, and we will be using its advanced version introduced in Section 3.14 called the Adam optimizer. Adam is an improved version of gradient descent that provides more stable training. It has three important hyperparameters, namely the learning rate , β_1 (which represents the weight of past gradients), and β_2 (which indicates the weight of the square of past gradients). By default, these are set to $(\alpha, \beta_1, \beta_2) = (0.001, 0.9, 0.999)$. If we do not specify any values, these default values will be used.

Next, we need to specify a loss function. As we learned in Section 3.13, the optimal choice for maximizing likelihood in multi-class cases is cross entropy. We also need to specify a performance metric that we will use to evaluate the model during training and testing. The accuracy metric is commonly used for this purpose. All of these can be set using the compile method.

```
model.compile(optimizer='adam',
              loss='sparse_categorical_crossentropy',
              metrics=['acc'])
```

To manually choose the hyperparameters of the Adam optimizer, we can define:

```
opt=tensorflow.keras.optimizers.Adam(
                learning_rate=0.01,
                beta_1 = 0.92,
                beta_2 = 0.992)
```

Next, we replace the previous option with optimizer=opt. As for the loss option in the compile method, we will use 'sparse_categorical_crossentropy' instead, which is suitable for cross entropy loss in cases beyond binary classification.

With these settings, we can now train the model on MNIST data. During the training process, we use a portion of the total examples to compute the gradient of the loss function, which is called a *batch*. Two more terms are used in this context: a *step* refers to the process of computing the loss for the examples in a single batch, while an *epoch* refers to the entire process associated with all the examples. For our experiment, we use a batch size of 64 and train the model for 20 epochs.

```
history = model.fit(X_train, y_train, batch_size=64, epochs=20)
```

```
Epoch 1/20
938/938 [===] - 2s 2ms/step - loss: 0.0025 - acc: 0.9992
Epoch 2/20
938/938 [===] - 2s 2ms/step - loss: 0.0059 - acc: 0.9981
Epoch 3/20
938/938 [===] - 2s 2ms/step - loss: 0.0031 - acc: 0.9990
Epoch 4/20
938/938 [===] - 2s 2ms/step - loss: 0.0074 - acc: 0.9976
Epoch 5/20
938/938 [===] - 2s 2ms/step - loss: 0.0025 - acc: 0.9993
Epoch 6/20
938/938 [===] - 2s 2ms/step - loss: 0.0043 - acc: 0.9984
Epoch 7/20
938/938 [===] - 2s 2ms/step - loss: 0.0044 - acc: 0.9984
Epoch 8/20
938/938 [===] - 2s 2ms/step - loss: 0.0010 - acc: 0.9998
Epoch 9/20
938/938 [===] - 2s 2ms/step - loss: 1.2813e-04 - acc: 1.0
Epoch 10/20
938/938 [===] - 2s 2ms/step - loss: 3.5169e-05 - acc: 1.0
Epoch 11/20
938/938 [===] - 2s 2ms/step - loss: 2.1899e-05 - acc: 1.0
Epoch 12/20
938/938 [===] - 2s 2ms/step - loss: 1.6756e-05 - acc: 1.0
Epoch 13/20
938/938 [===] - 2s 2ms/step - loss: 1.2778e-05 - acc: 1.0
```

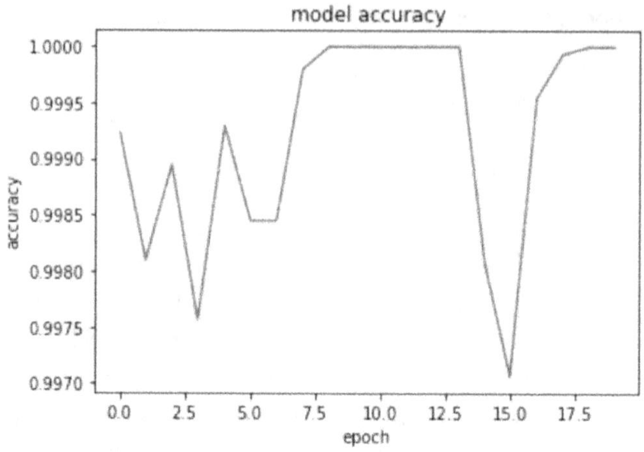

Figure B.6. Accuracy as a function of epochs.

```
Epoch 14/20
938/938 [===] - 2s 2ms/step - loss: 9.8947e-06 - acc: 1.0
Epoch 15/20
938/938 [===] - 2s 2ms/step - loss: 0.0082 - acc: 0.9981
Epoch 16/20
938/938 [===] - 2s 2ms/step - loss: 0.0090 - acc: 0.9971
Epoch 17/20
938/938 [===] - 2s 2ms/step - loss: 0.0016 - acc: 0.9995
Epoch 18/20
938/938 [===] - 2s 2ms/step - loss: 3.9583e-04 - acc: 0.9999
Epoch 19/20
938/938 [===] - 2s 2ms/step - loss: 7.6672e-05 - acc: 1.0
Epoch 20/20
938/938 [===] - 2s 2ms/step - loss: 2.4958e-05 - acc: 1.0
```

An advantage of using the fit() function is that it provides a dictionary of the metrics that were gathered during the training process. We can examine the metrics by running:

```
# list all data in history object
print(history.history.keys())
```

```
dict_keys(['loss', 'acc'])
```

We can create a plot of the accuracy as a function of epochs using the collected data.

```
plt.plot(history.history['acc'])
plt.title('model accuracy')
plt.xlabel('epoch')
plt.ylabel('accuracy')
```

Testing the trained model To conduct testing, we need to predict the model output using the predict() function in the following manner:

```
model.predict(X_test).argmax(1)
```

```
array([7, 2, 1, ..., 4, 5, 6], dtype=int64)
```

The function argmax(1) retrieves the class with the highest softmax output among the 10 available classes. In order to assess the accuracy of the test set, we utilize the evaluate() function:

```
model.evaluate(X_test, y_test)
```

```
313/313 [===] - 0s 751us/step - loss: 0.1001 - acc: 0.9847

[0.10007859766483307, 0.9847000241279602]
```

Saving and loading Saving and loading the trained model is a straightforward process, as shown below.

```
model.save('saved_classifier')
```

```
INFO:tensorflow:Assets written to: saved_classifier\assets
```

```
import tensorflow
loaded_model = tensorflow.keras.models.load_model(
'saved_classifier')
```

DOI: 10.1561/9781638281153.ch6

Appendix C

A Special Note on Research

In this appendix, we aim to provide some insights that could be valuable for your career advancement, specifically in the realm of conducting research. We will cover two key topics related to research. Firstly, we will provide advice on what aspects to concentrate on while conducting research. Secondly, we will outline a methodology for reading research papers.

C.1 Power of Fundamentals

An advice The main message we want to convey through this book is the importance of fundamental concepts and tools such as phase transitions, entropy, mutual information, and the KL divergence. One piece of advice we would like to offer is to focus on strengthening your understanding of these fundamentals, particularly as they relate to modern technologies. To further explain what we mean, let's examine how fundamentals have evolved from the past to the present day.

Fundamentals in old days Technologies in the past were shaped by the 1st, 2nd, and 3rd industrial revolutions, which were made possible by groundbreaking inventions inspired by scientific discoveries. The steam engine was the key invention of the 1st industrial revolution, based on the principles of thermodynamics in which physics and chemistry played foundational roles. The 2nd revolution was triggered by the invention of electricity, which is based on electromagnetism, again with physics as the underlying theory. The 3rd revolution was brought about by the computer, which is based on the invention of the semiconductor, with physics and chemistry providing the foundation. Although these fundamentals are crucial, they are not the ones we are referring to in this book. Our emphasis is on modern-day technologies.

Fundamentals in modern days The 4th industrial revolution is currently driving modern day technologies. It is widely accepted that the main focus of this revolution is Artificial Intelligence (AI). It is important to note that machine learning and deep learning, which are key methodologies for achieving machine intelligence, rely on optimization techniques that fall under the umbrella of mathematics. Therefore, mathematics is a significant driving force for the development of AI.

Four fundamentals in mathematics The 4th industrial revolution, centered around Artificial Intelligence (AI), relies heavily on mathematics. In particular, four branches of mathematics play foundational roles in AI: *optimization*, *linear algebra*, *probability*, and *information theory*. Optimization, a branch of mathematics that deals with finding the optimal solution to a problem, is a key methodology in achieving machine intelligence. Linear algebra provides instrumental tools for obtaining simple and tractable formulas of the objective function and/or constraints. Probability is used to deal with random quantities, and information theory sheds optimal architectural insights into machine learning models. These fundamentals are essential in the 4th industrial revolution. Therefore, our advice is to be strong in these fundamentals. However, it is worth noting that it is easier to build these fundamentals when you are younger and in school, as you may not have enough time or stamina to develop a deep understanding of these principles after graduation.

Programming skills Do fundamentals alone suffice? Unfortunately, the answer is negative. There is another vital skill that you must possess. Remember that the ultimate product of machine learning is "algorithms." In other words, it is a set of instructions that typically require extensive computation. Manual computation is virtually impossible, and therefore, it must be performed on a computer. This is where *programming tools*, such as Python and TensorFlow, that we have utilized throughout this book, become essential. We strongly suggest that you become proficient in this tool as well. Swift implementations through exceptional programming abilities will assist you in realizing and advancing your concepts. One caveat to note is that programming tools change over time. This is due to the rapid evolution of computational resources and capabilities, which influence the efficiency of programming languages. As a result, you must stay up to date with such changes.

C.2 How to Read Papers?

We would like to share with you our thoughts on a methodology for conducting research, specifically on how to approach reading papers. We believe this is a crucial skill for students, yet it is often not explicitly taught in university curriculums. Many students may approach reading papers without a clear strategy or purpose, which can be challenging even for experienced researchers.
Disclaimer: The following recommendations are based on our own opinions.

Two strategies We present two approaches: a passive strategy that is useful when you are trying to grasp the basics and trends of a field you are interested in (if you are not planning to write a paper), and an active strategy that is more suitable for those who want to become experts in a trending but not fully developed field, or for those who need to write papers on such topics.

Passive approach The concept is to depend on the expertise of professionals. This means waiting until the experts can provide instruction in a very understandable way, and then reading papers written by these professionals. Ideally, these papers should be of the survey type. Two questions that come to mind are: (i) how can you determine who the experts are?; and (ii) how can you read well-written papers effectively?

How to figure out experts? We suggest three practices that can help you find experts in the field. The first practice is to approach professors, seniors, and peers who are already working in the field. In most cases, well-known figures in the field can be easily recognized, so you can simply ask them for recommendations. The second practice is to use Google search with appropriate keywords. Nowadays, many relevant blogs are available on websites. These blogs may provide references or pointers to experts in the field. Once you have identified an expert, you can check their track record, such as their Google citations. The third practice is to look for organizers and keynote speakers in flagship conferences and workshops. These individuals are likely to be experts in the field. However, you may feel unsure when you identify scholars who are too young and have weak track records. In such cases, you may want to check their advisors, as great advisors usually produce great students.

How to read well-written papers? After identifying the experts, the next step is to read their well-written papers. Survey papers that have been cited frequently are generally a good choice. Here are some guidelines on how to read them.

First and foremost, it is crucial to read the papers intensively rather than simply skimming through them. It is best to read the paper while taking notes and attempting to translate the authors' words into your own understanding. It is recommended to ponder on the theorems before reading their proofs. The main messages and their implications are usually conveyed through theorems, and comprehending them is the most important aspect. Attempting to read technical proofs prior to grasping what the theorems mean will be unproductive, and you may quickly lose interest or become burnt out. Once you understand the main messages, you can delve into the technical details.

Secondly, it is recommended to read the paper multiple times until you have completely absorbed its contents. A well-written paper is like a bible, and repeated readings can provide a deep and diverse understanding of its contents. It is also important to be familiar with any proof techniques used in the paper. A good paper typically contains simple and insightful proof techniques. A short proof may not necessarily be good unless it is insightful. We believe that a good proof comes with insights, even if it is lengthy.

Finally, it is essential to respect the experts' notations, logical flows, mindsets, and writing styles. Expert writers usually spend a lot of time selecting their notations, developing a storyline, and refining their writing. Their choices are well-considered and carefully crafted. You may have your own preferred style, but if you find their methods to be better, don't hesitate to adopt them. If you don't have a preference or are unable to judge, simply follow theirs. It is worth emulating their style.

Active approach Let's now discuss the second approach, which is an active approach designed for individuals who wish to write papers in a relatively new or emerging field. This approach is suitable for those who cannot wait for experts to become available to teach the field. The spirit of the approach is based on trial and error. It involves reading numerous recent papers and going back and forth until you can grasp the big picture. This approach is not easy and requires two key skills: (i) the ability to quickly identify relevant and high-quality papers; and (ii) the ability to quickly comprehend the main ideas presented in these papers.

How to identify relevant/good papers quickly? We recommend two approaches. The first is similar to the one suggested in the passive approach, which is to ask experts in the field, such as professors, seniors, or peers. However, we emphasize the importance of seeking out experts who are familiar with the field, as identifying good papers in a new field can be challenging. Non-experts may not be able to identify relevant papers easily, and even if they can, they may not be motivated to invest a significant amount of time searching for papers on behalf of others.

The second approach is to search for papers in relevant conferences and workshops, but in a strategic manner. With so many conferences and workshops, it can be overwhelming to go through all the papers. To streamline the process, we suggest a two-step filtering method. First, look at the title and only consider papers that contain relevant keywords, are grammatically correct, and are well-written. If a paper fails to meet these criteria, remove it from the list. Second, read the abstract of the surviving papers. If the abstract is well-written and relevant, add the paper to a shortlist. Authors typically spend a significant amount of time crafting and refining the abstract, so if it is poorly written, the main body is likely to be even worse, which can make it challenging to understand the paper's main idea. Therefore, it is best to exclude such papers from the shortlist.

Further short-list papers if needed After applying the aforementioned guidelines, you might be left with a considerable number of papers. If the number of selected papers is approximately 10 or more, we advise you to refine the list by delving deeper into each paper. Here's how to go about it: determine your objective and what you hope to gain from reading each paper, then begin by reading the "introduction" section with your objective in mind. If the paper's introduction is well-written and relevant to your goal, it should remain on your list. If not, it should be removed from the list.

Figure out main ideas of the short-listed papers At this point, the list should only contain a few papers. If it still contains too many, repeat the previous step with stricter criteria until you have only a few papers left. Once you've narrowed down the list, search for the main idea sentence in the introduction of each paper. If you can't find it or don't understand it, read other sections of the paper until you do. Once you understand it, rephrase the main idea sentence in your own words and write it down in the heading on the first page of the paper. If you get frustrated during the process, it's okay to stop reading and move on to the next paper.

Do back-and-forth Continue the iterative process of summarizing the main ideas of the short-listed papers. Two important considerations to keep in mind are: first, do not invest too much time reading "related works." These sections are often dense and not critical to the main story of the paper. They exist mainly to avoid criticism from non-cited authors. You may choose to ignore them entirely. Second, when investigating other references, read them quickly and stay focused on your goal. It's best not to get sidetracked by reading other references or spending too much time on them.

You can end the iterative process when you either: (a) gain a complete understanding of the topic, or (b) find a well-written "anchor paper" that presents a clear overview of the subject. If you find an anchor paper, follow the passive approach to read it thoroughly.

Do your own research After figuring out the big picture, it's best to concentrate on your own research and avoid getting sidetracked by reading more papers. Follow these steps to get started:

1. Select a challenge that you want to tackle;
2. Define a specific and tangible problem that can address the challenge;
3. Attempt to solve the problem using conventional wisdom and first-principle thinking.

It's possible to encounter difficulties during the third step, especially if you're stuck. In that case, consider discussing the issue with your advisors, seniors, or peers.

Two final remarks We have two final remarks to make. Firstly, we advise you not to give up during the research process. Research is a complex task and it is natural to feel discouraged or overwhelmed at times. However, persistence and patience will ultimately pay off. Secondly, regarding communication skills, it is crucial to quickly grasp the writing quality and main idea of a paper as outlined in the guidelines. We strongly encourage you to work on improving your reading comprehension and grammar skills.

References

Abbe, E. (2017). "Community detection and stochastic block models: recent developments". *The Journal of Machine Learning Research*. 18(1): 6446–6531.

Ahn, K., K. Lee, H. Cha, and C. Suh. (2018). "Binary rating estimation with graph side information". *Advances in neural information processing systems*. 31.

Angwin, J., J. Larson, S. Mattu, and L. Kirchner. (2020). "There's software used across the country to predict future criminals and it's biased against blacks. 2016".

Arikan, E. (2009). "Channel polarization: A method for constructing capacity-achieving codes for symmetric binary-input memoryless channels". *IEEE Transactions on information Theory*. 55(7): 3051–3073.

Arora, S., R. Ge, Y. Liang, T. Ma, and Y. Zhang. (2017). "Generalization and equilibrium in generative adversarial nets (gans)". In: *International Conference on Machine Learning*. PMLR. 224–232.

Bansal, N., A. Blum, and S. Chawla. (2004). "Correlation clustering". *Machine learning*. 56(1): 89–113.

Beauchamp, K. (2001). *History of telegraphy*. No. 26. Iet. Iet.

Bertsekas, D. and J. N. Tsitsiklis. (2008). *Introduction to probability*. Vol. 1. Athena Scientific.

Bondyopadhyay, P. K. (1995). "Guglielmo Marconi-The father of long distance radio communication-An engineer's tribute". In: *1995 25th European Microwave Conference*. Vol. 2. IEEE. 879–885.

Boyd, S. P. and L. Vandenberghe. (2004). *Convex optimization*. Cambridge university press.

Bradley, R. A. and M. E. Terry. (1952). "Rank analysis of incomplete block designs: I. The method of paired comparisons". *Biometrika*. 39(3/4): 324–345.

Browning, S. R. and B. L. Browning. (2011). "Haplotype phasing: existing methods and new developments". *Nature Reviews Genetics*. 12(10): 703–714.

Candes, E. and B. Recht. (2012). "Exact matrix completion via convex optimization". *Communications of the ACM*. 55(6): 111–119.

Candès, E. J. and T. Tao. (2010). "The power of convex relaxation: Near-optimal matrix completion". *IEEE Transactions on Information Theory*. 56(5): 2053–2080.

Chartrand, G. (1977). *Introductory graph theory*. Courier Corporation.

Chen, J. and B. Yuan. (2006). "Detecting functional modules in the yeast protein–protein interaction network". *Bioinformatics*. 22(18): 2283–2290.

Chen, Y., G. Kamath, C. Suh, and D. Tse. (2016a). "Community recovery in graphs with locality". In: *International conference on machine learning*. PMLR. 689–698.

Chen, Y. and C. Suh. (2015). "Spectral MLE: Top-K rank aggregation from pairwise comparisons". In: *International Conference on Machine Learning*. PMLR. 371–380.

Chen, Y., C. Suh, and A. J. Goldsmith. (2016b). "Information recovery from pairwise measurements". *IEEE Transactions on Information Theory*. 62(10): 5881–5905.

Cho, J., G. Hwang, and C. Suh. (2020). "A fair classifier using mutual information". In: *2020 IEEE International Symposium on Information Theory (ISIT)*. IEEE. 2521–2526.

Coe, L. (1995). *The telephone and its several inventors: A history*. McFarland.

Cover, T. and A. T. Joy. (2006). *Elements of information theory*. Wiley-Interscience.

Cover, T. M. (1999). *Elements of information theory*. John Wiley & Sons.

Csiszár, I. and J. Körner. (2011). *Information theory: coding theorems for discrete memoryless systems*. Cambridge University Press.

Das, S. and H. Vikalo. (2015). "SDhaP: haplotype assembly for diploids and polyploids via semi-definite programming". *BMC genomics*. 16(1): 1–16.

El Gamal, A. and Y.-H. Kim. (2011). *Network information theory*. Cambridge university press.

Elmahdy, A., J. Ahn, C. Suh, and S. Mohajer. (2020). "Matrix completion with hierarchical graph side information". *Advances in neural information processing systems*. 33: 9061–9074.

Erdős, P., A. Rényi, *et al.* (1960). "On the evolution of random graphs". *Publ. Math. Inst. Hung. Acad. Sci.* 5(1): 17–60.

Fortunato, S. (2010). "Community detection in graphs". *Physics reports*. 486(3–5): 75–174.

Freedman, D., R. Pisani, and R. Purves. (2007). *Statistics*. W.W. Norton & Co.

Gallager, R. (1962). "Low-density parity-check codes". *IRE Transactions on information theory*. 8(1): 21–28.

Gallager, R. G. (1968). *Information theory and reliable communication*. Vol. 588. Springer.

Gallager, R. G. (2013). *Stochastic processes: theory for applications.* Cambridge University Press.

Garnier, J.-G. and A. Quetelet. (1838). *Correspondance mathématique et physique.* Vol. 10. Impr. d'H. Vandekerckhove.

Girvan, M. and M. E. Newman. (2002). "Community structure in social and biological networks". *Proceedings of the national academy of sciences.* 99(12): 7821–7826.

Gleick, J. (2011). *The information: A history, a theory, a flood.* Vintage.

Glorot, X., A. Bordes, and Y. Bengio. (2011). "Deep sparse rectifier neural networks". In: *Proceedings of the fourteenth international conference on artificial intelligence and statistics.* JMLR Workshop and Conference Proceedings. 315–323.

Golub, G. H. and C. F. Van Loan. (2013). *Matrix computations.* JHU press.

Goodfellow, I., J. Pouget-Abadie, M. Mirza, B. Xu, D. Warde-Farley, S. Ozair, A. Courville, and Y. Bengio. (2014). "Generative adversarial nets". *Advances in neural information processing systems.* 27.

Gray, R. M. (2011). *Entropy and information theory.* Springer Science & Business Media.

Grimmett, G. and D. Stirzaker. (2020). *Probability and random processes.* Oxford university press.

Gutmann, M. and A. Hyvärinen. (2010). "Noise-contrastive estimation: A new estimation principle for unnormalized statistical models". In: *Proceedings of the thirteenth international conference on artificial intelligence and statistics.* JMLR Workshop and Conference Proceedings. 297–304.

Hamming, R. W. (1950). "Error detecting and error correcting codes". *The Bell system technical journal.* 29(2): 147–160.

Hinton, G., N. Srivastava, and K. Swersky. (2012). "Neural networks for machine learning lecture 6a overview of mini-batch gradient descent". *Cited on.* 14(8): 2.

Huffman, D. A. (1952). "A method for the construction of minimum-redundancy codes". *Proceedings of the IRE.* 40(9): 1098–1101.

Ioffe, S. and C. Szegedy. (2015). "Batch normalization: Accelerating deep network training by reducing internal covariate shift". In: *International conference on machine learning.* PMLR. 448–456.

Ivakhnenko, A. G. (1971). "Polynomial theory of complex systems". *IEEE transactions on Systems, Man, and Cybernetics.* (4): 364–378.

Jalali, A., Y. Chen, S. Sanghavi, and H. Xu. (2011). "Clustering partially observed graphs via convex optimization". In: *ICML.*

Karush, W. (1939). "Minima of functions of several variables with inequalities as side constraints". *M. Sc. Dissertation. Dept. of Mathematics, Univ. of Chicago.*

Keshavan, R. H., A. Montanari, and S. Oh. (2010). "Matrix completion from a few entries". *IEEE transactions on information theory*. 56(6): 2980–2998.

Kingma, D. P. and J. Ba. (2014). "Adam: A method for stochastic optimization". *arXiv preprint arXiv:1412.6980*.

Krizhevsky, A., I. Sutskever, and G. E. Hinton. (2012). "Imagenet classification with deep convolutional neural networks". *Advances in neural information processing systems*. 25.

Kuhn, H. W. and A. W. Tucker. (2014). "Nonlinear programming". In: *Traces and emergence of nonlinear programming*. Springer. 247–258.

Larson, J., S. Mattu, L. Kirchner, and J. Angwin. (2016). "How we analyzed the COMPAS recidivism algorithm".

LeCun, Y., L. Bottou, Y. Bengio, and P. Haffner. (1998). "Gradient-based learning applied to document recognition". *Proceedings of the IEEE*. 86(11): 2278–2324.

MacKay, D. J. (2003). *Information theory, inference and learning algorithms*. Cambridge university press.

Meta. (2022). "Investor earnings report for 3Q 2022".

Negahban, S., S. Oh, and D. Shah. (2012). "Iterative ranking from pair-wise comparisons". *Advances in neural information processing systems*. 25.

News, B. (2016). "Artificial Intelligence: Google's AlphaGo Beats Go Master Lee Se-Dol".

Nielsen, R., J. S. Paul, A. Albrechtsen, and Y. S. Song. (2011). "Genotype and SNP calling from next-generation sequencing data". *Nature Reviews Genetics*. 12(6): 443–451.

Page, L., S. Brin, R. Motwani, and T. Winograd. (1999). "The PageRank citation ranking: Bringing order to the web." *Tech. rep.* Stanford InfoLab.

Pierce, J. R. (2012). *An introduction to information theory: symbols, signals and noise*. Courier Corporation.

Polyak, B. T. (1964). "Some methods of speeding up the convergence of iteration methods". *Ussr computational mathematics and mathematical physics*. 4(5): 1–17.

Roh, Y., K. Lee, S. Whang, and C. Suh. (2020). "Fr-train: A mutual information-based approach to fair and robust training". In: *International Conference on Machine Learning*. PMLR. 8147–8157.

Rosenblatt, F. (1958). "The perceptron: a probabilistic model for information storage and organization in the brain." *Psychological review*. 65(6): 386.

Salomon, D. (2004). *Data compression: the complete reference*. Springer Science & Business Media.

Samuel, A. L. (1967). "Some studies in machine learning using the game of checkers. II—Recent progress". *IBM Journal of research and development.* 11(6): 601–617.

Shannon, C. (1956). "The zero error capacity of a noisy channel". *IRE Transactions on Information Theory.* 2(3): 8–19.

Shannon, C. E. (1938). "A symbolic analysis of relay and switching circuits". *Electrical Engineering.* 57(12): 713–723.

Shannon, C. E. (2001). "A mathematical theory of communication". *ACM SIGMOBILE mobile computing and communications review.* 5(1): 3–55.

Shen, J., T. Tang, and L.-L. Wang. (2011). *Spectral methods: algorithms, analysis and applications.* Vol. 41. Springer Science & Business Media.

Si, H., H. Vikalo, and S. Vishwanath. (2014). "Haplotype assembly: An information theoretic view". In: *2014 IEEE Information Theory Workshop (ITW 2014).* IEEE. 182–186.

Silver, D., A. Huang, C. J. Maddison, A. Guez, L. Sifre, G. Van Den Driessche, J. Schrittwieser, I. Antonoglou, V. Panneershelvam, M. Lanctot, *et al.* (2016). "Mastering the game of Go with deep neural networks and tree search". *Nature.* 529(7587): 484–489.

Stewart, J. (2015). *Calculus.* Cengage Learning.

Suh, C. (2022). *Convex optimization for machine learning.* Now Publishers.

Suh, C. (2023). *Communication principles for data science.* Springer.

Wilde, M. M. (2013). *Quantum information theory.* Cambridge University Press.

Yeung, R. W. (2008). *Information theory and network coding.* Springer Science & Business Media.

Zafar, M. B., I. Valera, M. G. Rogriguez, and K. P. Gummadi. (2017). "Fairness constraints: Mechanisms for fair classification". In: *Artificial Intelligence and Statistics.* PMLR. 962–970.

Zhang, Q., V. Y. Tan, and C. Suh. (2021). "Community detection and matrix completion with social and item similarity graphs". *IEEE Transactions on Signal Processing.* 69: 917–931.

Ziv, J. and A. Lempel. (1977). "A universal algorithm for sequential data compression". *IEEE Transactions on information theory.* 23(3): 337–343.

Ziv, J. and A. Lempel. (1978). "Compression of individual sequences via variable-rate coding". *IEEE transactions on Information Theory.* 24(5): 530–536.

Index

About the Author

Dr. Changho Suh is an Associate Professor of Electrical Engineering at KAIST. He received the B.S. and M.S. degrees in Electrical Engineering from KAIST in 2000 and 2002 respectively, and the Ph.D. degree in Electrical Engineering and Computer Sciences from UC Berkeley in 2011. From 2011 to 2012, he was a postdoctoral associate at the Research Laboratory of Electronics in MIT. From 2002 to 2006, he was with Samsung Electronics.

Prof. Suh is a recipient of numerous awards in research and teaching: the 2022 Google Research Award, the 2021 James L. Massey Research & Teaching Award for Young Scholars from the IEEE Information Theory Society, the 2020 LINKGENESIS Best Teacher Award (the campus-wide Grand Prize in Teaching), the 2019 AFOSR Grant, the 2019 Google Education Grant, the 2018 IEIE/IEEE Joint Award, the 2015 IEIE Haedong Young Engineer Award, the 2015 Bell Labs Prize finalist, the 2013 IEEE Communications Society Stephen O. Rice Prize, the 2011 David J. Sakrison Memorial Prize (the best dissertation award in UC Berkeley EECS), the 2009 IEEE ISIT Best Student Paper Award, and the five Department Teaching Awards (2013, 2019, 2020, 2021, 2022). Dr. Suh is an IEEE Fellow, a Distinguished Lecturer of the IEEE Information Theory Society from 2020 to 2022, the General Chair of the Inaugural IEEE East Asian School of Information Theory 2021, an Associate Head of the KAIST AI Institute from 2021 to 2022, and a Member of the Young Korean Academy of Science and Technology. He is also an Associate Editor of Machine Learning for IEEE TRANSACTIONS ON INFORMATION THEORY, a Guest Editor for the IEEE

JOURNAL ON SELECTED AREAS IN INFORMATION THEORY, the Editor for IEEE INFORMATION THEORY NEWSLETTER, an Area Editor for IEEE BITS the Information Theory Magazine, an Area Chair of NeurIPS 2021–2022 and a Senior Program Committee of IJCAI 2019–2021.